Complex Variables

K. A. Stroud
Formerly Principal Lecturer
Department of Mathematics
Coventry University

Dexter J. Booth
Principal Lecturer
School of Computing and Engineering
University of Huddersfield

INDUSTRIAL PRESS, INC.
NEW YORK

Library of Congress Cataloging-in-Publication Data

Stroud, K. A.
Complex variables / K.A. Stroud and Dexter Booth.
p. cm.
ISBN 978–0–8311–3266–8
1. Functions of complex variables—Programmed instruction. I. Booth, Dexter J.
II. Title.
QA331.7.S77 2006
515′.9—dc22

2006041833

Published in North America under license from Palgrave Publishers Ltd, Houndmills, Basingstoke, Hants RG21 6XS, United Kingdom

Industrial Press, Inc.
989 Avenue of Americas, 19th Floor
New York, NY 10018

10 9 8 7 6 5 4 3 2 1

Contents

Hints on using this book

This book contains lessons called *Programs*. Each Program has been written in such a way as to make learning more effective and more interesting. It is like having a personal tutor because you proceed at your own rate of learning and any difficulties you may have are cleared before you have the chance to practise incorrect ideas or techniques.

You will find that each Program is divided into numbered sections called *frames*. When you start a Program, begin at Frame 1. Read each frame carefully and carry out any instructions or exercise that you are asked to do. In almost every frame, you are required to make a response of some kind, testing your understanding of the information in the frame, and you can immediately compare your answer with the correct answer given in the next frame. To obtain the greatest benefit, you are strongly advised to cover up the following frame until you have made your response. When a series of dots occurs, you are expected to supply the missing word, phrase, number or mathematical expression. At every stage you will be guided along the right path. There is no need to hurry: read the frames carefully and follow the directions exactly. In this way, you must learn.

Each Program opens with a list of **Learning outcomes** which specify exactly what you will learn by studying the contents of the Program. The Program ends with a matching checklist of **Can You?** questions that enables you to rate your success in having achieved the **Learning outcomes**. If you feel sufficiently confident then tackle the short **Test exercise** which follows. This is set directly on what you have learned in the Program: the questions are straightforward and contain no tricks. To provide you with the necessary practice, a set of **Further problems** is also included: do as many of these problems as you can. Remember, that in mathematics, as in many other situations, practice makes perfect – or more nearly so.

Useful background information

Symbols used in the text

Symbol	Meaning	Symbol	Meaning
$=$	is equal to	$\rightarrow$	tends to
$\approx$	is approximately equal to	$\neq$	is not equal to
$>$	is greater than	$\equiv$	is identical to
$\geq$	is greater than or equal to	$<$	is less than
$n!$	factorial $n = 1 \times 2 \times 3 \times \ldots \times n$	$\leq$	is less than or equal to
$\lvert k \rvert$	modulus of k, i.e. size of k irrespective of sign	∞	infinity
$\sum$	summation	$\lim_{n\to\infty}$	limiting value as $n \rightarrow \infty$

Useful mathematical information

1 Algebraic identities

$(a+b)^2 = a^2 + 2ab + b^2 \qquad (a+b)^3 = a^3 + 3a^2b + 3ab^2 + b^3$

$(a-b)^2 = a^2 - 2ab + b^2 \qquad (a-b)^3 = a^3 - 3a^2b + 3ab^2 - b^3$

$(a+b)^4 = a^4 + 4a^3b + 6a^2b^2 + 4ab^3 + b^4$

$(a-b)^4 = a^4 - 4a^3b + 6a^2b^2 - 4ab^3 + b^4$

$a^2 - b^2 = (a-b)(a+b) \qquad a^3 - b^3 = (a-b)(a^2 + ab + b^2)$

$a^3 + b^3 = (a+b)(a^2 - ab + b^2)$

2 Trigonometrical identities

(a) $\sin^2\theta + \cos^2\theta = 1$; $\sec^2\theta = 1 + \tan^2\theta$; $\operatorname{cosec}^2\theta = 1 + \cot^2\theta$

(b) $\sin(A+B) = \sin A\cos B + \cos A\sin B$

$\sin(A-B) = \sin A\cos B - \cos A\sin B$

$\cos(A+B) = \cos A\cos B - \sin A\sin B$

$\cos(A-B) = \cos A\cos B + \sin A\sin B$

$$\tan(A+B) = \frac{\tan A + \tan B}{1 - \tan A\tan B}$$

$$\tan(A-B) = \frac{\tan A - \tan B}{1 + \tan A\tan B}$$

(c) Let $A = B = \theta$ $\therefore$ $\sin 2\theta = 2\sin\theta\cos\theta$

$\cos 2\theta = \cos^2\theta - \sin^2\theta = 1 - 2\sin^2\theta = 2\cos^2\theta - 1$

$$\tan 2\theta = \frac{2\tan\theta}{1 - \tan^2\theta}$$

(d) Let $\theta = \dfrac{\phi}{2}$ $\therefore$ $\sin\phi = 2\sin\dfrac{\phi}{2}\cos\dfrac{\phi}{2}$

$$\cos\phi = \cos^2\frac{\phi}{2} - \sin^2\frac{\phi}{2} = 1 - 2\sin^2\frac{\phi}{2} = 2\cos^2\frac{\phi}{2} - 1$$

$$\tan\phi = \frac{2\tan\dfrac{\phi}{2}}{1 - 2\tan^2\dfrac{\phi}{2}}$$

(e) $\sin C + \sin D = 2\sin\dfrac{C+D}{2}\cos\dfrac{C-D}{2}$

$\sin C - \sin D = 2\cos\dfrac{C+D}{2}\sin\dfrac{C-D}{2}$

$\cos C + \cos D = 2\cos\dfrac{C+D}{2}\cos\dfrac{C-D}{2}$

$\cos D - \cos C = 2\sin\dfrac{C+D}{2}\sin\dfrac{C-D}{2}$

(f) $2\sin A\cos B = \sin(A+B) + \sin(A-B)$

$2\cos A\sin B = \sin(A+B) - \sin(A-B)$

$2\cos A\cos B = \cos(A+B) + \cos(A-B)$

$2\sin A\sin B = \cos(A-B) - \cos(A+B)$

(g) Negative angles: $\sin(-\theta) = -\sin\theta$

$\cos(-\theta) = \cos\theta$

$\tan(-\theta) = -\tan\theta$

(h) Angles having the same trigonometrical ratios:

(i) Same sine: θ and $(180^\circ - \theta)$

(ii) Same cosine: θ and $(360^\circ - \theta)$, i.e. $(-\theta)$

(iii) Same tangent: θ and $(180^\circ + \theta)$

(i) $a\sin\theta + b\cos\theta = A\sin(\theta + \alpha)$

$a\sin\theta - b\cos\theta = A\sin(\theta - \alpha)$

$a\cos\theta + b\sin\theta = A\cos(\theta - \alpha)$

$a\cos\theta - b\sin\theta = A\cos(\theta + \alpha)$

where $\begin{cases} A = \sqrt{a^2 + b^2} \\ \alpha = \tan^{-1}\dfrac{b}{a} \quad (0^\circ < \alpha < 90^\circ) \end{cases}$

3 Standard curves

(a) *Straight line*

Slope, $m = \dfrac{dy}{dx} = \dfrac{y_2 - y_1}{x_2 - x_1}$

Angle between two lines, $\tan\theta = \dfrac{m_2 - m_1}{1 + m_1 m_2}$

For parallel lines, $m_2 = m_1$

For perpendicular lines, $m_1 m_2 = -1$

Equation of a straight line (slope $= m$)

(i) Intercept c on real y-axis: $y = mx + c$

(ii) Passing through (x_1, y_1): $y - y_1 = m(x - x_1)$

(iii) Joining (x_1, y_1) and (x_2, y_2): $\dfrac{y - y_1}{y_2 - y_1} = \dfrac{x - x_1}{x_2 - x_1}$

(b) *Circle*

Centre at origin, radius r: $x^2 + y^2 = r^2$

Centre (h, k), radius r: $(x - h)^2 + (y - k)^2 = r^2$

General equation: $x^2 + y^2 + 2gx + 2fy + c = 0$

with centre $(-g, -f)$: radius $= \sqrt{g^2 + f^2 - c}$

Parametric equations: $x = r\cos\theta$, $y = r\sin\theta$

(c) *Parabola*

Vertex at origin, focus $(a, 0)$: $y^2 = 4ax$

Parametric equations : $x = at^2$, $y = 2at$

(d) *Ellipse*

Centre at origin, foci $\left(\pm\sqrt{a^2 + b^2}, 0\right)$: $\dfrac{x^2}{a^2} + \dfrac{y^2}{b^2} = 1$

where $a =$ semi-major axis, $b =$ semi-minor axis

Parametric equations: $x = a\cos\theta$, $y = b\sin\theta$

(e) *Hyperbola*

Centre at origin, foci $\left(\pm\sqrt{a^2 + b^2}, 0\right)$: $\dfrac{x^2}{a^2} - \dfrac{y^2}{b^2} = 1$

Parametric equations: $x = a\sec\theta$, $y = b\tan\theta$

Rectangular hyperbola:

Centre at origin, vertex $\pm\left(\dfrac{a}{\sqrt{2}}, \dfrac{a}{\sqrt{2}}\right)$: $xy = \dfrac{a^2}{2} = c^2$

i.e. $xy = c^2$ where $c = \dfrac{a}{\sqrt{2}}$

Parametric equations: $x = ct$, $y = c/t$

4 Laws of mathematics

(a) *Associative laws* – for addition and multiplication

$a + (b + c) = (a + b) + c$

$a(bc) = (ab)c$

(b) *Commutative laws* – for addition and multiplication

$a + b = b + a$

$ab = ba$

(c) *Distributive laws* – for multiplication and division

$a(b + c) = ab + ac$

$\dfrac{b + c}{a} = \dfrac{b}{a} + \dfrac{c}{a}$ (provided $a \neq 0$)

Preface

It is now nearly 40 years since Ken Stroud first developed his approach to personalized learning with his classic text *Engineering Mathematics*, now in its sixth edition. That unique and hugely successful programmed learning style is exemplified in this text and I am delighted to have been asked to contribute to it. I have endeavored to retain the very essence of his style that has contributed to so many students' mathematical abilities over the years, particularly the time-tested Stroud format with its close attention to technique development throughout.

The first Program deals with a review of the elements of functions of a real variable and the next two Programs introduce complex numbers and their polar form. Program 4 introduces the hyperbolic trigonometric functions and ends by displaying their relationships to the circular trigonometric functions via the complex domain. Program 5 discusses the concept of a complex mapping and considers translations, magnifications and rotations; the Program culminates in the bilinear transformation. Programs 6 and 7 develop the elements of partial differentiation and multiple integrals necessary to begin the discussion of the complex calculus that is introduced in Program 8 and which culminates in Program 9 with the calculus of residues.

To give the student as much assistance as possible in organizing their study there are specific **Learning outcomes** at the beginning and **Can You?** checklists at the end of each Program. In this way, the learning experience is made more explicit and the student is given greater confidence in what has been learned. Test exercises and Further problems follow, in which the student can consolidate their newly-found knowledge.

This is a further opportunity that I have had to work on the Stroud books, having made additions to both the *Engineering Mathematics* and *Advanced Engineering Mathematics* texts. It is as ever a challenge and an honor to be able to work with Ken Stroud's material. Ken had an understanding of his students and their learning and thinking processes which was second to none, and this is reflected in every page of this book. As always, my thanks go to the Stroud family for their continuing support for and encouragement of new projects and ideas which are allowing Ken's work an ever wider public.

Huddersfield
June 2007

Dexter J Booth

Functions

Learning outcomes

When you have completed this Program you will be able to:

- Identify a function as a rule and recognize rules that are not functions
- Determine the domain and range of a function
- Construct the inverse of a function and draw its graph
- Construct compositions of functions and de-construct them into their component functions
- Develop the trigonometric functions from the trigonometric ratios
- Find the period, amplitude and phase of a periodic function
- Distinguish between the inverse of a trigonometric function and the inverse trigonometric function
- Solve trigonometric equations using the inverse trigonometric functions and trigonometric identities
- Recognize that the exponential function and the natural logarithmic function are mutual inverses and solve indicial and logarithmic equations
- Find the even and odd parts of a function when they exist
- Construct the hyperbolic functions from the odd and even parts of the exponential function
- Evaluate limits of simple functions

Processing numbers

1

The equation that states that y *is equal to some expression in* x, written as:

$y = f(x)$

has been described with the words 'y *is a function of* x'. Despite being widely used and commonly accepted, this description is not strictly correct as will be seen in Frame 3. Put simply, for all the functions that you have considered so far, both x and y are *numbers*.

Take out your calculator and enter the number:

5 this is x, the *input* number

Now press the x^2 key and the display changes to:

25 this is y, the *output* number where $y = x^2$

The *function* is a *rule* embodied in a *set of instructions* within the calculator that changed the 5 to 25, activated by you pressing the x^2 key. A diagram can be constructed to represent this:

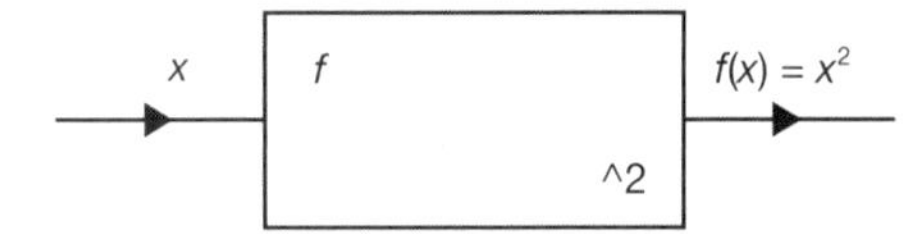

The box labelled f represents the function. The notation ^2 inside the box means *raising to the power 2* and describes the rule – what the set of instructions will do when activated. The diagram tells you that the input number x is *processed* by the function f to produce the output number $y = f(x)$. So that $y = f(x)$ is the *result* of function f acting on x.

So, use diagrams and describe the functions appropriate to each of the following equations:

(a) $y = \dfrac{1}{x}$ (b) $y = x - 6$

(c) $y = 4x$ (d) $y = \sin x$

Just follow the reasoning above, the answers are in the next frame

2

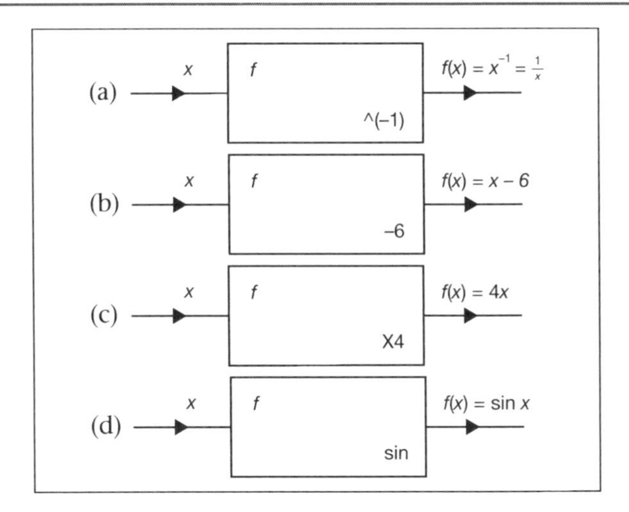

(a) Function f produces the reciprocal of the input
(b) Function f subtracts 6 from the input
(c) Function f multiplies the input by 4
(d) Function f produces the sine of the input

Let's now expand this idea

Functions are rules but not all rules are functions

3

A function of a variable x is a *rule* that describes how a value of the variable x is manipulated to generate a value of the variable y. The rule is often expressed in the form of an equation $y = f(x)$ with the proviso that for any input x there is a unique value for y. Different outputs are associated with different inputs – the function is said to be *single valued* because for a given input there is only one output. For example, the equation:

$$y = 2x + 3$$

expresses the rule '*multiply the value of x and add three*' and this rule is the function. On the other hand, the equation:

$y = x^{\frac{1}{2}}$ which is the same as $y = \pm\sqrt{x}$

expresses the rule '*take the positive and negative square roots of the value of x*'. This rule is not a function because to each value of the input $x > 0$ there are two different values of output y.

▶

The graph of $y = \pm\sqrt{x}$ illustrates this quite clearly:

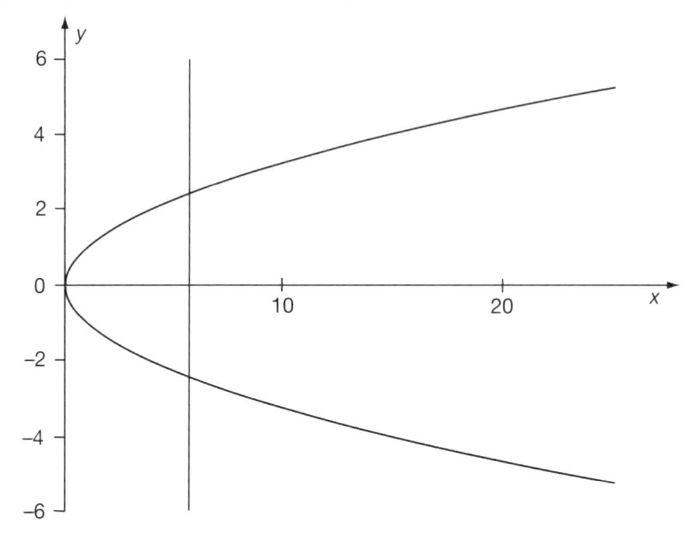

If a vertical line is drawn through the x-axis (for $x > 0$) it intersects the graph at more than one point. The fact that for $x = 0$ the vertical line intersects the graph at only *one* point does not matter – that there are other points where the vertical line intersects the graph in more than one point is sufficient to bar this from being the graph of a function. Notice that $y = x^{\frac{1}{2}}$ has no real values of y for $x < 0$.

Also note that your calculator only gives a single answer to $x^{\frac{1}{2}}$ because it is, in fact, calculating $\sqrt{x}$.

So, which of the following equations express rules that are functions?

(a) $y = 5x^2 + 2x^{-\frac{1}{4}}$

(b) $y = 7x^{\frac{1}{3}} - 3x^{-1}$

Next frame

4

(a) $y = 5x^2 + 2x^{-\frac{1}{4}}$ does not
(b) $y = 7x^{\frac{1}{3}} - 3x^{-1}$ does

(a) $y = 5x^2 + 2x^{-\frac{1}{4}}$ does not express a function because to each value of x $(x > 0)$ there are two values of $x^{-\frac{1}{4}}$, positive and negative because $x^{-\frac{1}{4}} \equiv (x^{-\frac{1}{2}})^{\frac{1}{2}} \equiv \pm\sqrt{x^{-\frac{1}{2}}}$. Indeed, *any* even root produces two values.

(b) $y = 7x^{\frac{1}{3}} - 3x^{-1}$ does express a function because to each value of x $(x \neq 0)$ there is just one value of y.

All the input numbers x that a function can process are collectively called the function's *domain*. The complete collection of numbers y that correspond to the numbers in the domain is called the *range* (or *co-domain*) of the function. For example, if:

$y = \sqrt{1 - x^2}$ where both x and y are real numbers

the domain is $-1 \leq x \leq 1$ because these are the only values of x for which y has a real value. The range is $0 \leq y \leq 1$ because 0 and 1 are the minimum and maximum values of y over the domain. Other functions may, for some purpose or other, be defined on a restricted domain. For example, if we specify:

$y = x^3, \quad -2 \leq x < 3$ (the function is defined only for the restricted set of x-values given)

the domain is given as $-2 \leq x < 3$ and the range as $-8 \leq y < 27$ because -8 and 27 are the minimum and maximum values of y over the domain.
So the domains and ranges of each of the following are:

(a) $y = x^3 \quad -5 \leq x < 4$ (b) $y = x^4$ (c) $y = \dfrac{1}{(x-1)(x+2)} \quad 0 \leq x \leq 6$

The answers are in the next frame

5

(a) $y = x^3 \quad -5 \leq x < 4$
domain $-5 \leq x < 4$, range $-125 \leq y < 64$
(b) $y = x^4$
domain $-\infty < x < \infty$, range $0 \leq y < \infty$
(c) $y = \dfrac{1}{(x-1)(x+2)}, \quad 0 \leq x \leq 6$
domain $0 \leq x < 1$ and $1 < x \leq 6$,
range $-\infty < y \leq -0.5$, $0.025 \leq y < \infty$

Because

(a) The domain is given as $-5 \leq x < 4$ and the range as $-125 \leq y < 64$ because -125 and 64 are the minimum and maximum values of y over the domain.
(b) The domain is not given and is assumed to consist of all finite values of x, that is, $-\infty < x < \infty$. The range values are all positive because of the even power.
(c) The domain is $0 \leq x < 1$ and $1 < x \leq 6$ since y is not defined when $x = 1$ where there is a vertical asymptote. To the left of the asymptote $(0 \leq x < 1)$ the y-values range from $y = -0.5$ when $x = 0$ and increase negatively towards $-\infty$ as $x \to 1$. To the right of the asymptote $1 < y \leq 6$ the y-values range from infinitely large and positive to 0.025 when $x = 6$. If you plot the graph on your spreadsheet this will be evident.

Next frame

6 Functions and the arithmetic operations

Functions can be combined under the action of the arithmetic operators provided care is taken over their common domains. For example:

If $f(x) = x^2 - 1, \quad -2 \le x < 4$ and $g(x) = \dfrac{2}{x+3}, \quad 0 < x \le 5$ then, for example

(a) $h(x) = f(x) + g(x) = x^2 - 1 + \dfrac{2}{x+3}, \quad 0 < x < 4$

because $g(x)$ is not defined for $-2 \le x \le 0$ and $f(x)$ is not defined for $4 \le x \le 5$ so $0 < x < 4$ is the common domain between them.

(b) $k(x) = \dfrac{g(x)}{f(x)} = \dfrac{2}{(x+3)(x^2-1)}, \quad 0 < x < 4$ and $x \ne 1$

because $g(x)$ is not defined for $-2 \le x \le 0$, $f(x)$ is not defined for $4 \le x \le 5$ and $k(x)$ is not defined when $x = 1$.

So if:

$f(x) = \dfrac{2x}{x^3 - 1}$, where $-3 < x < 3$ and $x \ne 1$ and

$g(x) = \dfrac{4x - 8}{x + 5}, 0 < x \le 6$ then $h(x) = \dfrac{f(x)}{g(x)}$ is ....

The answer is in the next frame

7

$$h(x) = \frac{f(x)}{g(x)} = \frac{2x(x+5)}{(x^3-1)(4x-8)} \text{ where } 0 < x < 3,\ x \ne 1 \text{ and } x \ne 2$$

Because when $x = 1$ or 2, $h(x)$ is not defined; when $-3 < x \le 0$, $g(x)$ is not defined; and when $3 \le x \le 6$, $f(x)$ is not defined.

8 Inverses of functions

The process of generating the output of a function is assumed to be reversible so that what has been constructed can be de-constructed. The effect can be described by reversing the flow of information through the diagram so that, for example, if:

$y = f(x) = x + 5$

x
f
+5
f(x) = x + 5

the flow is reversed by making the output the input and *retrieving the original input as the new output*:

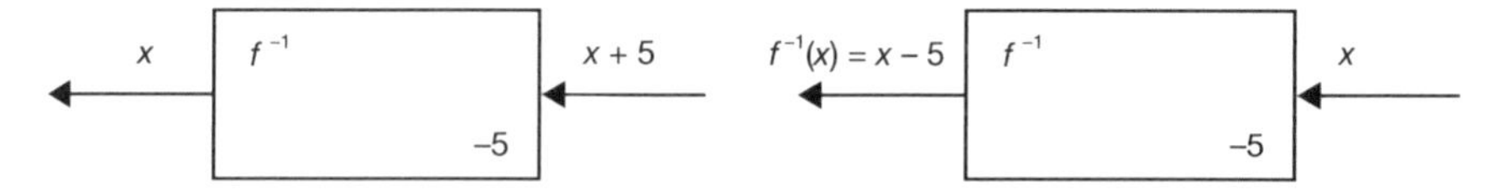

The reverse process is different because instead of adding 5 to the input, 5 is now subtracted from the input. The rule that describes the reversed process is called the *inverse of the function* which is labelled as either f^{-1} or *arcf*. That is:

$$f^{-1}(x) = x - 5$$

The notation f^{-1} is very commonly used but care must be taken to remember that the -1 does not mean that it is in any way related to the reciprocal of f.

Try some. Find $f^{-1}(x)$ in each of the following cases:

(a) $f(x) = 6x$ (b) $f(x) = x^3$ (c) $f(x) = \dfrac{x}{2}$

Draw the diagram, reverse the flow and find the inverse of the function in each case

---

9

(a) $f^{-1}(x) = \dfrac{x}{6}$

(b) $f^{-1}(x) = x^{\frac{1}{3}}$

(c) $f^{-1}(x) = 2x$

Because

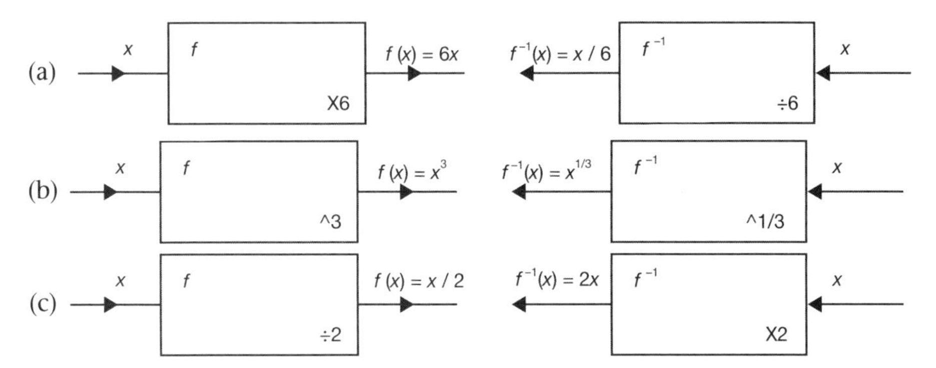

The inverses of the arithmetic operations are just as you would expect:

addition and subtraction are inverses of each other
multiplication and division are inverses of each other
raising to the power k and raising to a power 1/k are inverses of each other

Now, can you think of two functions that are each identical to their inverse?

Think carefully

10

$f(x) = x$ and $f(x) = \dfrac{1}{x}$

Because the function with output $f(x) = x$ does not alter the input at all so the inverse does not either, and the function with output $f(x) = \dfrac{1}{x}$ is its own

▶

inverse because the reciprocal of the reciprocal of a number is the number:

$$\frac{1}{1/x} = x$$

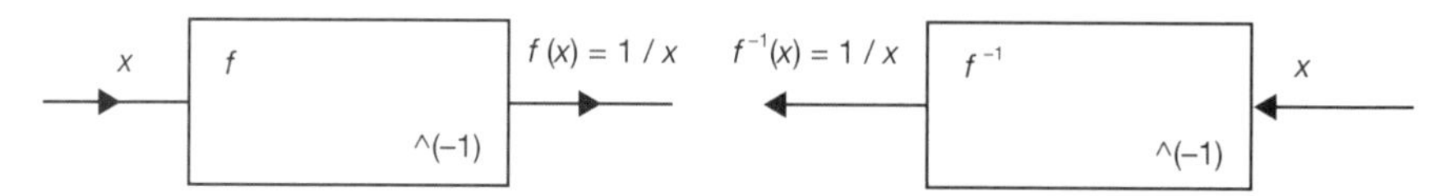

Let's progress

11 Graphs of inverses

The diagram of the inverse of a function can be drawn by reversing the flow of information and this is the same as interchanging the contents of each ordered pair generated by the function. As a result, when the ordered pairs generated by the inverse of a function are plotted, the graph takes up the shape of the original function but reflected in the line $y = x$. Let's try it. Use your spreadsheet to plot $y = x^3$ and the inverse $y = x^{\frac{1}{3}}$. The spreadsheet that is used in this text is the Microsoft Excel spreadsheet which is a component of Microsoft Office. The version used is Excel 2002 and it is assumed that later versions will exhibit the same features exemplified here.

What you are about to do is a little involved, so follow the instructions to the letter and take it slowly and carefully

12 The graph of $y = x^3$

Open up your spreadsheet

Enter −1.1 in cell **A1**
Highlight **A1** to **A24**
Click **Edit-Fill-Series** and enter the **step value as** 0.1

The cells **A1** to **A24** then fill with the numbers −1.1 to 1.2.

In cell **B1** enter the formula **=A1^3** and press **Enter**

Cell **B1** now contains the cube of the contents of cell **A1**

Make **B1** the active cell
Click **Edit-Copy** — This copies the contents of B1 to the Clipboard
Highlight **B2** to **B24**
Click **Edit-Paste** — This pastes the contents of the Clipboard to B2 to B24

Each of the cells **B1** to **B24** contains the cube of the contents of the adjacent cell in the **A** column.

Highlight the block of cells **A1** to **B24**
Click the *Chart Wizard* button to create an **XY (Scatter)** graph with joined-up points

▶

The graph you obtain will look like that depicted below:

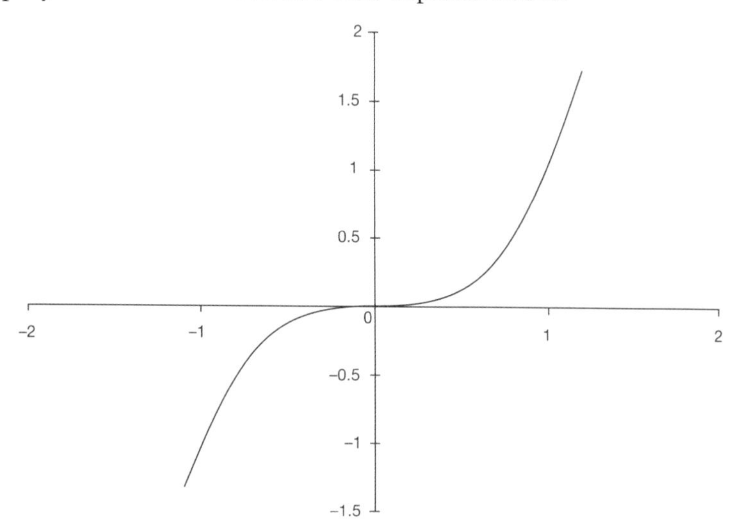

Now for the graph of $y = x^{\frac{1}{3}}$

The graph of $y = x^{\frac{1}{3}}$ 13

Keep the data you already have on the spreadsheet, you are going to use it:

Highlight cells **A1** to **A24**
Click **Edit-Copy** This copies the contents of A1 to A24 to the Clipboard
Place the cursor in cell **B26**
Click **Edit-Paste** This pastes the contents of the Clipboard from B26 to B49

The cells **B26** to **B49** then fill with the same values as those in cells **A1** to **A24**

Highlight cells **B1** to **B24**
Click **Edit-Copy** This copies the contents of B1 to B24 to the Clipboard
Place the cursor in cell **A26**
Click **Edit-Paste Special**
In the *Paste Special* window select **Values** and click **OK**

The cells **A26** to **A49** then fill with the same values as those in cells **B1** to **B24**. Because the cells **B1** to **B24** contain formulas, using **Paste Special** rather than simply **Paste** ensures that you copy the values rather than the formulas.

What you now have are the original ordered pairs for the first function reversed in readiness to draw the graph of the inverse of the function.

Notice that row 25 is empty. This is essential because later on you are going to obtain a plot of two curves on the same graph.

▶

For now you must first clear away the old graph:

Click the boundary of the graph to display the handles
Click **Edit-Clear-All**

and the graph disappears. Now, to draw the new graph:

Highlight the block of cells **A26** to **B49**
Click the *Chart Wizard* button to create an **XY (Scatter)** graph with joined-up points

The graph you obtain will look like that depicted to the right:

Same shape as the previous one but a different orientation.

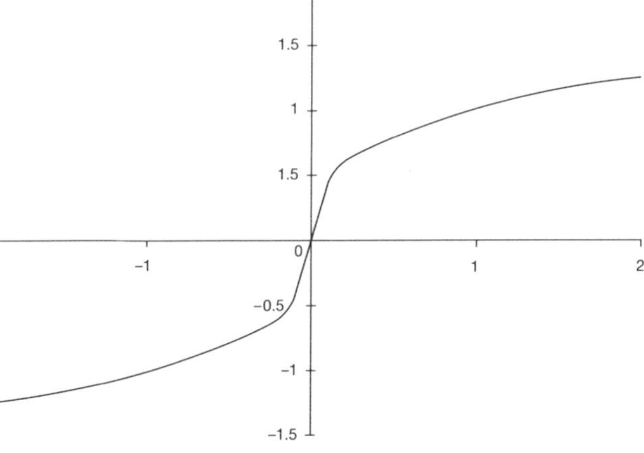

Now for both the graph of $y = x^3$ and $y = x^{\frac{1}{3}}$ together

14 The graphs of $y = x^3$ and $y = x^{\frac{1}{3}}$ plotted together

Clear away the graph you have just drawn. Then:

Highlight the block of cells **A1** to **B49**
Click the *Chart Wizard* button to create an **XY (Scatter)** graph with joined-up points

The graph you obtain will look like that depicted to the right:

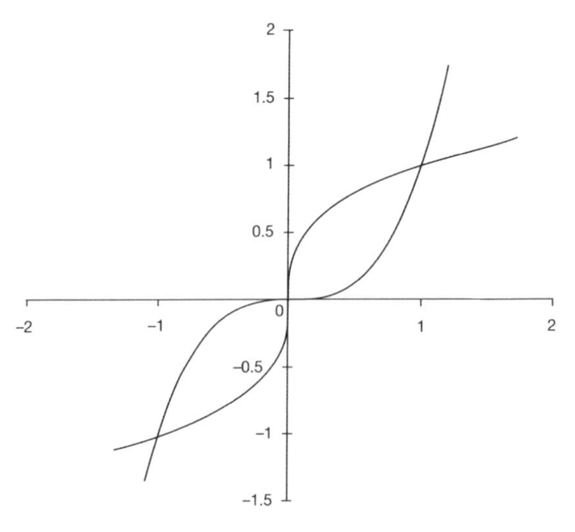

Now you can see that the two graphs are each a reflection of the other in the line $y = x$. To firmly convince yourself of this:

Place the cursor in cell **A51** and enter the number -1.1
Enter the number -1.1 in cell **B51**
Enter the number 1.2 in cell **A52**
Enter the number 1.2 in cell **B52**

You now have two points with which to plot the straight line $y = x$. *Notice again, row 50 this time is empty.*

Clear away the last graph. Then:

Highlight the block of cells **A1** to **B52**
Click the *Chart Wizard* button to create an **XY (Scatter)** graph with joined-up points

The graph you obtain will look like that depicted below:

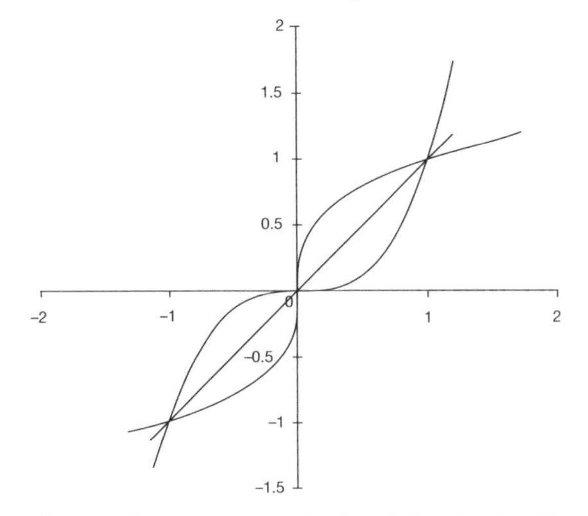

As you can see, the graphs are symmetric about the sloping line $y = x$. We say they are **reflection symmetric about $y = x$** because each one could be considered as a reflection of the other in a double-sided mirror lying along this line.

Now you try one. Use the spreadsheet to plot the graphs of $y = x^2$ and its inverse $y = x^{\frac{1}{2}}$. You do not need to start from scratch, just used the sheet you have already used and change the contents of cell **B1** to the formula **=A1^2**, copy this down the **B** column to **B24** and then **Paste Special** these values into cells **A26** to **A49**.

15

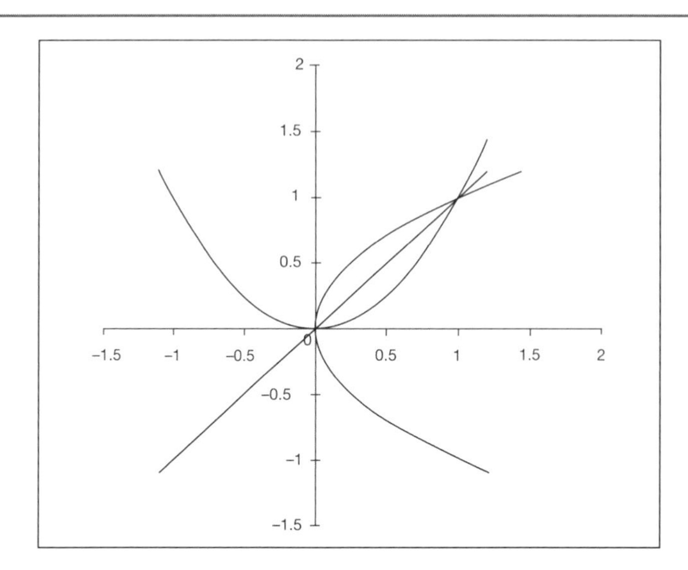

The graph of the inverse of the square function is a parabola on its side. However, as you have seen earlier, this is not a graph of a function. If, however, the bottom branch of this graph is removed, what is left is the graph of the function expressed by $y = \sqrt{x}$ which is called the *inverse function* because it is single valued.

Plot the graph of $y = x^4 - x^2 + 1$ by simply changing the formula in **B1** and copying it into cells **B2** to **B24**. So:

(a) What does the inverse of the function look like?

(b) Is the inverse of the function the inverse function?

16

(a)

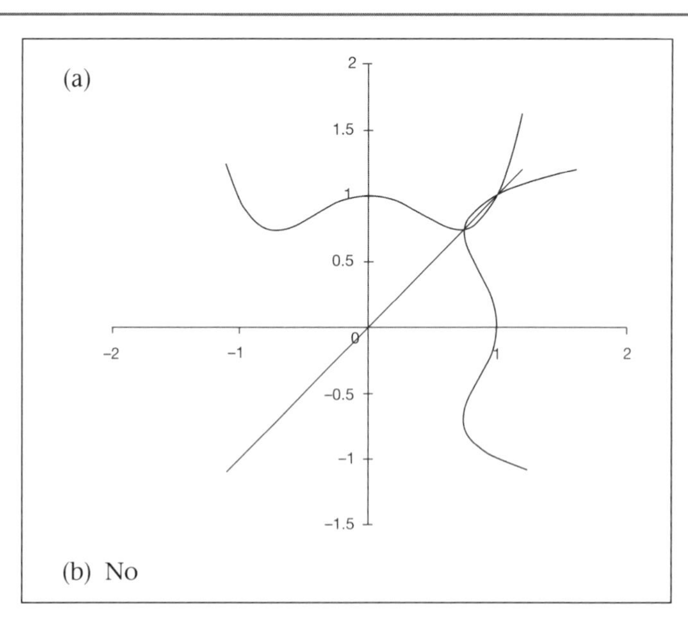

(b) No

Because

(b) The inverse of the function is not single valued so it cannot be a function. The inverse function would have to be obtained by removing parts of the inverse of the function to obtain a function that was single valued.

At this point let us pause and summarize the main facts so far on functions and their inverses

Summary

1 A function is a rule expressed in the form $y = f(x)$ with the proviso that for each value of x there is a unique value of y.

2 The collection of permitted input values to a function is called the *domain* of the function and the collection of corresponding output values is called the *range*.

3 The inverse of a function is a rule that associates range values to domain values of the original function.

17

Review exercise

1 Which of the following equations expresses a rule that is a function:

(a) $y = 6x - 2$

(b) $y = \sqrt{x^3}$

(c) $y = \left(\dfrac{3x}{x^2 + 3}\right)^{\frac{5}{2}}$

18

2 Given the two functions f and g expressed by:

$$f(x) = 2x - 1 \text{ for } -2 < x < 4 \text{ and } g(x) = \frac{4}{x-2} \text{ for } 3 < x < 5,$$

find the domain and range of:

(a) $h(x) = f(x) - g(x)$

(b) $k(x) = -\dfrac{2f(x)}{g(x)}$

3 Use your spreadsheet to draw each of the following and their inverses. Is the inverse a function?

(a) $y = x^6$ Use the data from the text and just change the formula

(b) $y = -3x$ Use the data from the text and just change the formula

(c) $y = \sqrt{x^3}$ Enter 0 in cell **A1** and **Edit-Fill-Series** with *step value* 0.1

19

1 (a) $y = 6x - 2$ expresses a rule that is a function because to each value of x there is only one value of y.

(b) $y = \sqrt{x^3}$ expresses a rule that is a function because to each value of x there is only one value of y. The surd sign $\sqrt{\ }$ stands for the positive square root.

(c) $y = \left(\frac{3x}{x^2+3}\right)^{\frac{5}{2}}$ expresses a rule that is not a function because to each positive value of the bracket there are two values of y. The power 5/2 represents raising to the power 5 and taking the square root, and there are always two square roots to each positive number.

2 (a) $h(x) = f(x) - g(x) = 2x - 1 - \frac{4}{x-2}$ for $3 < x < 4$ because $g(x)$ is not defined for $-2 < x \leq 3$ and $f(x)$ is not defined for $4 \leq x < 5$. Range $1 < h(x) < 5$.

(b) $k(x) = -\frac{2f(x)}{g(x)} = -\frac{(2x-1)(x-2)}{2}$ for $3 < x < 4$.

Range $-7 < k(x) < -5/2$.

3 (a) $y = x^6$ has an inverse $y = x^{\frac{1}{6}}$. This does not express a function because there are always two values to an even root (see Frame 4).

(b) $y = -3x$ has an inverse $y = -\frac{x}{3}$. This does express a function because there is only one value of y to each value of x.

(c) $y = \sqrt{x^3}$ has an inverse $y = x^{\frac{2}{3}}$ because $\sqrt{x^3}$ represents the positive value of $y = x^{\frac{3}{2}}$. The inverse does express a function.

Now let's move on

Composition – 'function of a function'

20

Chains of functions can be built up where the output from one function forms the input to the next function in the chain. Take out your calculator again and this time enter the number:

[4]

Now press the [$\frac{1}{x}$] key – the reciprocal key – and the display changes to:

0.25 the reciprocal of 4

Now press the [x^2] key and the display changes to:

0.0625 the square of 0.25

▶

Here, the number 4 was the input to the reciprocal function and the number 0.25 was the output. This same number 0.25 was then the input to the squaring function with output 0.0625. This can be represented by the following diagram:

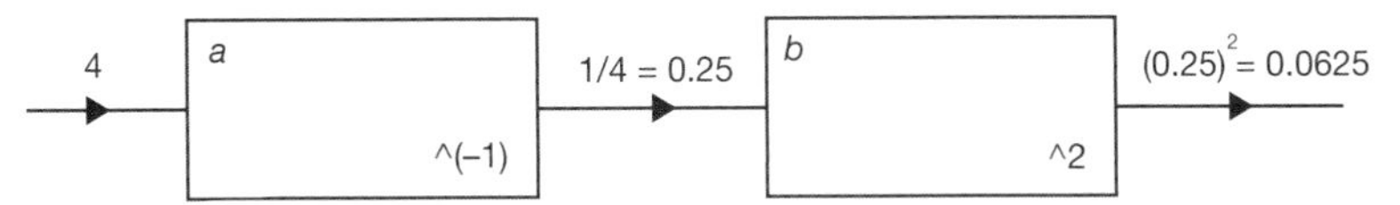

Notice that the two functions have been named a and b, but any letter can be used to label a function.

At the same time the *total* processing by f could be said to be that the number 4 was input and the number 0.0625 was output:

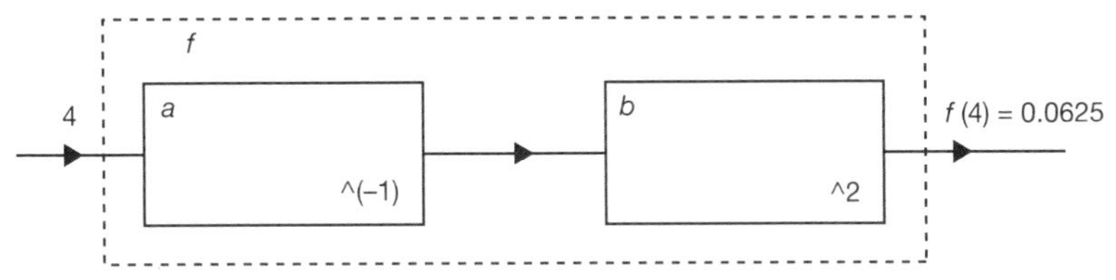

So the function f is *composed* of the two functions a and b where $a(x) = \frac{1}{x}$, $b(x) = x^2$ and $f(x) = \left(\frac{1}{x}\right)^2$. It is said that f is the *composition* of a and b, written as:

$$f = b \circ a$$

and read as *b of a*. Notice that the functions a and b are written down algebraically in the reverse order from the order in which they are given in the diagram. This is because in the diagram the input to the composition enters on the left, whereas algebraically the input is placed to the right:

$$f(x) = b \circ a(x)$$

So that $f(x) = b \circ a(x)$, which is read as *f of x equals b of a of x*. An alternative notation, more commonly used, is:

$$f(x) = b[a(x)]$$

and f is described as being a *function of a function*.

Now you try. Given that $a(x) = x + 3$, $b(x) = 4x$ find the functions f and g where:

(a) $f(x) = b[a(x)]$
(b) $g(x) = a[b(x)]$

Stick with what you know, draw the boxes and see what you find

21

(a) $f(x) = 4x + 12$
(b) $g(x) = 4x + 3$

Because

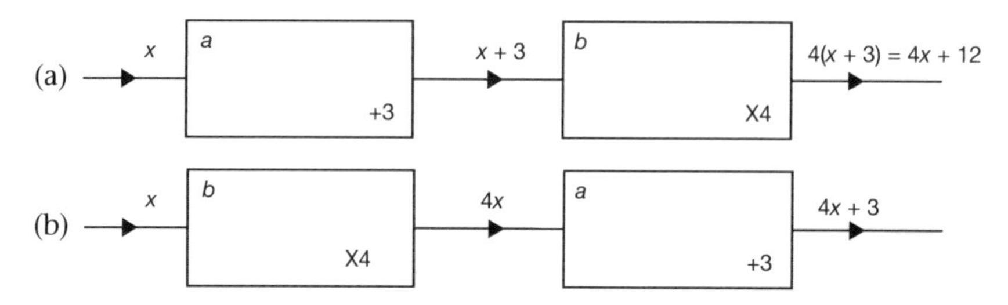

Notice how these two examples show that $b[a(x)]$ is different from $a[b(x)]$. That is, the order of composition matters.

Now, how about something a little more complicated? Given the three functions a, b and c where $a(x) = x^3$, $b(x) = 2x$ and $c(x) = \tan x$, find each of the following as expressions in x:

(a) $f(x) = a(b[c(x)])$ (b) $g(x) = c(a[b(x)])$ (c) $h(x) = a(a[c(x)])$

Remember, draw the boxes and follow the logic

22

(a) $f(x) = 8\tan^3 x$
(b) $g(x) = \tan(8x^3)$
(c) $h(x) = \tan^9 x$

Because

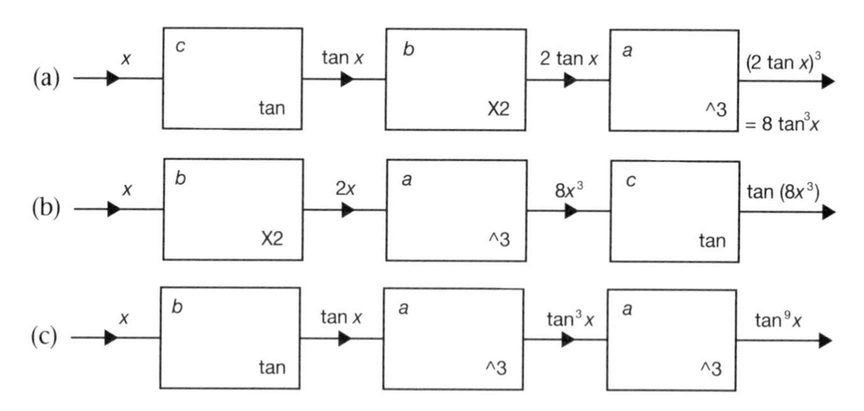

How about working the other way? Given the expression $f(x)$ for the output from a composition of functions, how do you decompose it into its component functions? This is particularly easy because you already know how to do it even though you may not yet realize it.

Let's look at a specific example first ▶

Given the output from a composition of functions as $f(x) = 6x - 4$ ask yourself how, given a calculator, would you find the value of $f(2)$? You would:

enter the number 2	the input	x	
multiply by 6 to give 12	the first function	$a(x) = 6x$	input times 6
subtract 4 to give 8	the second function	$b(x) = 6x - 4$	input minus 4

so that $f(x) = b[a(x)]$. The very act of using a calculator to enumerate the output from a composition requires you to decompose the composition automatically as you go.

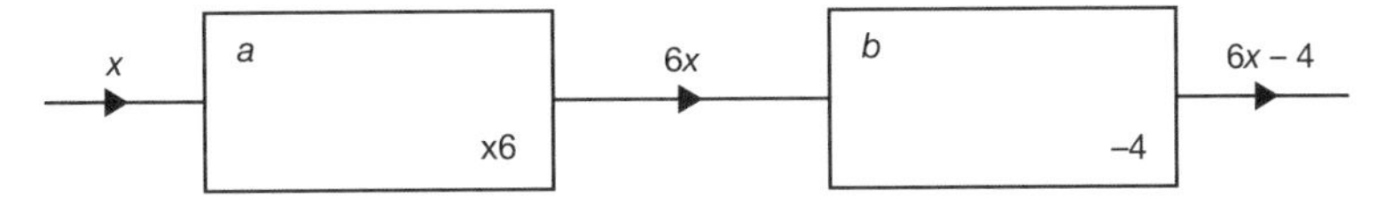

Try it yourself. Decompose the composition with output $f(x) = (x + 5)^4$.

Get your calculator out and find the output for a specific input

---

23

$f(x) = b[a(x)]$ where $a(x) = x + 5$ and $b(x) = x^4$

Because

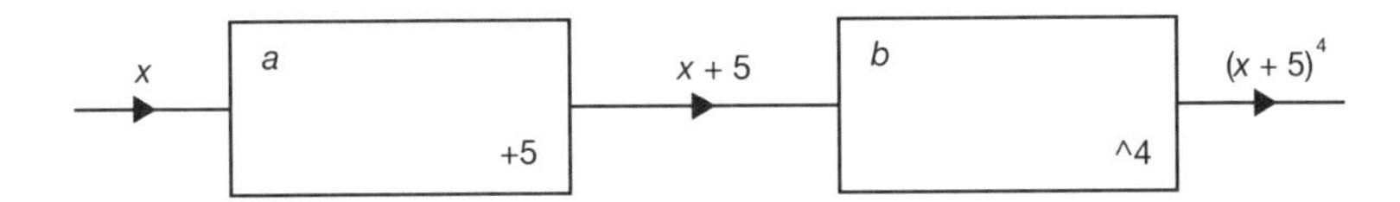

Notice that this decomposition is not unique. You could have defined $b(x) = x^2$ in which case the composition would have been $f(x) = b(b[a(x)])$.

Just to make sure you are clear about this, decompose the composition with output $f(x) = 3\sin(2x + 7)$.

Use your calculator and take it steady, there are four functions here

---

24

$f(x) = d[c(b[a(x)])]$ where $a(x) = 2x$
$b(x) = x + 7$, $c(x) = \sin x$ and $d(x) = 3x$

Because

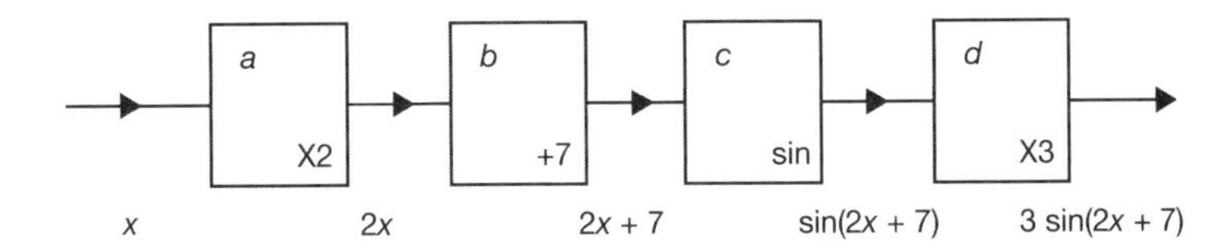

Let's keep going

---

25 Inverses of compositions

As has been stated before, the diagram of the inverse of a function can be drawn as the function with the information flowing through it in the reverse direction:

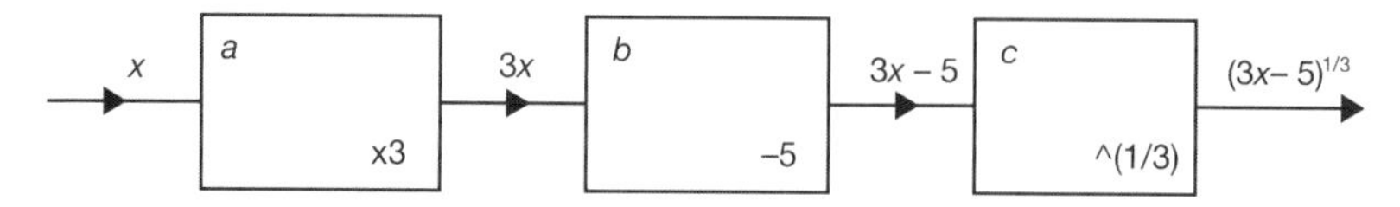

The diagram permits the inverse of a composition of functions to be found. For example, consider the function f with output $f(x) = (3x - 5)^{\frac{1}{3}}$. By decomposing f you find that:

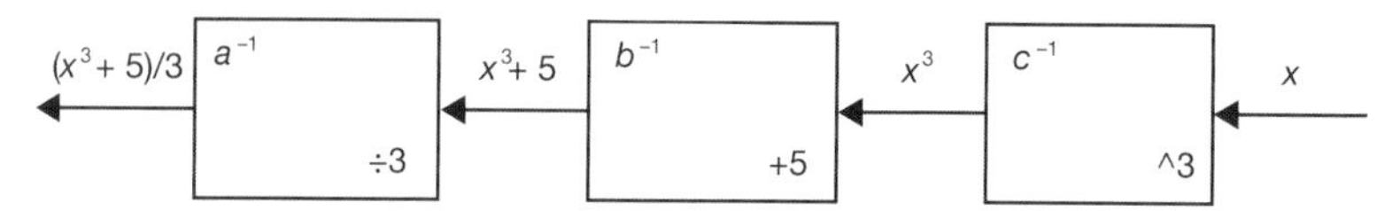

$f(x) = c(b[a(x)])$ where $a(x) = 3x$, $b(x) = x - 5$ and $c(x) = x^{\frac{1}{3}}$. Each of the three functions a, b and c has its respective inverse as:

$$a^{-1}(x) = \frac{x}{3}$$
$$b^{-1}(x) = x + 5$$
$$c^{-1}(x) = x^3$$

and from the diagram you can see that $f^{-1}(x) = a^{-1}(b^{-1}[c^{-1}(x)]) = (x^3 + 5)/3$.

Notice the *reversal* of the order of the components:

$$f(x) = c(b[a(x)]), \qquad f^{-1}(x) = a^{-1}(b^{-1}[c^{-1}(x)])$$

Now you try this one. Find the inverse of the function f with output $f(x) = \left(\dfrac{x+2}{4}\right)^5$.

Answer in the next frame

26

$$\boxed{f^{-1}(x) = 4x^{\frac{1}{5}} - 2}$$

Because

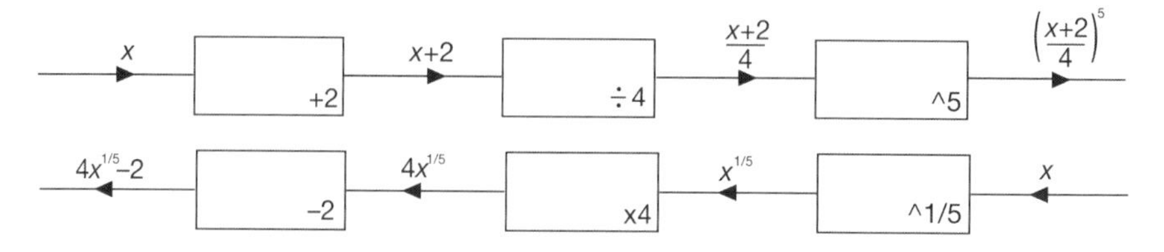

Now try the following exercises

27

1 Given that $a(x) = 4x$, $b(x) = x^2$, $c(x) = x - 5$ and $d(x) = \sqrt{x}$ find:

(a) $f(x) = a[b(c[d(x)])]$
(b) $f(x) = a(a[d(x)])$
(c) $f(x) = b[c(b[c(x)])]$

2 Given that $f(x) = (2x - 3)^3 - 3$, decompose f into its component functions and find its inverse. Is the inverse a function?

Take it steady – you will find the solutions in the next frame

28

1 (a) $f(x) = a[b(c[d(x)])] = 4(\sqrt{x} - 5)^2$
(b) $f(x) = a(a[d(x)]) = 16\sqrt{x}$
(c) $f(x) = b[c(b[c(x)])] = ((x - 5)^2 - 5)^2 = x^4 - 20x^3 + 140x^2 - 400x + 400$

2 $f = b \circ c \circ b \circ a$ so that $f(x) = b[c(b[a(x)])]$ where $a(x) = 2x, b(x) = x - 3$ and $c(x) = x^3$. The inverse is $f^{-1}(x) = a^{-1}[b^{-1}(c^{-1}[b^{-1}(x)])]$ so that:

$$f^{-1}(x) = a^{-1}[b^{-1}(c^{-1}[b^{-1}(x)])] = \frac{(x+3)^{\frac{1}{3}} + 3}{2} \text{ where } a^{-1}(x) = x/2,$$

$b^{-1}(x) = x + 3$ and $c^{-1}(x) = x^{\frac{1}{3}}$. The inverse is a function.

So far our work on functions has centered around *algebraic functions*. This is just one category of function. We shall now move on and consider other types of function and their specific properties.

Next frame

Trigonometric functions

Rotation

29

The trigonometric ratios are initially defined for the two acute angles in a right-angled triangle. These definitions can be extended to form *trigonometric functions* that are valid for *any* angle and yet retain all the properties of the original ratios. Start with the circle generated by the end point A of a straight line OA of unit length rotating anticlockwise about the end O as shown in the diagram over the page:

▶

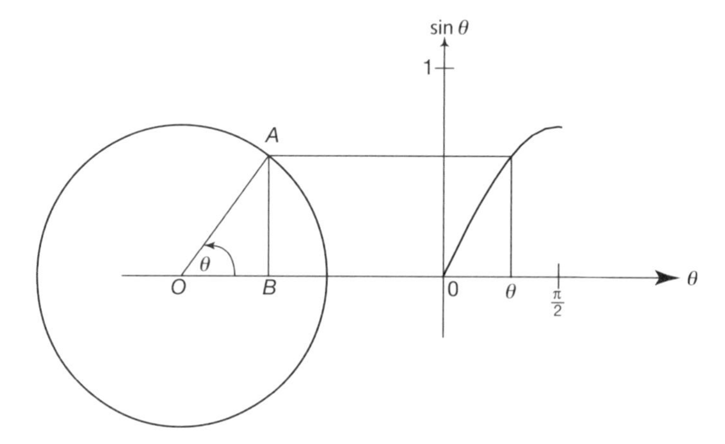

For angles θ where $0 < \theta < \pi/2$ radians you already know that:

$$\sin\theta = \frac{AB}{OA} = AB \text{ since } OA = 1$$

That is, the value of the trigonometric ratio $\sin\theta$ is equal to the height of A above B. The *sine* function with output $\sin\theta$ is now defined as the height of A above B *for any angle* θ $(0 \leq \theta < \infty)$.

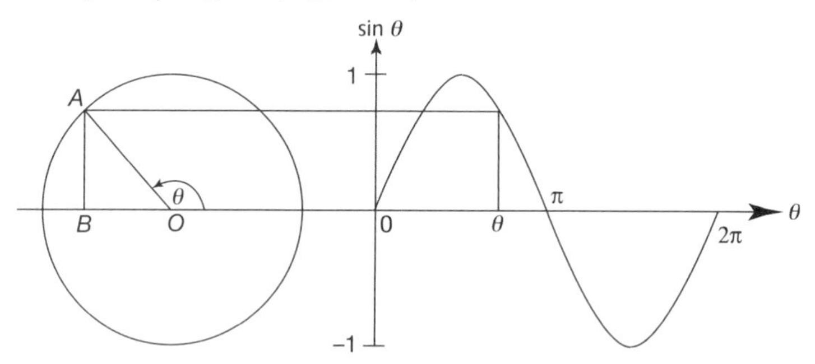

Notice that when A is *below* B the height is *negative*. The definition of the sine function can be further extended by taking into account negative angles, which represent a clockwise rotation of the line OA giving the complete graph of the sine function as in the diagram below:

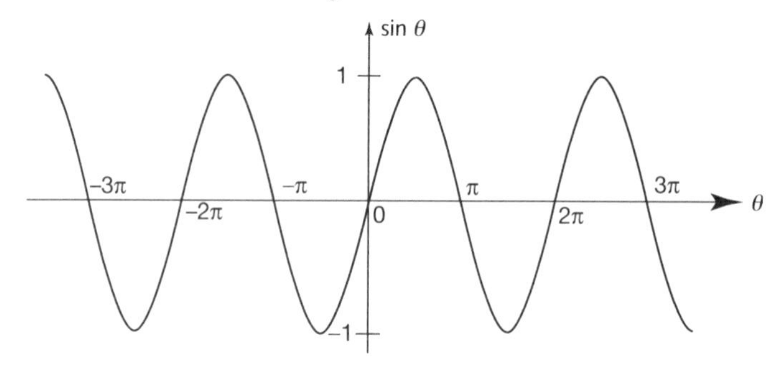

As you can see from this diagram, the value of $\sin\theta$ ranges from $+1$ to -1 depending upon the value of θ. You can reproduce this graph using a spreadsheet – in cells **A1** to **A21** enter the numbers -10 to 10 in steps of 1 and in cell **B1** enter the formula **=sin(A1)** and copy this into cells **B2** to **B21**. Use the *Chart Wizard* to draw the graph.

Just as before, you can use a calculator to find the values of the sine of an angle. So the sine of $153°$ is ....

Remember to put your calculator in degree mode

30

0.4540 to 4 dp

and the sine of $-\pi/4$ radians is ....

Remember to put your calculator in radian mode

31

-0.7071 to 4 dp

By the same reasoning, referring back to the first diagram in Frame 29, for angles θ where $0 < \theta < \pi/2$ radians:

$$\cos\theta = \frac{OB}{OA} = OB \text{ since } OA = 1$$

This time, the value of the trigonometric ratio $\cos\theta$ is equal to the distance from O to B. The *cosine* function with output $\cos\theta$ is now defined as the distance from O to B *for any angle* θ $(-\infty < \theta < \infty)$.

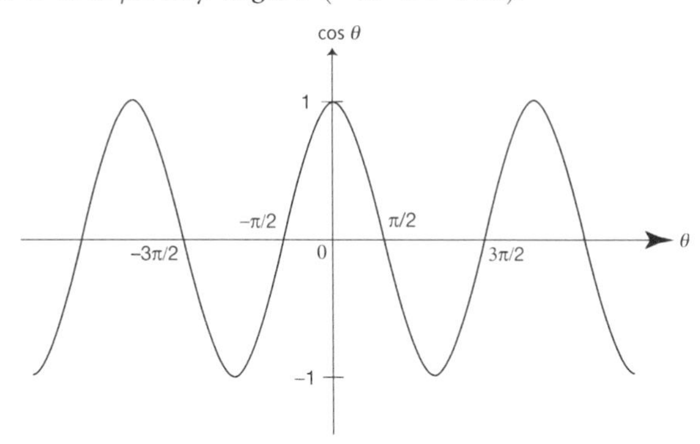

Notice that when B is to the left of O the distance from O to B is negative.

Again, you can reproduce this graph using a spreadsheet – in cells **A1** to **A21** enter the numbers -10 to 10 in steps of 1 and in cell **B1** enter the formula **=cos (A1)** and copy this into cells **B2** to **B21**. Use the *Chart Wizard* to draw the graph.

A calculator is used to find the values of the cosine of an angle. So the cosine of $-272°$ is ....

Remember to put your calculator in degree mode

32

0.0349 to 4 dp

and the cosine of $2\pi/3$ radians is ....

Remember to put your calculator in radian mode

33

-0.5

Now to put these two functions together

34 The tangent

The third basic trigonometric function, the tangent, is defined as the ratio of the sine to the cosine:

$$\tan\theta = \frac{\sin\theta}{\cos\theta}$$

Because $\cos\theta = 0$ whenever θ is an odd multiple of $\pi/2$, the tangent is not defined at these points. Instead the graph has vertical asymptotes as seen below:

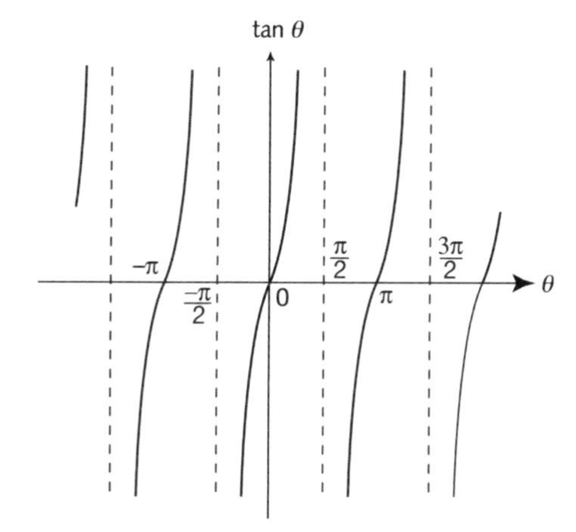

You can plot a single branch of the tangent function using your spreadsheet. Enter -1.5 in cell **A1** and **Edit-Fill-Series** down to **A21** with *step value* 0.15, then use the function **=tan (A1)** in cell **B1** and copy down to **B21**. Use the *Chart Wizard* to create the graph.

Make a note of the diagram in this frame and the diagrams in Frames 29 and 31. It is essential that you are able to draw sketch graphs of these functions.

For the sine and cosine functions the repeated sinusoidal wave pattern is easily remembered, all you then have to remember is that each rises and falls between $+1$ and -1 and crosses the horizontal axis:

(a) every whole multiple of π for the sine function

(b) every odd multiple of $\pi/2$ for the cosine function

▶

The repeated *branch* pattern of the tangent function is also easily remembered, all you then have to remember is that it rises from $-\infty$ to $+\infty$, crosses the horizontal axis every even multiple of $\pi/2$ and has a vertical asymptote every odd multiple of $\pi/2$.

Just as before, you can use a calculator to find the values of the tangent of an angle. So the tangent of 333° is ....

Remember to put your calculator in degree mode

35

-0.5095 to 4 dp

and the tangent of $-6\pi/5$ radians is ....

Remember to put your calculator in radian mode

36

-0.7265 to 4 dp

Now to look at some common properties of these trigonometric functions

---

37

Period

Any function whose output repeats itself over a regular interval of the input is called a *periodic function*, the regular interval of the input being called the *period* of the function. From the graphs of the trigonometric functions you can see that:

Both the sine and cosine functions repeat themselves every 2π radians so both are periodic with period 2π radians. The tangent function repeats itself every π radians so it is periodic with period π radians. Finding the periods of trigonometric functions with more involved outputs requires some manipulation. For example, to find the period of $\sin 3\theta$ note that:

$$\sin 3\theta = \sin(3\theta + 2\pi)$$

It is tempting to say that the period is, therefore, 2π but this is not the case because there is a smaller interval of θ over which the basic sinusoidal shape repeats itself:

$$\sin 3\theta = \sin(3\theta + 2\pi) = \sin 3\left(\theta + \frac{2\pi}{3}\right) \text{ so the period is } \frac{2\pi}{3}$$

$\sin 3\theta$ certainly repeats itself over 2π but within 2π the basic sinusoidal shape is repeated three times.

So the period of $\cos 4\theta$ is ....

Answer in the next frame

38

$\dfrac{\pi}{2}$

Because

$$\cos 4\theta = \cos(4\theta + 2\pi) = \cos 4\left(\theta + \frac{2\pi}{4}\right) = \cos 4\left(\theta + \frac{\pi}{2}\right)$$

And the period of $\tan 5\theta = \ldots\ldots\ldots\ldots$

Answer in the next frame

39

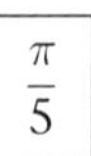

Because

$$\tan 5\theta = \tan(5\theta + \pi) = \tan 5(\theta + \pi/5)$$

Now, try another one. The period of $\sin(\theta/3) = \ldots\ldots\ldots\ldots$

Just follow the same procedure. The answer may surprise you

40

Because

$$\sin(\theta/3) = \sin(\theta/3 + 2\pi) = \sin\frac{1}{3}(\theta + 6\pi)$$

The answer is not 2π because the basic sinusoidal shape is only completed over the interval of 6π radians. If you are still not convinced of all this, use the spreadsheet to plot their graphs. Just one more before moving on.

The period of $\cos(\theta/2 + \pi/3) = \ldots\ldots\ldots\ldots$

Just follow the procedure

41

Because

$$\cos(\theta/2 + \pi/3) = \cos(\theta/2 + \pi/3 + 2\pi) = \cos\left(\frac{1}{2}[\theta + 4\pi] + \pi/3\right)$$

The $\pi/3$ has no effect on the period, it just shifts the basic sinusoidal shape $\pi/3$ radians to the left.

Move on

Amplitude

42

Every periodic function possesses an *amplitude* that is given as *the difference between the maximum value and the average value of the output taken over a single period*. For example, the average value of the output from the cosine function is zero (it ranges between +1 and −1) and the maximum value of the output is 1, so the amplitude is $1 - 0 = 1$.

So the amplitude of $4\cos(2\theta - 3)$ is . . . .

Next frame

43

4

Because

The maximum and minimum values of the cosine function are +1 and −1 respectively, so the output here ranges from +4 to −4 with an average of zero. The maximum value is 4 so that the amplitude is $4 - 0 = 4$.

Periodic functions are not always trigonometric functions. For example, the function with the graph shown in the diagram below is also periodic:

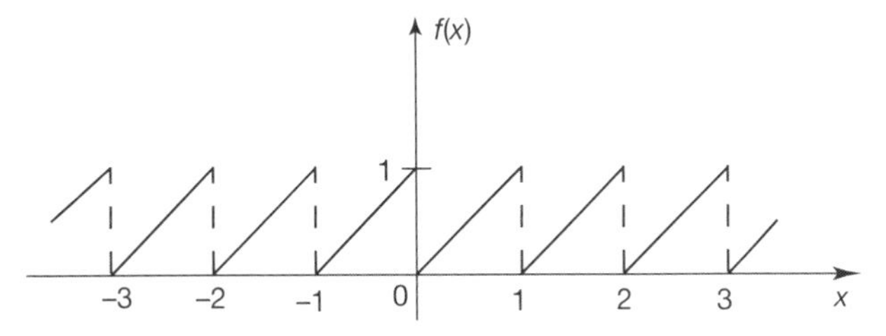

The straight line branch between $x = 0$ and $x = 1$ repeats itself indefinitely. For $0 \le x < 1$ the output from f is given as $f(x) = x$. The output from f for $1 \le x < 2$ matches the output for $0 \le x < 1$. That is:

$$f(x + 1) = f(x) \text{ for } 0 \le x < 1$$

So for example, $f(1.5) = f(0.5 + 1) = f(0.5) = 0.5$

The output from f for $2 \le x < 3$ also matches the output for $0 \le x < 1$. That is:

$$f(x + 2) = f(x) \text{ for } 0 \le x < 1$$

So that, for example, $f(2.5) = f(0.5 + 2) = f(0.5) = 0.5$

This means that we can give the prescription for the function as:

$$f(x) = x \text{ for } 0 \le x < 1$$

$$f(x + n) = f(x) \text{ for any integer } n$$

▶

For a periodic function of this type with period P where the first branch of the function is given for $a \leq x < a + P$ we can say that:

$$f(x) = \text{ some expression in } x \text{ for } a \leq x < a + P$$
$$f(x + nP) = f(x)$$

Because of its shape, the specific function we have considered is called a *sawtooth wave*.

The amplitude of this sawtooth wave is ....

Remember the definition of amplitude

44

$$\boxed{\frac{1}{2}}$$

Because

The amplitude is given as the *difference between the maximum value and the average value of the output taken over a single period*. Here the maximum value of the output is 1 and the average output is $\frac{1}{2}$, so the amplitude is $1 - 1/2 = 1/2$.

Next frame

45 Phase difference

The phase difference of a periodic function is the interval of the input by which the output leads or lags behind the *reference function*. For example, the plots of $y = \sin x$ and $y = \sin(x + \pi/4)$ on the same graph are shown below:

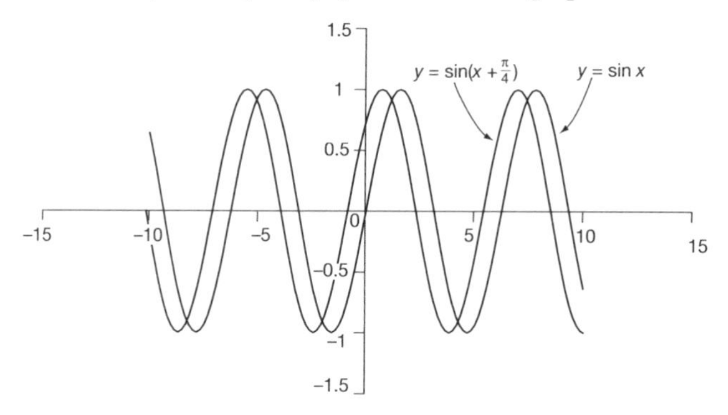

The diagram shows that $y = \sin(x + \pi/4)$ has the identical shape to $y = \sin x$ but is *leading* it by $\pi/4$ radians. It might appear to lag behind when you look at the diagram but it is, in fact, leading because when $x = 0$ then $\sin(x + \pi/4)$ already has the value $\sin \pi/4$, whereas $\sin x$ only has the value $\sin 0$. It is said that $y = \sin(x + \pi/4)$ leads with a *phase difference* of $\pi/4$ radians relative to the reference function $y = \sin x$. A function with a negative phase difference is said to *lag behind* the reference function. So that $y = \sin(x - \pi/4)$ lags behind $y = \sin x$ with a phase difference of $-\pi/4$.

So the phase difference of $y = \sin(x - \pi/6)$ relative to $y = \sin x$ is ....

Next frame

46

$-\pi/6$ radians

Because

The graph of $y = \sin(x - \pi/6)$ *lags behind* $y = \sin x$ by $\pi/6$ radians.

The phase difference of $y = \cos x$ relative to the reference function $y = \sin x$ is . . . .

Think how $\cos x$ *relates to* $\sin x$

47

$\pi/2$ radians

Because

$\cos x = \sin(x + \pi/2)$ and $\sin(x + \pi/2)$ *leads* $\sin x$ by $\pi/2$ radians.

Finally, try this. The phase difference of $y = \sin(3x + \pi/8)$ relative to $y = \sin 3x$ is . . . .

Take care to compare like with like – plot the graph if necessary

48

$\pi/24$ radians

Because

$\sin(3x + \pi/8) = \sin(3[x + \pi/24])$ and $\sin(3[x + \pi/24])$ *leads* $\sin 3x$ by $\pi/24$ radians.

Now for inverse trigonometric functions

49

Inverse trigonometric functions

If the graph of $y = \sin x$ is reflected in the line $y = x$, the graph of the inverse of the sine function is what results:

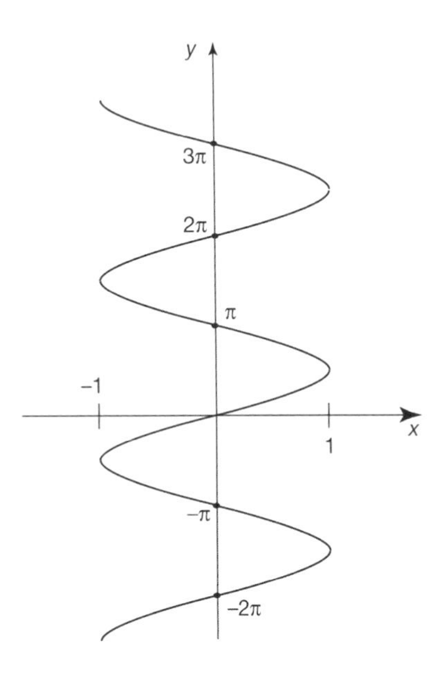

▶

However, as you can see, this is not a function because there is more than one value of y corresponding to a given value of x. If you cut off the upper and lower parts of the graph you obtain a single-valued function and it is this that is the *inverse sine function*:

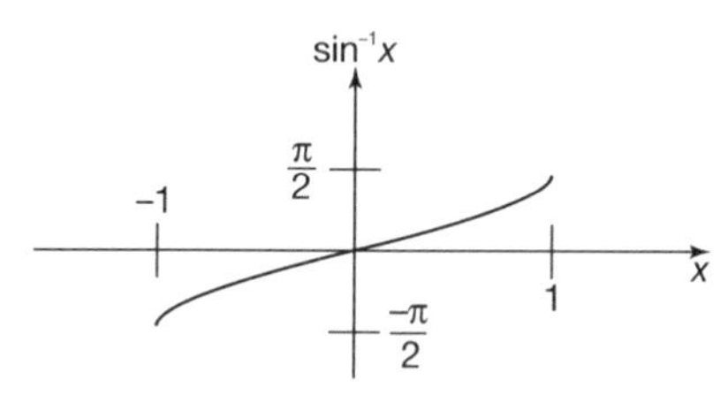

In a similar manner you can obtain the *inverse cosine function* and the *inverse tangent function*:

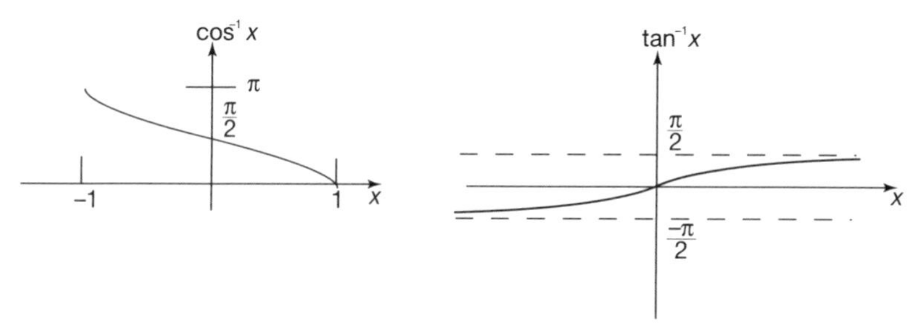

As in the case of the trigonometric functions, the values of these inverse functions are found using a calculator.

So that:

(a) $\sin^{-1}(0.5) = \ldots\ldots\ldots$

(b) $\tan^{-1}(-3.5) = \ldots\ldots\ldots$

(c) $\sec^{-1}(10) = \ldots\ldots\ldots \quad (\sec\theta = 1/\cos\theta)$

Next frame for the answer

50

(a) $30°$	(b) $-74.05°$	(c) $84.26°$

Because

(c) If $\sec^{-1}(10) = \theta$ then $\sec\theta = 10 = \dfrac{1}{\cos\theta}$ so that $\cos\theta = 0.1$ and $\theta = \cos^{-1}(0.1) = 84.26°$

So remember $\sec^{-1}\theta = \cos^{-1}\dfrac{1}{\theta}$.

Similar results are obtained for $\text{cosec}^{-1}\theta$ and $\cot^{-1}\theta$:

$$\text{cosec}^{-1}\theta = \sin^{-1}\frac{1}{\theta} \text{ and } \cot^{-1}\theta = \tan^{-1}\frac{1}{\theta}$$

Now to use these functions and their inverses to solve equations

Trigonometric equations

51

A simple trigonometric equation is one that involves just a single trigonometric expression. For example, the equation:

$\sin 3x = 0$ is a simple trigonometric equation

The solution of this equation can be found from inspecting the graph of the sine function $\sin\theta$ which crosses the θ-axis whenever θ is an integer multiple of π. That is, $\sin n\pi = 0$ where n is an integer. This means that the solutions to $\sin 3x = 0$ are found when:

$$3x = n\pi \text{ so that } x = \frac{n\pi}{3},\ n = 0,\ \pm 1,\ \pm 2,\ \ldots$$

So the values of x that satisfy the simple trigonometric equation:

$\cos 2x = 1$ are ....

Next frame

52

$$\boxed{x = n\pi,\ n = 0,\ \pm 1,\ \pm 2,\ \ldots}$$

Because

From the graph of the cosine function you can see that it rises to its maximum $\cos\theta = 1$ whenever θ is an even multiple of π, that is $\theta = 0,\ \pm 2\pi,\ \pm 4\pi$. Consequently, $\cos 2x = 1$ when $2x = 2n\pi$ so that $x = n\pi,\ n = 0,\ \pm 1,\ \pm 2,\ \ldots$

Just look at another. Consider the equation:

$$2\sin 3x = \sqrt{2}$$

This can be rewritten as $\sin 3x = \frac{\sqrt{2}}{2} = \frac{1}{\sqrt{2}}$. From what you know about the right-angled isosceles triangle, you can say that when $\theta = \frac{\pi}{4}$ then $\sin\theta = \frac{1}{\sqrt{2}}$. However if you look at the graph of the sine function you can see that between $\theta = 0$ and $\theta = 2\pi$ there are two values of θ where $\sin\theta = \frac{1}{\sqrt{2}}$, namely $\theta = \frac{\pi}{4}$ and $\frac{3\pi}{4}$.

▶

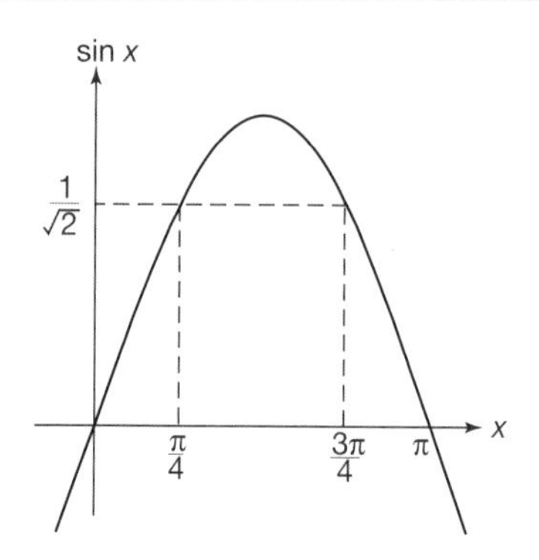

Consequently:

$$\sin 3x = \frac{1}{\sqrt{2}} \text{ when } 3x = \frac{\pi}{4} \pm 2n\pi \text{ and when } 3x = \frac{3\pi}{4} \pm 2n\pi,\ n = 0,\ \pm 1,\ \pm 2,\ \ldots$$

So the values of x that satisfy $2\sin 3x = \sqrt{2}$ are:

$$x = \frac{\pi}{12} \pm \frac{2n\pi}{3} \text{ and } x = \frac{\pi}{4} \pm \frac{2n\pi}{3} \text{ where } n = 0,\ \pm 1,\ \pm 2,\ \ldots$$

So, the values of x that satisfy $\cos 4x = \frac{1}{2}$ are ....

Next frame

53

$$\boxed{x = \frac{\pi}{12} \pm \frac{n\pi}{2},\text{ and } \frac{5\pi}{12} \pm \frac{n\pi}{2},\ n = 0,\ \pm 1,\ \pm 2,\ \ldots}$$

Because

From the graph of the cosine function you see that when $\cos\theta = \frac{1}{2}$

$$\theta = \frac{\pi}{3} \pm 2n\pi \text{ or } \theta = \frac{5\pi}{3} \pm 2n\pi \text{ where } n = 0,\ \pm 1,\ \pm 2,\ \ldots$$

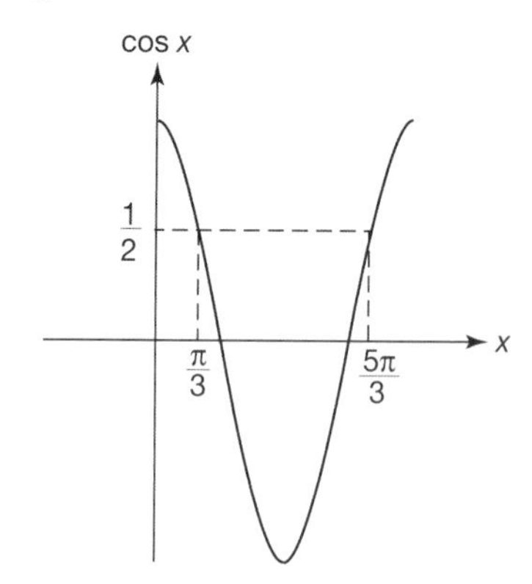

So that when $\cos 4x = \frac{1}{2}$, $4x = \frac{\pi}{3} \pm 2n\pi$ or $4x = \frac{5\pi}{3} \pm 2n\pi$. That is:

$$x = \frac{\pi}{12} \pm \frac{n\pi}{2} \text{ and } x = \frac{5\pi}{12} \pm \frac{n\pi}{2} \text{ where } n = 0,\ \pm 1,\ \pm 2,\ \ldots$$

54

Equations of the form $a\cos x + b\sin x = c$

A plot of $f(x) = a\cos x + b\sin x$ against x will produce a sinusoidal graph. Try it. Use your spreadsheet to plot:

$f(x) = 3\cos x + 4\sin x$ against x for $-10 \leq x \leq 10$ with *step value* 1. The result is shown below:

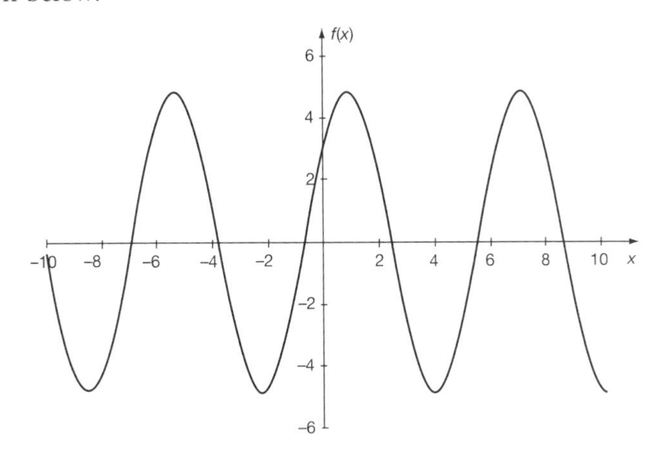

This sinusoidal shape possesses an amplitude and a phase, so its equation must be of the form:

$f(x) = R\sin(x + \theta)$ or $f(x) = R\cos(x + \phi)$

Either form will suffice – we shall select the first one to find solutions to the equation:

$3\cos x + 4\sin x = 5$

That is:

$R\sin(x + \theta) = 5$

The left-hand side can be expanded to give:

$R\sin\theta\cos x + R\cos\theta\sin x = 5$

Comparing this equation with the equation $3\cos x + 4\sin x = 5$ enables us to say that:

$3 = R\sin\theta$ and $4 = R\cos\theta$

Now:

$$R^2\sin^2\theta + R^2\cos^2\theta = R^2 = 3^2 + 4^2 = 25 = 5^2$$

so that $R = 5$. This means that $5\sin(\theta + x) = 5$ so that:

$\sin(\theta + x) = 1$ with solution $\theta + x = \dfrac{\pi}{2} \pm 2n\pi$

Thus:

$$x = \frac{\pi}{2} - \theta \pm 2n\pi$$

▶

Now, $\frac{R\sin\theta}{R\cos\theta} = \tan\theta = \frac{3}{4}$ so that $\theta = \tan^{-1}\left(\frac{3}{4}\right) = 0.64$ rad. This gives the solution to the original equation as:

$$x = \frac{\pi}{2} - 0.64 \pm 2n\pi = 0.93 \pm 2n\pi \text{ radians to 2 dp}$$

Notice that if we had assumed a form $f(x) = R\cos(x + \phi)$ the end result would have been the same.

55

Try this one yourself.

The solutions to the equation $\sin x - \sqrt{2}\cos x = 1$ are ....

Next frame for the answer

56

$$\boxed{x = 1.571 \pm 2n\pi \text{ radians}}$$

Because

Letting $R\sin(x + \theta) = 1$ we find, by expanding the sine, that:

$$R\sin x\cos\theta + R\sin\theta\cos x = 1$$

Comparing this equation with $\sin x - \sqrt{2}\cos x = 1$ enables us to say that:

$$R\cos\theta = 1 \text{ and } R\sin\theta = -\sqrt{2}$$

where $R^2\cos^2\theta + R^2\sin^2\theta = R^2 = 1^2 + \left(-\sqrt{2}\right)^2 = 3$ so that $R = \sqrt{3}$.

This means that $\sqrt{3}\sin(x + \theta) = 1$ so that $\sin(x + \theta) = \frac{1}{\sqrt{3}}$ giving:

$$x + \theta = \sin^{-1}\left(\frac{1}{\sqrt{3}}\right) = 0.6155 \pm 2n\pi \text{ radians}$$

That is, $x = 0.6155 \pm 2n\pi - \theta$ rad. Now:

$$\frac{R\sin\theta}{R\cos\theta} = \tan\theta = -\sqrt{2} \text{ so that } \theta = \tan^{-1}\left(-\sqrt{2}\right) = -0.9553 \text{ rad,}$$

giving the final solution as $x = 1.571 \pm 2n\pi$ radians.

At this point let us pause and summarize the main facts so far on trigonometric functions and equations

Summary

57

1. The definitions of the trigonometric ratios, valid for angles greater than 0° and less than 90°, can be extended to the trigonometric functions valid for any angle.
2. The trigonometric functions possess periods, amplitudes and phases.
3. The inverse trigonometric functions have restricted ranges.

 Review exercise

58

1 Use a calculator to find the value of each of the following (take care to ensure that your calculator is in the correct mode):

(a) $\sin(3\pi/4)$ (b) $\text{cosec}(-\pi/13)$ (c) $\tan(125^\circ)$
(d) $\cot(-30^\circ)$ (e) $\cos(-5\pi/7)$ (f) $\sec(18\pi/11)$

2 Find the period, amplitude and phase (in radians) of each of the following:

(a) $f(\theta) = 3\sin 9\theta$ (b) $f(\theta) = -7\cos(5\theta - 3)$
(c) $f(\theta) = \tan(2 - \theta)$ (d) $f(\theta) = -\cot(3\theta - 4)$

3 A function is defined by the following prescription:

$$f(x) = -x + 4, \quad 0 \le x < 3, \quad f(x + 3) = f(x)$$

Plot a graph of this function for $-9 \le x < 9$ and find:

(a) the period

(b) the amplitude

(c) the phase of $f(x) + 2$ with respect to $f(x)$

4 Solve the following trigonometric equations:

(a) $\tan 4x = 1$

(b) $\sin(x + 2\pi) + \sin(x - 2\pi) = \dfrac{1}{2}$

(c) $2\cot\theta + 3\cot\phi = 1.4$
$\cot\theta - \cot\phi = 0.2$

(d) $18\cos^2 x + 3\cos x - 1 = 0$

(e) $3\sin^2 x - \cos^2 x = \sin 2x$

(f) $\cos x + \sqrt{3}\sin x = \sqrt{2}$ for $0 \le x \le 2\pi$

59

1 Using your calculator you will find:

(a) 0.7071 (b) − 4.1786 (c) − 1.4281
(d) − 1.7321 (e) − 0.6235 (f) 2.4072

2 (a) $3\sin 9\theta = 3\sin(9\theta + 2\pi) = 3\sin 9(\theta + 2\pi/9)$ so the period of $f(\theta)$ is $2\pi/9$ and the phase is 0. The maximum value of $f(\theta)$ is 3 and the average value is 0, so the amplitude of $f(\theta)$ is 3.

(b) $-7\cos(5\theta - 3) = -7\cos(5\theta - 3 + 2\pi) = -7\sin 5(\theta + 2\pi/5 - 3/5)$ so the period of $f(\theta)$ is $2\pi/5$, the phase is $-3/5$ and the amplitude is 7.

(c) $\tan(2 - \theta) = \tan(2 - \theta + \pi) = \tan(-\theta + 2 + \pi)$ so the period of $f(\theta)$ is π and $f(\theta)$ leads $\tan(-\theta)$ by the phase 2 with an infinite amplitude.

(d) $-\cot(3\theta - 4) = \cot(-3\theta + 4 + \pi) = \cot 3(-\theta + 4/3 + \pi/3)$ so the period of $f(\theta)$ is $\pi/3$ and $f(\theta)$ leads $\cot(-3\theta)$ by the phase 4/3 with an infinite amplitude.

3 (a) 3 (b) 1.5 (c) 0

▶

4 (a) If $\tan\theta = 1$ then $\theta = \dfrac{\pi}{4} \pm n\pi$ radians. Since $\tan 4x = 1$ then $4x = \dfrac{\pi}{4} \pm n\pi$ so that $x = \dfrac{\pi}{16} \pm n\dfrac{\pi}{4}$.

(b)
$$\begin{aligned}\text{LHS} &= (\sin x\cos 2\pi + \sin 2\pi\cos x) + (\sin x\cos 2\pi - \sin 2\pi\cos x)\\ &= 2\sin x\cos 2\pi\\ &= 2\sin x \qquad \text{because } \cos 2\pi = 1\\ &= 1/2 \qquad \text{right-hand side}\end{aligned}$$

Therefore $\sin x = 1/4$ so $x = \sin^{-1}\left(\dfrac{1}{4}\right) = 0.2526 \pm 2n\pi$ radians.

(c) Multiplying the second equation by 3 yields:

$2\cot\theta + 3\cot\phi = 1.4$

$3\cot\theta - 3\cot\phi = 0.6$ adding yields $5\cot\theta = 2.0$ so that

$\theta = \cot^{-1}\left(\dfrac{2}{5}\right) = 1.1903 \pm n\pi$ radians.

Also, substituting $\cot\theta = 0.4$ into the first equation gives

$\cot\phi = \dfrac{1.4 - 0.8}{3} = 0.2$ so that $\phi = \cot^{-1}(0.2) = 1.3734 \pm n\pi$ radians.

(d) This equation factorizes as $(6\cos x - 1)(3\cos x + 1) = 0$ so that:

$\cos x = 1/6$ or $-1/3$. Thus $x = \pm 1.4033 \pm 2n\pi$ radians or

$x = \pm 1.9106 \pm 2n\pi$ radians.

(e) This equation can be written as $3\sin^2 x - \cos^2 x = 2\sin x\cos x$.

That is: $3\sin^2 x - 2\sin x\cos x - \cos^2 x = 0$.

That is $(3\sin x + \cos x)(\sin x - \cos x) = 0$

so that $3\sin x + \cos x = 0$ or $\sin x - \cos x = 0$.

If $3\sin x + \cos x = 0$ then $\tan x = -1/3$ and so $x = -0.3218 \pm n\pi$

and if $\sin x - \cos x = 0$ then $\tan x = 1$ and so $x = \pi/4 \pm n\pi$.

(f) To solve $\cos x + \sqrt{3}\sin x = \sqrt{2}$, write $R\sin\theta = 1$ and $R\cos\theta = \sqrt{3}$. The equation then becomes:

$$R\sin\theta\cos x + R\cos\theta\sin x = \sqrt{2}$$

That is:

$$R\sin(\theta + x) = \sqrt{2}$$

Now:

$$R^2\sin^2\theta + R^2\cos^2\theta = R^2 = 1^2 + (\sqrt{3})^2 = 4$$

so that $R = 2$. This means that $2\sin(\theta + x) = \sqrt{2}$ so that:

$$\sin(\theta + x) = \frac{1}{\sqrt{2}} \text{ with solution } \theta + x = \frac{\pi}{4} \pm 2n\pi \text{ or } \theta + x = \frac{3\pi}{4} \pm 2n\pi$$

Thus:

$$\theta = \frac{\pi}{4} - x \pm 2n\pi \text{ or } \theta = \frac{3\pi}{4} - x \pm 2n\pi$$

Now $\frac{R\sin\theta}{R\cos\theta} = \tan\theta = \frac{1}{\sqrt{3}}$ so that $\theta = \tan^{-1}\left(\frac{1}{\sqrt{3}}\right) = \frac{\pi}{6}$ rad. This gives the solution to the original equation as:

$x = \frac{\pi}{4} - \frac{\pi}{6} = \frac{\pi}{12}$ or $x = \frac{3\pi}{4} - \frac{\pi}{6} = \frac{7\pi}{12}$ within the range $0 \le x \le 2\pi$.

Exponential and logarithmic functions

Exponential functions

60

The exponential function is expressed by the equation:

$y = e^x$ or $y = \exp(x)$

where e is the exponential number 2.7182818... . The graph of this function lies entirely above the x-axis as does the graph of its reciprocal $y = e^{-x}$, as can be seen in the diagram:

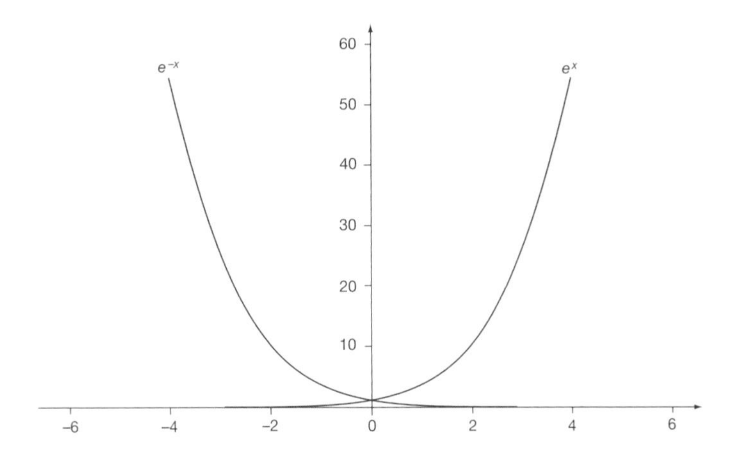

The value of e^x can be found to any level of precision desired from the series expansion:

$$e^x = 1 + x + \frac{x^2}{2!} + \frac{x^3}{3!} + \frac{x^4}{4!} + \dots$$

In practice a calculator is used. The general exponential function is given by:

$y = a^x$ where $a > 0$

and because $a = e^{\ln a}$ the general exponential function can be written in the form:

$y = e^{x \ln a}$

Because $\ln a < 1$ when $a < e$ you can see that the graph increases less quickly than the graph of e^x and if $a > e$ it grows faster.

▶

The inverse exponential function is the logarithmic function expressed by the equation:

$$y = \log_a x$$

with the graph shown in the diagram:

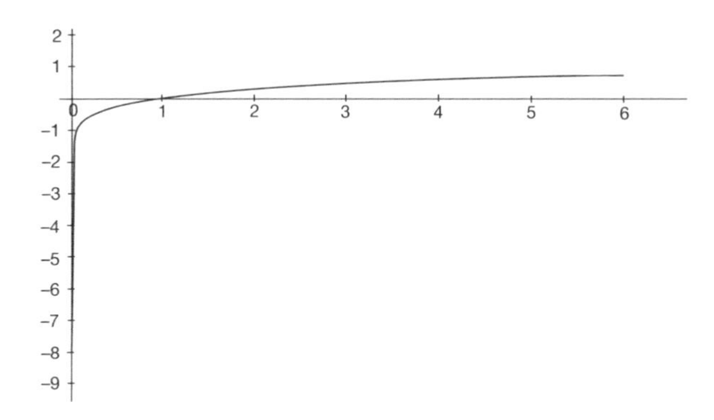

When the base a of the logarithmic function takes on the value of the exponential number e the notation $y = \ln x$ is used.

Indicial equations

An *indicial equation* is an equation where the variable appears as an index and the solution of such an equation requires the application of logarithms.

Example 1

Here is a simple case. We have to find the value of x, give that $12^{2x} = 35.4$.

Taking logs of both sides – and using $\log(A^n) = n \log A$ we have

$$(2x) \log 12 = \log 35.4$$

i.e. $$(2x)1.0792 = 1.5490$$

$$2.1584x = 1.5490$$

$$\therefore\ x = \frac{1.5490}{2.1584} = 0.71766$$

$$\therefore\ x = 0.7177 \text{ to 4 sig fig}$$

Example 2

Solve the equation $4^{3x-2} = 26^{x+1}$

The first line in this solution is ....

61

$$(3x-2)\log 4=(x+1)\log 26$$

$\therefore\ (3x-2)0.6021=(x+1)1.4150$

Multiplying out and collecting up, we eventually get

$x=$ .... to 4 sig fig

62

$$6.694$$

Because we have

$$\begin{aligned}(3x-2)0.6021&=(x+1)1.4150\\1.8063x-1.2042&=1.4150x+1.4150\\(1.8063-1.4150)x&=(1.4150+1.2042)\\0.3913x&=2.6192\\\therefore\ x&=\frac{2.6192}{0.3913}=6.6936\\\therefore\ x&=6.694\text{ to 4 sig fig}\end{aligned}$$

Care must be taken to apply the rules of logarithms rigidly.

Now we will deal with another example

63

Example 3

Solve the equation $5.4^{x+3}\times 8.2^{2x-1}=4.8^{3x}$

We recall that $\log(A\times B)=\log A+\log B$

Therefore, we have $\log\{5.4^{x+3}\}+\log\{8.2^{2x-1}\}=\log\{4.8^{3x}\}$

i.e. $(x+3)\log 5.4+(2x-1)\log 8.2=3x\log 4.8$

You can finish it off, finally getting

$x=$ .... to 4 sig fig

64

$$2.485$$

Here is the working as a check:

$$\begin{aligned}&(x+3)0.7324+(2x-1)0.9138=(3x)0.6812\\&0.7324x+2.1972+1.8276x-0.9138=2.0436x\\&(0.7324+1.8276)x+(2.1972-0.9138)=2.0436x\\&\therefore\ 2.5600x+1.2834=2.0436x\\&(2.5600-2.0436)x=-1.2834\\&0.5164x=-1.2834\\&x=-\frac{1.2834}{0.5164}=-2.4853\quad\therefore\ x=-2.485\text{ to 4 sig fig}\end{aligned}$$

Finally, here is one to do all on your own. ▶

Example 4

Solve the equation $7(14.3^{x+5}) \times 6.4^{2x} = 294$

Work right through it, giving the result to 4 sig fig.

$x = \ldots\ldots\ldots\ldots$

65

$\boxed{-1.501}$

Check the working:

$$7(14.3^{x+5}) \times 6.4^{2x} = 294$$

$$\therefore \log 7 + (x+5)\log 14.3 + (2x)\log 6.4 = \log 294$$

$$0.8451 + (x+5)1.1553 + (2x)0.8062 = 2.4683$$

$$0.8451 + 1.1553x + 5.7765 + 1.6124x = 2.4683$$

$$(1.1553 + 1.6124)x + (0.8451 + 5.7765) = 2.4683$$

$$2.7677x + 6.6216 = 2.4683$$

$$2.7677x = 2.4683 - 6.6216 = -4.1533$$

$$\therefore x = -\frac{4.1533}{2.7677} = -1.5006$$

$$x = -1.501 \text{ to 4 sig fig}$$

66

Some indicial equations may need a little manipulation before the rules of logarithms can be applied.

Example 5

Solve the equation $2^{2x} - 6 \times 2^x + 8 = 0$

Because $2^{2x} = (2^x)^2$ this is an equation that is quadratic in 2^x. We can, therefore, write $y = 2^x$ and substitute y for 2^x in the equation to give:

$$y^2 - 6y + 8 = 0$$

which factorizes to give:

$$(y-2)(y-4) = 0$$

so that $y = 2$ or $y = 4$. That is $2^x = 2$ or $2^x = 4$ so that $x = 1$ or $x = 2$.

Try this one. Solve $2 \times 3^{2x} - 6 \times 3^x + 4 = 0$.

The answer is in Frame 67

67

$$x = 0.631 \text{ to 3 dp or } x = 0$$

Because $3^{2x} = (3^x)^2$ this is an equation that is quadratic in 3^x. We can, therefore, write $y = 3^x$ and substitute y for 3^x in the equation to give:

$$2y^2 - 6y + 4 = 0$$

which factorizes to give:

$$(2y - 4)(y - 1) = 0$$

so that $y = 2$ or $y = 1$. That is $3^x = 2$ or $3^x = 1$ so that $x \log 3 = \log 2$ or $x = 0$. That is $x = \dfrac{\log 2}{\log 3} = 0.631$ to 3 dp or $x = 0$.

And now to the final topic of Part I

Odd and even functions

68

If, by replacing x by $-x$ in $f(x)$ the expression does not change its value, f is called an *even* function. For example, if:

$f(x) = x^2$ then $f(-x) = (-x)^2 = x^2 = f(x)$ so that f is an even function.

On the other hand, if $f(-x) = -f(x)$ then f is called an *odd* function. For example, if:

$f(x) = x^3$ then $f(-x) = (-x)^3 = -x^3 = -f(x)$ so that f is an odd function .

Because $\sin(-\theta) = -\sin\theta$ the sine function is an odd function and because $\cos(-\theta) = \cos\theta$ the cosine function is an even function. Notice how the graph of the cosine function is reflection symmetric about the vertical axis through $\theta = 0$ in the diagram in Frame 31. All even functions possess this type of symmetry. The graph of the sine function is rotation symmetric about the origin as it goes into itself under a rotation of 180° about this point as can be seen from the third diagram in Frame 29. All odd functions possess this type of antisymmetry. Notice also that $\tan(-\theta) = -\tan\theta$ so that the tangent function, like the sine function, is odd and has an antisymmetric graph (see the diagram in Frame 34).

69 Odd and even parts

Not every function is either even or odd but many can be written as the sum of an even part and an odd part. If, given $f(x)$ where $f(-x)$ is also defined then:

$$f_e(x) = \frac{f(x) + f(-x)}{2} \text{ is even and } f_o(x) = \frac{f(x) - f(-x)}{2} \text{ is odd.}$$

Furthermore $f_e(x)$ is called the *even part* of $f(x)$ and

$f_o(x)$ is called the *odd part* of $f(x)$.

For example, if $f(x) = 3x^2 - 2x + 1$ then

$$f(-x) = 3(-x)^2 - 2(-x) + 1 = 3x^2 + 2x + 1$$

so that the even and odd parts of $f(x)$ are:

$$f_e(x) = \frac{(3x^2 - 2x + 1) + (3x^2 + 2x + 1)}{2} = 3x^2 + 1 \text{ and}$$

$$f_o(x) = \frac{(3x^2 - 2x + 1) - (3x^2 + 2x + 1)}{2} = -2x$$

So, the even and odd parts of $f(x) = x^3 - 2x^2 - 3x + 4$ are . . . .

Apply the two formulas; the answer is in the next frame

70

$$f_e(x) = -2x^2 + 4$$
$$f_o(x) = x^3 - 3x$$

Because

$$\begin{aligned} f_e(x) &= \frac{f(x) + f(-x)}{2} \\ &= \frac{(x^3 - 2x^2 - 3x + 4) + ((-x)^3 - 2(-x)^2 - 3(-x) + 4)}{2} \\ &= \frac{x^3 - 2x^2 - 3x + 4 - x^3 - 2x^2 + 3x + 4}{2} = \frac{-2x^2 + 4 - 2x^2 + 4}{2} \\ &= -2x^2 + 4 \end{aligned}$$

and

$$\begin{aligned} f_o(x) &= \frac{f(x) - f(-x)}{2} \\ &= \frac{(x^3 - 2x^2 - 3x + 4) - ((-x)^3 - 2(-x)^2 - 3(-x) + 4)}{2} \\ &= \frac{x^3 - 2x^2 - 3x + 4 + x^3 + 2x^2 - 3x - 4}{2} = \frac{x^3 - 3x + x^3 - 3x}{2} \\ &= x^3 - 3x \end{aligned}$$

Make a note here that *even polynomial functions consist of only even powers* and *odd polynomial functions consist of only odd powers.*

So the odd and even parts of $f(x) = x^3(x^2 - 3x + 5)$ are . . . .

Answer in next frame

71

$$f_e(x) = -3x^4 \qquad f_o(x) = x^5 + 5x^3$$

Because

Even polynomial functions consist of only even powers and odd polynomial functions consist of only odd powers. Consequently, the even part of $f(x)$ consists of even powers only and the odd part of $f(x)$ consists of odd powers only.

Now try this. The even and odd parts of $f(x) = \dfrac{1}{x-1}$ are ....

Next frame

72

$$f_e(x) = \frac{1}{(x-1)(x+1)}$$
$$f_o(x) = \frac{x}{(x-1)(x+1)}$$

Because

$$\begin{aligned}
f_e(x) &= \frac{f(x)+f(-x)}{2} \\
&= \frac{1}{2}\left(\frac{1}{(x-1)} + \frac{1}{(-x-1)}\right) = \frac{1}{2}\left(\frac{1}{(x-1)} - \frac{1}{(x+1)}\right) \\
&= \frac{1}{2}\frac{x+1-(x-1)}{(x-1)(x+1)} = \frac{1}{2}\frac{2}{(x-1)(x+1)} \\
&= \frac{1}{(x-1)(x+1)}
\end{aligned}$$

and

$$\begin{aligned}
f_o(x) &= \frac{f(x)-f(-x)}{2} \\
&= \frac{1}{2}\left(\frac{1}{(x-1)} - \frac{1}{(-x-1)}\right) = \frac{1}{2}\left(\frac{1}{(x-1)} + \frac{1}{(x+1)}\right) \\
&= \frac{1}{2}\frac{x+1+x-1}{(x-1)(x+1)} = \frac{1}{2}\frac{2x}{(x-1)(x+1)} \\
&= \frac{x}{(x-1)(x+1)}
\end{aligned}$$

Next frame

73 Odd and even parts of the exponential function

The exponential function is neither odd nor even but it can be written as a sum of an odd part and an even part.

That is, $\exp_e(x) = \dfrac{\exp(x) + \exp(-x)}{2}$ and $\exp_o(x) = \dfrac{\exp(x) - \exp(-x)}{2}$. These two functions are known as the *hyperbolic cosine* and the *hyperbolic sine* respectively:

$$\cosh x = \frac{e^x + e^{-x}}{2} \quad \text{and} \quad \sinh x = \frac{e^x - e^{-x}}{2}$$

Using these two functions the hyperbolic tangent can also be defined:

$$\tanh x = \frac{e^x - e^{-x}}{e^x + e^{-x}}$$

No more will be said about these hyperbolic trigonometric functions here. Instead they will be looked at in more detail in Program 4.

The logarithmic function $y = \log_a x$ is neither odd nor even and indeed does not possess even and odd parts because $\log_a(-x)$ is not defined.

74 Limits of functions

There are times when a function is not defined for a particular value of x, say $x = x_0$ but it is defined for values of x that are arbitrarily close to x_0. For example, the expression:

$$f(x) = \frac{x^2 - 1}{x - 1}$$

is not defined when $x = 1$ because at that point the denominator is zero and division by zero is not defined. However, we note that:

$$f(x) = \frac{x^2 - 1}{x - 1} = \frac{(x-1)(x+1)}{x - 1} = x + 1 \text{ provided that } x \neq 1$$

We still cannot permit the value $x = 1$ because to do so would mean that the cancellation of the $x - 1$ factor would be division by zero. But we can say that as the value of x approaches 1, the value of $f(x)$ approaches 2. Clearly the value of $f(x)$ never actually attains the value of 2 but it does get as close to it as you wish it to be by selecting a value of x sufficiently close to 1. We say that the limit of $\dfrac{x^2 - 1}{x - 1}$ as x approaches 1 is 2 and we write:

$$\operatorname*{Lim}_{x \to 1}\left(\frac{x^2 - 1}{x - 1}\right) = 2$$

You try one, $\operatorname*{Lim}_{x \to -3}\left(\dfrac{x^2 - 9}{x + 3}\right) = \ldots\ldots\ldots\ldots$

The answer is in the next frame

75

$$\boxed{-6}$$

Because

$$\begin{aligned} f(x) &= \frac{x^2 - 9}{x + 3} \\ &= \frac{(x + 3)(x - 3)}{x + 3} \\ &= x - 3 \text{ provided } x \neq -3 \end{aligned}$$

So that:

$$\underset{x \to -3}{Lim}\left(\frac{x^2 - 9}{x + 3}\right) = -6$$

The rules of limits

Listed here are a number of simple rules used for evaluating limits. They are what you would expect with no surprises.

If $\underset{x \to x_0}{Lim} f(x) = A$ and $\underset{x \to x_0}{Lim} g(x) = B$ then:

$$\underset{x \to x_0}{Lim}\,(f(x) \pm g(x)) = \underset{x \to x_0}{Lim} f(x) \pm \underset{x \to x_0}{Lim} g(x) = A \pm B$$

$$\underset{x \to x_0}{Lim}\,(f(x)g(x)) = \underset{x \to x_0}{Lim} f(x)\,\underset{x \to x_0}{Lim} g(x) = AB$$

$$\underset{x \to x_0}{Lim}\frac{f(x)}{g(x)} = \frac{\underset{x \to x_0}{Lim} f(x)}{\underset{x \to x_0}{Lim} g(x)} = \frac{A}{B} \text{ provided } \underset{x \to x_0}{Lim} g(x) = B \neq 0$$

$$\underset{x \to x_0}{Lim} f(g(x)) = f\left(\underset{x \to x_0}{Lim} g(x)\right) = f(B) \text{ provided } f(g(x)) \text{ is continuous at } x_0$$

So you try these for yourself:

(a) $\underset{x \to \pi}{Lim}\,(x^2 - \sin x)$

(b) $\underset{x \to \pi}{Lim}\,(x^2 \sin x)$

(c) $\underset{x \to \pi/4}{Lim} \dfrac{(\tan x)}{(\sin x)}$

(d) $\underset{x \to 1}{Lim} \cos(x^2 - 1)$

The answers are in the next frame

76

(a) π^2	(b) 0	(c) $\sqrt{2}$	(d) 1

Because:

(a) $\underset{x\to\pi}{Lim}\,(x^2 - \sin x) = \underset{x\to\pi}{Lim}\, x^2 - \underset{x\to\pi}{Lim}\sin x = \pi^2 - 0 = \pi^2$

(b) $\underset{x\to\pi}{Lim}\,(x^2 \sin x) = \underset{x\to\pi}{Lim}\,(x^2)\,\underset{x\to\pi}{Lim}\,(\sin x) = \pi^2 \times 0 = 0$

(c) $\underset{x\to\pi/4}{Lim}\,\dfrac{(\tan x)}{(\sin x)} = \dfrac{\underset{x\to\pi/4}{Lim}\,\tan x}{\underset{x\to\pi/4}{Lim}\,\sin x} = \dfrac{1}{1/\sqrt{2}} = \sqrt{2}$

(d) $\underset{x\to 1}{Lim}\cos(x^2 - 1) = \cos\left(\underset{x\to 1}{Lim}\,(x^2 - 1)\right) = \cos 0 = 1$

All fairly straightforward for these simple problems. Difficulties do occur for the limits of quotients when both the numerator and the denominator are simultaneously zero at the limit point but those problems we shall leave for later.

At this point let us pause and summarize the main facts so far on exponential and logarithmic functions as well as on odd and even functions

Summary

77

1 Any function of the form $f(x) = a^x$ is called an *exponential* function.

2 The exponential function $f(x) = a^x$ and the logarithmic function $g(x) = \log_a x$ are *mutual inverses*: $f^{-1}(x) = g(x)$ and $g^{-1}(x) = f(x)$.

3 If $f(-x) = f(x)$ then f is called an *even* function and if $f(-x) = -f(x)$ then f is called an *odd* function.

4 If $f_e(x) = \dfrac{f(x) + f(-x)}{2}$ can be defined it if called the *even* part of $f(x)$ and if $f_o(x) = \dfrac{f(x) - f(-x)}{2}$ can be defined it is called the *odd* part of $f(x)$.

5 The limit of a sum (difference) is equal to the sum (difference) of the limits.
The limit of a product is equal to the product of the limits.
The limit of a quotient is equal to the quotient of the limits provided the denominator limit is not zero.

Review exercise

78

1 Solve the following indicial equations giving the results to 4 dp:

(a) $13^{3x} = 8.4$ (b) $2.8^{2x+1} \times 9.4^{2x-1} = 6.3^{4x}$

(c) $7^{2x} - 9 \times 7^x + 14 = 0$

2 Evaluate (a) $\lim_{x \to -1} \left(\dfrac{x^2 + 2x + 1}{x^2 + 3x + 2} \right)$ (b) $\lim_{x \to \pi/4} (3x - \tan x)$

(c) $\lim_{x \to \pi/2} (2x^2 \cos[3x - \pi/2])$

79

1 (a) $13^{3x} = 8.4$

Taking logs of both sides gives: $3x \log 13 = \log 8.4$ so that:

$$x = \frac{\log 8.4}{3 \log 13} = \frac{0.9242\ldots}{3.3418\ldots} = 0.277 \text{ to 3 dp}$$

(b) $2.8^{2x+1} \times 9.4^{2x-1} = 6.3^{4x}$

Taking logs:

$$(2x + 1) \log 2.8 + (2x - 1) \log 9.4 = 4x \log 6.3$$

Factorizing x gives:

$$x(2 \log 2.8 + 2 \log 9.4 - 4 \log 6.3) = \log 9.4 - \log 2.8$$

So that:

$$x = \frac{\log 9.4 - \log 2.8}{2 \log 2.8 + 2 \log 9.4 - 4 \log 6.3}$$
$$= \frac{0.9731\ldots - 0.4471\ldots}{2(0.4471\ldots) + 2(0.9731\ldots) - 4(0.7993\ldots)} = -1.474 \text{ to 3 dp}$$

(c) $7^{2x} + 9 \times 7^x - 14 = 0$

This equation is quadratic in 7^x so let $y = 7^x$ and rewrite the equation as:

$y^2 - 9y + 14 = 0$ which factorizes to $(y - 2)(y - 7) = 0$
with solution: $y = 2$ or $y = 7$ that is $7^x = 2$ or $7^x = 7$

so that $x = \dfrac{\log 2}{\log 7} = 0.356$ to 3 dp or $x = 1$

▶

2 (a) $\lim\limits_{x\to -1}\left(\dfrac{x^2+2x+1}{x^2+3x+2}\right) = \lim\limits_{x\to -1}\left(\dfrac{(x+1)^2}{(x+1)(x+2)}\right)$

$= \lim\limits_{x\to -1}\left(\dfrac{x+1}{x+2}\right)$ The cancellation is permitted since $x \neq -1$

$= \dfrac{0}{1} = 0$

(b) $\lim\limits_{x\to \pi/4}(3x - \tan x) = \lim\limits_{x\to \pi/4}(3x) - \lim\limits_{x\to \pi/4}(\tan x)$

$= \dfrac{3\pi}{4} - \tan\dfrac{\pi}{4} = \dfrac{3\pi}{4} - 1$

(c) $\lim\limits_{x\to \pi/2}\left(2x^2\cos[3x - \pi/2]\right) = \lim\limits_{x\to \pi/2}\left(2x^2\right)\lim\limits_{x\to \pi/2}(\cos[3x - \pi/2])$

$= \dfrac{2\pi^2}{4} \times \cos \lim\limits_{x\to \pi/2}[3x - \pi/2]$

$= \dfrac{2\pi^2}{4} \times \cos[3\pi/2 - \pi/2]$

$= \dfrac{2\pi^2}{4} \times \cos\pi = -\dfrac{2\pi^2}{4}$

80

You have now come to the end of this Program. A list of **Can You?** questions follows for you to gauge your understanding of the material in the Program. These questions match the **Learning outcomes** listed at the beginning of the Program so try the **Test exercise**. *Work through the questions at your own pace, there is no need to hurry*. A set of **Further problems** provides additional valuable practice.

Can You?

Checklist 1

81

Check this list before and after you try the end of Program test.

On a scale of 1 to 5 how confident are you that you can: **Frames**

- Identify a function as a rule and recognize rules that are not functions? 1 to 4
  Yes ☐ ☐ ☐ ☐ ☐ *No*
- Determine the domain and range of a function? 4 to 7
  Yes ☐ ☐ ☐ ☐ ☐ *No*
- Construct the inverse of a function and draw its graph? 8 to 16
  Yes ☐ ☐ ☐ ☐ ☐ *No*
- Construct compositions of functions and de-construct them into their component functions? 20 to 28
  Yes ☐ ☐ ☐ ☐ ☐ *No*
- Develop the trigonometric functions from the trigonometric ratios? 29 to 36
  Yes ☐ ☐ ☐ ☐ ☐ *No*
- Find the period, amplitude and phase of a periodic function? 37 to 48
  Yes ☐ ☐ ☐ ☐ ☐ *No*
- Distinguish between the inverse of a trigonometric function and the inverse trigonometric function? 49 to 50
  Yes ☐ ☐ ☐ ☐ ☐ *No*
- Solve trigonometric equations using the inverse trigonometric functions? 51 to 56
  Yes ☐ ☐ ☐ ☐ ☐ *No*
- Recognize that the exponential function and the natural logarithmic function are mutual inverses and solve indicial and logarithmic equations? 60 to 67
  Yes ☐ ☐ ☐ ☐ ☐ *No*
- Find the even and odd parts of a function when they exist? 68 to 72
  Yes ☐ ☐ ☐ ☐ ☐ *No*
- Construct the hyperbolic functions from the odd and even parts of the exponential function? 73
  Yes ☐ ☐ ☐ ☐ ☐ *No*
- Evaluate limits of simple functions? 74 to 76
  Yes ☐ ☐ ☐ ☐ ☐ *No*

Test exercise 1

82

1 Which of the following equations expresses a rule that is a function?

(a) $y = -x^{\frac{3}{2}}$ (b) $y = x^2 + x + 1$ (c) $y = (\sqrt{x})^3$

2 Given the two functions f and g expressed by:

$f(x) = \dfrac{1}{x-2}$ for $2 < x \leq 4$ and $g(x) = x - 1$ for $0 \leq x < 3$

find the domain and range of functions h and k where:

(a) $h(x) = 2f(x) - 3g(x)$ (b) $k(x) = -\dfrac{3f(x)}{5g(x)}$

3 Use your spreadsheet to draw each of the following and their inverses. Is the inverse a function?

(a) $y = 3x^4$ (b) $y = -x^3$ (c) $y = 3 - x^2$

4 Given that $a(x) = 5x$, $b(x) = x^4$, $c(x) = x + 3$ and $d(x) = \sqrt{x}$ find:

(a) $f(x) = a[b(c[d(x)])]$ (b) $f(x) = a(a[d(x)])$

(c) $f(x) = b[c(b[c(x)])]$

5 Given that $f(x) = (5x - 4)^3 - 4$ decompose f into its component functions and find its inverse. Is the inverse a function?

6 Use a calculator to find the value of each of the following:

(a) $\sin(-320^\circ)$ (b) $\text{cosec}(\pi/11)$ (c) $\cot(-\pi/2)$

7 Find the period, amplitude and phase of each of the following:

(a) $f(\theta) = 4\cos 7\theta$ (b) $f(\theta) = -2\sin(2\theta - \pi/2)$

(c) $f(\theta) = \sec(3\theta + 4)$

8 A function is defined by the following prescription:

$f(x) = 9 - x^2, \quad 0 \leq x < 3$

$f(x + 3) = f(x)$

Plot a graph of this function for $-6 \leq x \leq 6$ and find:

(a) the period

(b) the amplitude

(c) the phase of $f(x) + 2$ with respect to $f(x)$

9 Solve the following trigonometric equations:

(a) $\tan(x + \pi) + \tan(x - \pi) = 0$

(b) $6\sin^2 x - 3\sin^2 2x + \cos^2 x = 0$

(c) $3\sin x - 5\cos x = 2$

10 Find the value of x corresponding to each of the following:

(a) $4^{-x} = 1$ (b) $\exp(3x) = e$

(c) $e^x = 54.32$ (d) $\log_5 x = 4$

(e) $\log_{10}(x - 3) = 0.101$ (f) $\ln 100 = x$

▶

11 Solve for x:

(a) $4^{x+2}5^{x+1} = 32{,}000$ (b) $e^{2x} - 5e^x + 6 = 0$
(c) $\log_x 36 = 2$ (d) $\frac{1}{2}\log(x^2) = 5\log 2 - 4\log 3$
(e) $\ln(x^{1/2}) = \ln x + \ln 3$

12 Find the even and odd parts of $f(x) = a^x$.

13 Evaluate:

(a) $\underset{x\to 4}{Lim}\left(\dfrac{x^2 - 8x + 16}{x^2 - 5x + 4}\right)$ (b) $\underset{x\to 1}{Lim}\,(6x^{-1} + \ln x)$

(c) $\underset{x\to 1/2}{Lim}\,[(3x^3 - 1)/(4 - 2x^2)]$ (d) $\underset{x\to -3}{Lim}\,\dfrac{(1 + 9x^{-2})}{(3x - x^2)}$

(e) $\underset{x\to 1}{Lim}\left(\tan^{-1}\left[\dfrac{x^2 - 1}{2(x - 1)}\right]\right)$

Further problems 1

83

1 Do the graphs of $f(x) = 3\log x$ and $g(x) = \log(3x)$ intersect?

2 Let $f(x) = \ln\left(\dfrac{1 + x}{1 - x}\right)$.

(a) Find the domain and range of f.

(b) Show that the new function g formed by replacing x in $f(x)$ by $\dfrac{2x}{1 + x^2}$ is given by $g(x) = 2f(x)$.

3 Describe the graph of $x^2 - 9y^2 = 0$.

4 Two functions C and S are defined as the even and odd parts of f where $f(x) = a^x$. Show that:

(a) $[C(x)]^2 - [S(x)]^2 = 1$
(b) $S(2x) = 2S(x)C(x)$

5 Show by using a diagram that, for functions f and g, $(f \circ g)^{-1} = g^{-1} \circ f^{-1}$.

6 Is it possible to find a value of x such that $\log_a(x) = a^x$ for $a > 1$?

7 Given the three functions a, b and c where $a(x) = 6x$, $b(x) = x - 2$ and $c(x) = x^3$ find the inverse of:

(a) $f(x) = a(b[c(x)])$ (b) $f = c \circ b \circ c$ (c) $f = b \circ c \circ a \circ b \circ c$

8 Use your spreadsheet to plot $\sin\theta$ and $\sin 2\theta$ on the same graph. Plot from $\theta = -5$ to $\theta = +5$ with a step value of 0.5.

9 The square sine wave with period 2 is given by the prescription:

$$f(x) = \begin{cases} 1 & 0 \le x < 1 \\ -1 & 1 \le x < 2 \end{cases}$$

▶

Plot this wave for $-4 \leq x \leq 4$ on a sheet of graph paper.

10 The absolute value of x is given as:

$$|x| = \begin{cases} x & \text{if } x \geq 0 \\ -x & \text{if } x < 0 \end{cases}$$

(a) Plot the graph of $y = |x|$ for $-2 \leq x \leq 2$.

(b) Find the derivative of y.

(c) Does the derivative exist at $x = 0$?

11 Use the spreadsheet to plot the rectified sine wave $f(x) = |\sin x|$ for $-10 \leq x \leq 10$ with step value 1.

12 Use your spreadsheet to plot $f(x) = \dfrac{\sin x}{x}$ for $-40 \leq x \leq 40$ with step value 4. (You will have to enter the value $f(0) = 1$ specifically in cell **B11**.)

13 Solve the following giving the results to 4 sig fig:

(a) $6\{8^{3x+2}\} = 5^{2x-7}$

(b) $4.5^{1-2x} \times 6.2^{3x+4} = 12.7^{5x}$

(c) $5\{17.2^{x+4}\} \times 3\{8.6^{2x}\} = 4.7^{x-1}$

14 Evaluate each of the following limits:

(a) $\displaystyle \mathop{Lim}_{x \to \pi/2} \left(\frac{(x^2 - \pi/4)\sin(\cos x)}{x - \pi/2} \right)$ (b) $\displaystyle \mathop{Lim}_{x \to -1} \ln\left(\exp\left[\frac{3x^2 + 2x - 1}{x + 1} \right] \right)$

(c) $\displaystyle \mathop{Lim}_{x \to 2+\sqrt{3}} \cos\left(\sin^{-1}\left(\frac{x - 2}{x - \sqrt{3}} \right) \right)$

Complex numbers

Learning outcomes

When you have completed this Program you will be able to:

- Recognize i as standing for $\sqrt{-1}$ and be able to reduce powers of i to $\pm i$ or ± 1
- Recognize that all complex numbers are in the form (real part) $+ \, i$(imaginary part)
- Add, subtract and multiply complex numbers
- Find the complex conjugate of a complex number
- Divide complex numbers
- State the conditions for the equality of two complex numbers
- Draw complex numbers and recognize the parallelogram law of addition
- Convert a complex number from Cartesian to polar form and vice versa
- Write a complex number in its exponential form
- Obtain the logarithm of a complex number

Introduction

1 Ideas and symbols

The numerals and their associated arithmetic symbols were devised so that we could make written calculations and records of quantities and measurements. When a grouping of symbols such as $\sqrt{-1}$ arises to which there is no corresponding quantity we must ask ourselves why? Why does the grouping of symbols occur if there is no quantity associated with it? Often, the only way to answer to such a question is to accept the grouping of symbols and carry on manipulating with it to see if any new ideas are forthcoming that will help answer the question. This we shall now do.

By the way, because $\sqrt{-1}$ can have no quantity associated with it we call it an **imaginary** number. This is to distinguish it from those numbers to which we can associate quantities; these we call **real** numbers. This is a most unfortunate choice of words because it gives the idea that those numbers we call 'real' are somehow more concrete and proper than their ethereal and elusive cousins that we call 'imaginary'. In fact all numbers are constructs of the human mind and imagination; no one type is more concrete or proper than any other and the numerals we use to describe numbers of any type are just symbols that permit us to manipulate numbers and so communicate ideas of number from one mind to another. When we put those numbers to use as in counting or measuring it is easy to confuse the idea of number and the real world to which that number is applied.

It may be thought that there are some quadratics that cannot be factorized into two linear factors. In fact all quadratics can be factorized into two linear factors but with the present state of your knowledge you may not always be able find them. This will now be remedied.

Read on and see how

The symbol *i*

2 Quadratic equations

The solutions of the quadratic equation:

$$x^2 - 1 = 0$$

are:

$x = \ldots\ldots\ldots\ldots$ and $x = \ldots\ldots\ldots\ldots$

The answers are in the next frame

3

$$x = 1 \text{ and } x = -1$$

Because:

$x^2 - 1 = 0$ can be written as $x^2 = 1$ by adding 1 to each side of the equation. The solution is then given as:

$$x = 1^{\frac{1}{2}} = \pm\sqrt{1} = 1 \text{ or } -1$$

Try this next one.

The solutions to the quadratic equation:

$$x^2 + 1 = 0$$

are:

$x = \ldots\ldots\ldots\ldots$ and $\ldots\ldots\ldots\ldots$

Again, the answers are in the next frame

4

$$x = \sqrt{-1} \text{ and } x = -\sqrt{-1}$$

Because

$x^2 + 1 = 0$ can be written as $x^2 = -1$ by subtracting 1 from each side of the equation. The solution is then given as:

$$x = (-1)^{\frac{1}{2}} = \pm\sqrt{-1} = \sqrt{-1} \text{ or } -\sqrt{-1}$$

This notation is rather clumsy so we denote $\sqrt{-1}$ by the symbol i. This means that we can write the solution of the equation $x^2 + 1 = 0$ as:

$\ldots\ldots\ldots\ldots$ and $\ldots\ldots\ldots\ldots$

Next frame

5

$$x = i \text{ and } x = -i$$

Because

$$i = \sqrt{-1} \text{ and so } -i = -\sqrt{-1}$$

Let's take a closer look at quadratic equations in the next frame

6

We have already seen that the general quadratic equation is of the form:

$$ax^2 + bx + c = 0 \text{ with solution } x = \frac{-b \pm \sqrt{b^2 - 4ac}}{2a}$$

For example, if:

$2x^2 + 9x + 7 = 0$, then we have $x = 2$, $b = 9$, and $c = 7$ and so:

$$x = \frac{-9 \pm \sqrt{81 - 56}}{4} = \frac{-9 \pm \sqrt{25}}{4} = \frac{-9 \pm 5}{4}$$

$$\therefore\ x = -\frac{4}{4} \text{ or } -\frac{14}{4}$$

$$\therefore\ x = -1 \text{ or } -3.5$$

That was straightforward enough, but if we solve the equation

$5x^2 - 6x + 5 = 0$ in the same way,

we get $x = \frac{6 \pm \sqrt{36 - 100}}{10} = \frac{6 \pm \sqrt{-64}}{10}$

and the next stage is now to determine the square root of -64.

Is it (a) 8
(b) -8
(c) neither?

Next frame

7

neither

It is, of course, neither, since $+8$ and -8 are the square roots of 64 and not of -64. In fact:

$$\sqrt{-64} = \sqrt{-1 \times 64} = \sqrt{-1} \times \sqrt{64} = 8i \text{ an } \textit{imaginary} \text{ number}$$

Similarly:

$$\sqrt{-36} = \sqrt{-1}\sqrt{36} = 6i$$

$$\sqrt{-7} = \sqrt{-1}\sqrt{7} = 2.646i$$

So $\sqrt{-25}$ can be written as ....

8

$5i$

We now have a way of finishing off the quadratic equation we started in Frame 6.

$$5x^2 - 6x + 5 = 0 \text{ therefore } x = \frac{6 \pm \sqrt{36 - 100}}{10} = \frac{6 \pm \sqrt{-64}}{10}$$

$$\therefore\ x = \frac{6 \pm 8i}{10} \quad \therefore\ x = 0.6 \pm 0.8i$$

$$\therefore\ x = 0.6 + 0.8i \text{ or } x = 0.6 - 0.8i$$

▶

Just try one yourself to make sure that you have followed all this. The solution of the quadratic $x^2 + x + 1 = 0$ equation is

$$x = \ldots\ldots\ldots\ldots \quad \text{and} \quad x = \ldots\ldots\ldots\ldots$$

Next frame

9

$$x = \frac{-1 + \sqrt{3}i}{2} \quad \text{and} \quad x = \frac{-1 - \sqrt{3}i}{2}$$

Because:

$x^2 + x + 1 = 0$ so that $a = 1$, $b = 1$ and $c = 1$

$$\text{Therefore } x = \frac{-1 \pm \sqrt{1-4}}{2} = \frac{-1 \pm \sqrt{3}i}{2},$$

$$\text{that is } x = \frac{-1 + \sqrt{3}i}{2} \quad \text{or} \quad x = \frac{-1 - \sqrt{3}i}{2}$$

Now let's look a little closer at that number i.

Move to the next frame

Powers of *i*

10

Positive integer powers

Because i represents $\sqrt{-1}$, it is clear that:

$$i^2 = -1$$
$$i^3 = (i^2)i = -i$$
$$i^4 = (i^2)^2 = (-1)^2 = 1$$
$$i^5 = (i^4)i = i$$

Note especially the third result: $i^4 = 1$. Every time a factor i^4 occurs, it can be replaced by the factor 1, so that i raised to a positive power is reduced to one of the four results $\pm i$ or ± 1. For example:

$$i^9 = (i^4)^2 i = (1)^2 i = 1 \cdot i = i$$
$$i^{20} = (i^4)^5 = (1)^5 = 1$$
$$i^{30} = (i^4)^7 i^2 = (1)^7(-1) = 1(-1) = -1$$
$$i^{15} = (i^4)^3 i^3 = (1)^3(-i) = -i$$

So, in the same way:

(a) $i^{42} = \ldots\ldots\ldots\ldots$

(b) $i^{12} = \ldots\ldots\ldots\ldots$

(c) $i^{11} = \ldots\ldots\ldots\ldots$

The answers are in the next frame

11

(a) $i^{42} = -1$
(b) $i^{12} = 1$
(c) $i^{11} = -i$

Because:

$$i^{42} = (i^4)^{10}\, i^2 = (1)^{10}(-1) = -1$$
$$i^{12} = (i^4)^3 = (1)^3 = 1$$
$$i^{11} = (i^4)^2\, i^3 = (1)^2(-i) = -i$$

Now for negative powers in the next frame

12 Negative integer powers

Negative powers follow from the reciprocal of i. Because

$$i^2 = -1$$

we see that by dividing both sides by i that $i = -\frac{1}{i} = -i^{-1}$ so that:

$$i^{-1} = -i$$
$$i^{-2} = (i^2)^{-1} = (-1)^{-1} = -1$$
$$i^{-3} = (i^{-2})\, i^{-1} = (-1)(-i) = i$$
$$i^{-4} = (i^{-2})^2 = (-1)^2 = 1$$

Note again the last result. Every time a factor i^{-4} occurs, it can be replaced by the number 1, so that i raised to a negative power is reduced to one of the four results $\pm i$ or ± 1. For example:

$$i^{-8} = (i^{-4})^2 = (1)^2 = 1$$
$$i^{-19} = (i^{-4})^4\, i^{-3} = (1)^4(i) = i$$
$$i^{-30} = (i^{-4})^7\, i^{-2} = (1)^7(-1) = -1$$
$$i^{-15} = (i^{-4})^3\, i^{-3} = (1)^3(-i) = -i$$

So, in the same way:

(a) $i^{-32} =$ ....
(b) $i^{-13} =$ ....
(c) $i^{-23} =$ ....

The answers are in the next frame

13

(a) $i^{-32} = 1$
(b) $i^{-13} = -i$
(c) $i^{-23} = i$

▶

Because:

$i^{-32} = (i^{-4})^8 = (1)^8 = 1$

$i^{-13} = (i^{-4})^3\, i^{-1} = (1)^3(-i) = -i$

$i^{-23} = (i^{-4})^5\, i^{-3} = (1)^5(i) = i$

Complex numbers

14

In Frame 8 we saw that the quadratic equation $5x^2 - 6x + 5 = 0$ had two solutions, one of which was $x = 0.6 + 0.8i$. This consists of two separate terms, 0.6 and $0.8i$ which cannot be combined any further – it is a **mixture** of the real number 0.6 and the imaginary number $0.8i$. This mixture is called a **complex** number.

The word 'complex' means either:

(a) complicated, or
(b) mixture

Answer in the next frame

15

mixture

A complex number is a mixture of a real number and an imaginary number. The symbol z is used to denote a complex number.

In the complex number $z = 3 + 5i$ the number 3 is called the *real part* of z and is denoted as $\text{Re}(z)$. The number 5 is called the *imaginary part* of z and is denoted $\text{Im}(z)$.

So $\text{Re}(3 + 5i) = 3$ and $\text{Im}(3 + 5i) = 5$

Notice: the imaginary part of a complex number is a real number so that:

$\text{Im}(3 + 5i) = 5$ and not $5i$

So, a complex number $z = \text{Re}(z) + \text{Im}(z)i$.

In the complex number $z = 2 + 7i$ the real and imaginary parts are

$\text{Re}(z) = \ldots\ldots\ldots\ldots$ and $\text{Im}(z) = \ldots\ldots\ldots\ldots$

Go to the next frame

16

$\text{Re}(z) = 2$ and $\text{Im}(z) = 7$ (*not* $7i$!)

Complex numbers have many applications in engineering and science. To use them, we must know how to carry out the usual arithmetical operations.

▶

1 Addition and subtraction of complex numbers

This is easy, as a few examples will show:

$$(4+5i)+(3-2i)$$

Although the real and imaginary parts cannot be combined, we can remove the brackets and total up terms of the same kind:

$$\begin{aligned}(4+5i)+(3-2i) &= 4+5i+3-2i=(4+3)+i(5-2)\\ &= 7+3i\end{aligned}$$

Another example:

$$\begin{aligned}(4+7i)-(2-5i) &= 4+7i-2+5i=(4-2)+i(7+5)\\ &= 2+12i\end{aligned}$$

So in general, $(a+ib)+(c+id)=(a+c)+i(b+d)$
Now you do this one:

$(5+7i)+(3-4i)-(6-3i)=$ ....

17

$2+6i$

since $(5+7i)+(3-4i)-(6-3i)$
$=5+7i+3-4i-6+3i$
$=(5+3-6)+i(7-4+3)=2+6i$

Now you do these in the same way:

(a) $(6+5i)-(4-3i)+(2-7i)=$ ....
and (b) $(3+5i)-(5-4i)-(-2-3i)=$ ....

18

(a) $4+i$ (b) $12i$

Here is the working:

(a) $(6+5i)-(4-3i)+(2-7i)$
$=6+5i-4+3i+2-7i$
$=(6-4+2)+i(5+3-7)=4+i$

(b) $(3+5i)-(5-4i)-(-2-3i)$
$=3+5i-5+4i+2+3i$ (Take care with signs!)
$=(3-5+2)+i(5+4+3)$
$=0+12i=12i$

This is very easy then, so long as you remember that the real and the imaginary parts must be treated quite separately – just like x's and y's in an algebraic expression.

On to Frame 19

19

2 Multiplication of complex numbers

Take as an example: $(3+4i)(2+5i)$

These are multiplied together in just the same way as you would determine the product $(3x+4y)(2x+5y)$.

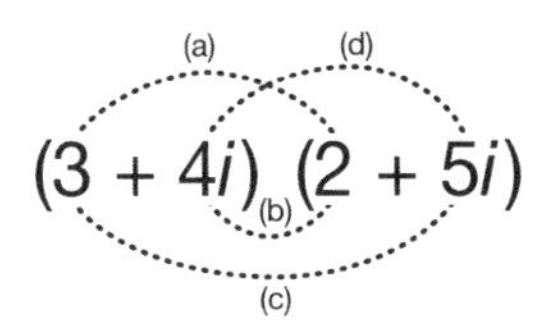

Form the product terms of

(a) the two left-hand terms
(b) the two inner terms
(c) the two outer terms
(d) the two right-hand terms

$$= 6 + 8i + 15i + 20i^2$$
$$= 6 + 23i - 20 \quad (\text{since } i^2 = -1)$$
$$= -14 + 23i$$

Likewise, $(4-5i)(3+2i)$ . . . .

20

$22 - 7i$

Because

$$(4-5i)(3+2i) = 12 - 15i + 8i - 10i^2$$
$$= 12 - 7i + 10 \quad (i^2 = -1)$$
$$= 22 - 7i$$

If the expression contains more than two factors, we multiply the factors together in stages:

$$(3+4i)(2-5i)(1-2i) = (6 + 8i - 15i - 20i^2)(1-2i)$$
$$= (6 - 7i + 20)(1-2i)$$
$$= (26 - 7i)(1-2i)$$
$$= \ldots\ldots\ldots\ldots$$

Finish it off

21

$12 - 59i$

Because

$$(26-7i)(1-2i)$$
$$= 26 - 7i - 52i + 14i^2$$
$$= 26 - 59i - 14 = 12 - 59i$$

Note that when we are dealing with complex numbers, the result of our calculations is also, in general, a complex number.

Now you do this one on your own.

$$(5+8i)(5-8i) = \ldots\ldots\ldots\ldots$$

22

89

Here it is:

$$(5+8i)(5-8i) = 25 + 40i - 40i - 64i^2$$
$$= 25 + 64$$
$$= 89$$

In spite of what we said above, here we have a result containing no i term. The result is therefore entirely real.

This is rather an exceptional case. Look at the two complex numbers we have just multiplied together. Can you find anything special about them? If so, what is it?

When you have decided, move on to the next frame

23

They are identical except for the middle sign in the brackets, i.e. $(5+8i)$ and $(5-8i)$

A pair of complex numbers like these are called *conjugate* complex numbers and *the product of two conjugate complex numbers is always entirely real.*

Look at it this way:

$(a+b)(a-b) = a^2 - b^2$ Difference of two squares

Similarly

$$(5+8i)(5-8i) = 5^2 - (8i)^2 = 5^2 - 8^2i^2$$
$$= 5^2 + 8^2 \quad \text{Sum of two squares}$$
$$= 25 + 64 = 89$$

Without actually working it out, will the product of $(7-6i)$ and $(4+3i)$ be

(a) a real number
(b) an imaginary number
(c) a complex number?

24

A complex number

since $(7-6i)(4+3i)$ is a product of two complex numbers which are *not* conjugate complex numbers or multiples of conjugates.

Remember: Conjugate complex numbers are identical except for the signs in the middle of the brackets.

	$(4+5i)$ and $(4-5i)$	*are*	conjugate complex numbers
	$(a+ib)$ and $(a-ib)$	*are*	conjugate complex numbers
but	$(6+2i)$ and $(2+6i)$	*are not*	conjugate complex numbers
	$(5-3i)$ and $(-5+3i)$	*are not*	conjugate complex numbers

So what must we multiply $(3-2i)$ by, to produce a result that is entirely real?

25

(3 + 2*i*) or a multiple of it.

because the conjugate of $(3 - 2i)$ is identical to it, except for the middle sign, i.e. $(3 + 2i)$, and we know that the product of two *conjugate* complex numbers is always real.

Here are two examples:

$$(3 - 2i)(3 + 2i) = 3^2 - (2i)^2 = 9 - 4i^2$$
$$= 9 + 4 = 13$$
$$(2 + 7i)(2 - 7i) = 2^2 - (7i)^2 = 4 - 49i^2$$
$$= 4 + 49 = 53$$

. . . and so on.

Complex numbers of the form $(a + ib)$ and $(a - ib)$ are called .... complex numbers.*

26

conjugate

Now you should have no trouble with these:

(a) Write down the following products

(i) $(4 - 3i)(4 + 3i)$ (ii) $(4 + 7i)(4 - 7i)$

(iii) $(a + ib)(a - ib)$ (iv) $(x - iy)(x + iy)$

(b) Multiply $(3 - 5i)$ by a suitable factor to give a product that is entirely real.

When you have finished, move on to Frame 27

27

Here are the results in detail:

(a) (i) $(4 - 3i)(4 + 3i) = 4^2 - 3^2i^2 = 16 + 9 = \boxed{25}$

(ii) $(4 + 7i)(4 - 7i) = 4^2 - 7^2i^2 = 16 + 49 = \boxed{65}$

(iii) $(a + ib)(a - ib) = a^2 - i^2b^2 = \boxed{a^2 + b^2}$

(iv) $(x - iy)(x + iy) = x^2 - i^2y^2 = \boxed{x^2 + y^2}$

(b) To obtain a real product, we can multiply $(3 - 5i)$ by its conjugate, i.e. $(3 + 5i)$, giving:

$$(3 - 5i)(3 + 5i) = 3^2 - 5^2i^2 = 9 + 25 = \boxed{34}$$

Now move on to the next frame for a short review exercise

*Note: By convention numerals are written before *i* whereas other symbols are written after.

Review exercise

28

1 Simplify (a) i^{12} (b) i^{10} (c) i^{-7}

2 Simplify:

(a) $(5 - 9i) - (2 - 6i) + (3 - 4i)$

(b) $(6 - 3i)(2 + 5i)(6 - 2i)$

(c) $(4 - 3i)^2$

(d) $(5 - 4i)(5 + 4i)$

3 Multiply $(4 - 3i)$ by an appropriate factor to give a product that is entirely real. What is the result?

When you have completed this exercise, move on to Frame 29

29

Here are the results. Check yours.

1 (a) $i^{12} = (i^4)^3 = 1^3 = \boxed{1}$

(b) $i^{10} = (i^4)^2 i^2 = 1^2(-1) = \boxed{-1}$

(c) $i^{-7} = (i^{-4})^2 i = (1)^2 i = \boxed{i}$

2 (a) $(5 - 9i) - (2 - 6i) + (3 - 4i)$

$= 5 - 9i - 2 + 6i + 3 - 4i$

$= (5 - 2 + 3) + i(6 - 9 - 4) = \boxed{6 - 7i}$

(b) $(6 - 3i)(2 + 5i)(6 - 2i)$

$= (12 - 6i + 30i - 15i^2)(6 - 2i)$

$= (27 + 24i)(6 - 2i)$

$= 162 + 144i - 54i + 48 = \boxed{210 + 90i}$

(c) $(4 - 3i)^2 = 16 - 24i - 9$

$= \boxed{7 - 24i}$

(d) $(5 - 4i)(5 + 4i)$

$= 25 - 16i^2 = 25 + 16 = \boxed{41}$

3 A suitable factor is the conjugate of the given complex number:

$(4 - 3i)(4 + 3i) = 16 + 9 = \boxed{25}$

All correct? Right.

Now move on to the next frame to continue the Program

30

Now let us deal with division.

Division of a complex number by a real number is easy enough:

$$\frac{5-4i}{3}=\frac{5}{3}-\frac{4}{3}i=1.67-1.33i \text{ (to 2 dp)}$$

But how do we manage with $\dfrac{7-4i}{4+3i}$?

If we could, somehow, convert the denominator into a real number, we could divide out as in the example above. So our problem is really, how can we convert $(4+3i)$ into a completely real denominator – and this is where our last piece of work comes in.

We know that we can convert $(4+3i)$ into a completely real number by multiplying it by its ....

31

conjugate

i.e. the same complex number but with the opposite sign in the middle, in this case $(4-3i)$.

But if we multiply the denominator by $(4-3i)$, we must also multiply the numerator by the same factor:

$$\frac{7-4i}{4+3i}=\frac{(7-4i)(4-3i)}{(4+3i)(4-3i)}=\frac{28-37i-12}{16+9}=\frac{16-37i}{25}$$

$$\frac{16}{25}-\frac{37}{25}i=0.64-1.48i$$

and the job is done. To divide one complex number by another, therefore, we multiply numerator and denominator by the conjugate of the denominator. This will convert the denominator into a real number and the final step can then be completed.

Thus, to simplify $\dfrac{4-5i}{1+2i}$, we shall multiply top and bottom by ....

32

the conjugate of the denominator, i.e. $(1-2i)$

If we do that, we get:

$$\frac{4-5i}{1+2i}=\frac{(4-5i)(1-2i)}{(1+2i)(1-2i)}=\frac{4-13i-10}{1+4}$$

$$=\frac{-6-13i}{5}=\frac{-6}{5}-\frac{13}{5}i=-1.2-2.6i$$

Now here is one for you to do: Simplify $\dfrac{3+2i}{1-3i}$

When you have done it, move on to the next frame

33

$$\boxed{-0.3 + 1.1i}$$

Because

$$\frac{3+2i}{1-3i} = \frac{(3+2i)(1+3i)}{(1-3i)(1+3i)} = \frac{3+11i-6}{1+9} = \frac{-3+11i}{10} = -0.3 + 1.1i$$

Now do these in the same way:

(a) $\dfrac{4-5i}{2-i}$ (b) $\dfrac{3+5i}{5-3i}$ (c) $\dfrac{(2+3i)(1-2i)}{3+4i}$

When you have worked these, move on to Frame 34 to check your results

34

Here are the solutions in detail:

(a) $\dfrac{4-5i}{2-i} = \dfrac{(4-5i)(2+i)}{(2-i)(2+i)} = \dfrac{8-6i+5}{4+1} = \dfrac{13-6i}{5} = \boxed{2.6 - 1.2i}$

(b) $\dfrac{3+5i}{5-3i} = \dfrac{(3+5i)(5+3i)}{(5-3i)(5+3i)} = \dfrac{15+34i-15}{25+9} = \dfrac{34i}{34} = \boxed{i}$

(c) $\dfrac{(2+3i)(1-2i)}{(3+4i)} = \dfrac{2-i+6}{3+4i} = \dfrac{8-i}{3+4i}$

$$= \frac{(8-i)(3-4i)}{(3+4i)(3-4i)}$$

$$= \frac{24-35i-4}{9+16} = \frac{20-35i}{25}$$

$$= \boxed{0.8 - 1.4i}$$

And now you know how to add, subtract, multiply and divide complex numbers

Equal complex numbers

35

Now let us see what we can find out about two complex numbers which we are told are equal.

Let the numbers be	$a + ib$ and $c + id$
Then we have	$a + ib = c + id$
Rearranging terms, we get	$a - c = i(d - b)$

In this last statement, the quantity on the left-hand side is entirely real, while that on the right-hand side is entirely imaginary, i.e. a real quantity equals an imaginary quantity! This seems contradictory and in general it just cannot be true. But there is *one* special case for which the statement can be true. That is when …………

36

each side is zero

$a - c = i(d - b)$

can be true only if $a - c = 0$, i.e. $a = c$
and if $d - b = 0$, i.e. $b = d$

So we get this important result:
If two complex numbers are equal

(a) the two real parts are equal
(b) the two imaginary parts are equal

For example, if $x + iy = 5 + 4i$, then we know $x = 5$ and $y = 4$
and if $a + ib = 6 - 3i$, then $a =$ .... and $b =$ ....

37

$a = 6$ and $b = -3$

Be careful to include the sign!

Now what about this one?

If $(a + b) + i(a - b) = 7 + 2i$, find the values of a and b.

Well now, following our rule about two equal complex numbers, what can we say about $(a + b)$ and $(a - b)$?

38

$a + b = 7$ and $a - b = 2$

since the two real parts are equal and the two imaginary parts are equal.

This gives you two simultaneous equations, from which you can determine the values of a and b.

So what are they?

39

$a = 4.5$; $b = 2.5$

For $\left.\begin{array}{l} a + b = 7 \\ a - b = 2 \end{array}\right\} \begin{array}{ll} 2a = 9 & \therefore\ a = 4.5 \\ 2b = 5 & \therefore\ b = 2.5 \end{array}$

We see then that an equation involving complex numbers leads to a pair of simultaneous equations by putting

(a) the two real parts equal
(b) the two imaginary parts equal

This is quite an important point to remember.

Graphical representation of a complex number

40 Argand diagram

So far we have looked at the algebra of complex numbers, now we want to look at the geometry. A real number can be graphically represented as a point on a line – the *real line*. By using a Cartesian coordinate system a pair of real numbers can be graphically represented by a point in the plane. In 1806, the French mathematician Jean-Robert Argand devised a means of representing a complex number using the same Cartesian coordinate system. He plotted the pair of real numbers (a, b) of $z = a + ib$ as a point in the plane and then joined that point to the origin with a straight line.

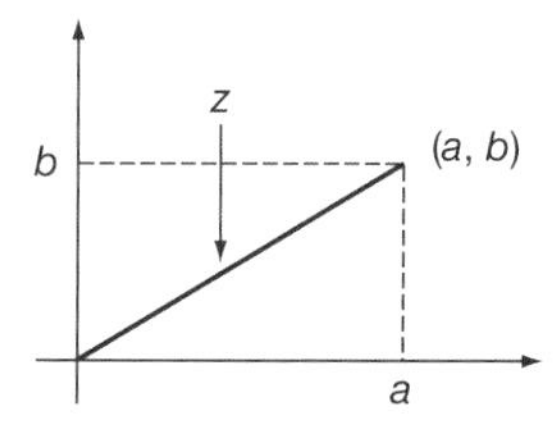

This straight line is then the graphical representation of the complex number $z = a + ib$ and the plane it is plotted against is referred to as the **complex plane**. The entire diagram is called an Argand diagram. The two axes are referred to as the real and imaginary axes respectively but note that it is *real numbers* that are plotted on both axes.

Draw an Argand diagram to represent the complex numbers:

(a) $z_1 = 2 + 3i$ (b) $z_2 = -3 + 2i$

(c) $z_3 = 4 - 3i$ (d) $z_4 = -4 - 5i$

Label each one clearly

41

Here they are. Check yours.

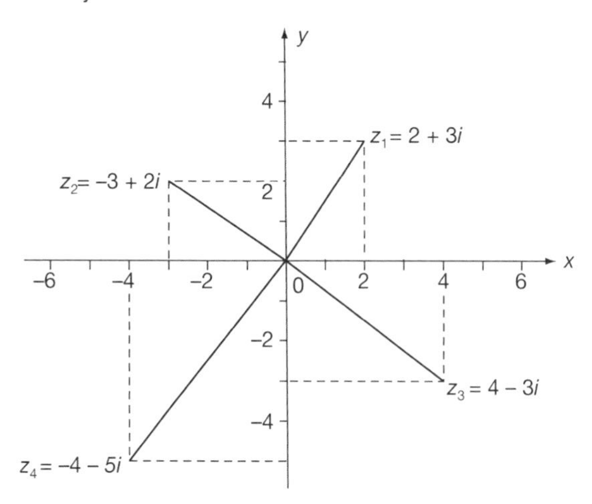

Note once again that the end of each complex number is plotted very much like plotting x and y coordinates.

The real part corresponds to the x-value.

The imaginary part corresponds to the y-value.

Move on to Frame 42

Graphical addition of complex numbers

42

If we draw the two complex numbers $z_1 = 5 + 2i$ and $z_2 = 2 + 3i$ on an Argand diagram we can then construct the parallelogram formed from the two adjacent sides z_1 and z_2 as shown.

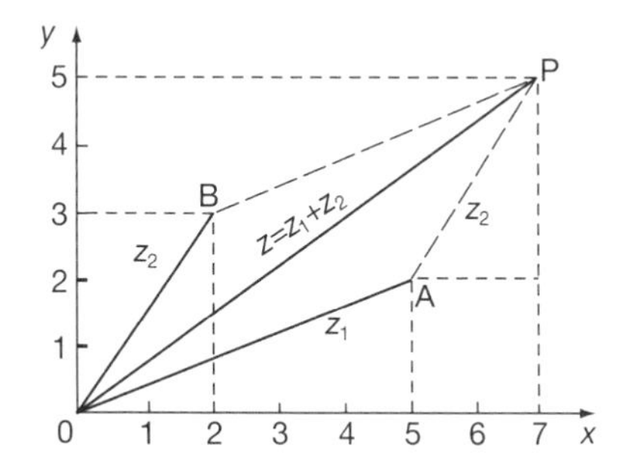

If the diagonal of the parallelogram OP represents the complex number $z = a + ib$, the values of a and b are, from the diagram:

$a = \dots\dots\dots\dots$

$b = \dots\dots\dots\dots$

Next frame

43

$a = 5 + 2 = 7 \qquad b = 2 + 3 = 5$

Therefore $OP = z = 7 + 5i$.

$$\begin{aligned} \text{Since } z_1 + z_2 &= (5 + 2i) + (2 + 3i) \\ &= 7 + 5i \\ &= z \end{aligned}$$

we can see that on an Argand diagram the sum of two complex numbers $z_1 + z_2$ is given by the .... of the parallelogram formed from the two adjacent sides z_1 and z_2.

Next frame

44

diagonal

How do we do subtraction by similar means? We do this rather craftily without learning any new methods. The trick is simply this:

$$z_1 - z_2 = z_1 + (-z_2)$$

That is, we draw the lines representing z_1 and the *negative* of z_2 and add them as before. The negative of z_2 is simply a line with the same magnitude (or length) as z_2 but in the opposite direction.

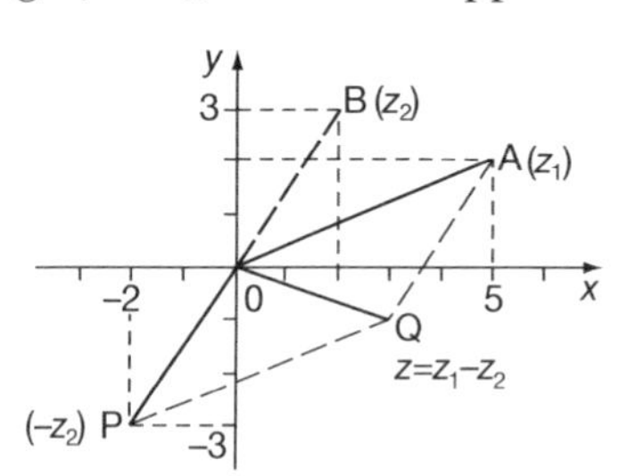

e.g. If $z_1 = 5 + 2i$ and $z_2 = 2 + 3i$

$$\begin{aligned} OA &= z_1 = 5 + 2i \\ OP &= -z_2 = -(2 + 3i) \\ \text{Then } OQ &= z_1 + (-z_2) \\ &= z_1 - z_2 \end{aligned}$$

Determine on an Argand diagram

$$(4 + 2i) + (-2 + 3i) - (-1 + 6i)$$

45

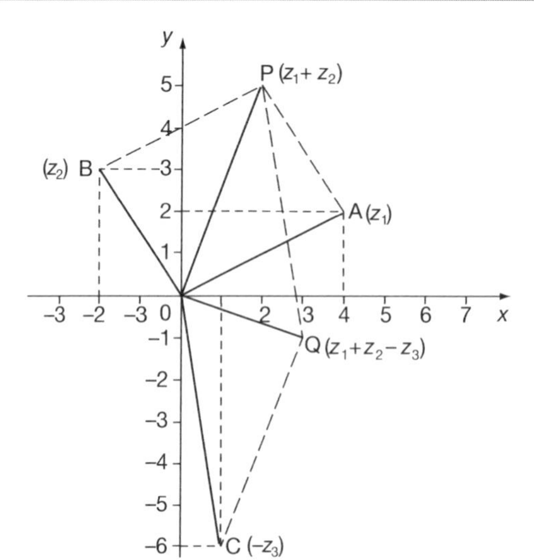

$$\begin{aligned} OA &= z_1 = 4 + 2i \\ OB &= z_2 = -2 + 3i \\ OC &= -z_3 = 1 - 6i \end{aligned}$$

$$\begin{aligned} \text{Then } OP &= z_1 + z_2 \\ OQ &= z_1 + z_2 - z_3 = 3 - i \end{aligned}$$

Polar form of a complex number

46

It is convenient sometimes to express a complex number $a + ib$ in a different form. On an Argand diagram, let OP be a complex number $a + ib$. Let r = length of the complex number and θ the angle made with OX.

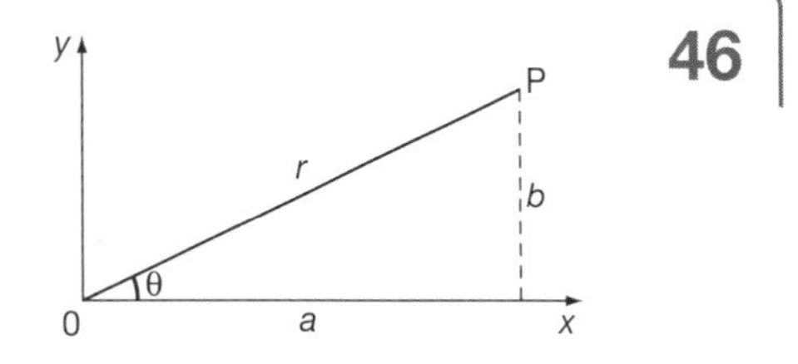

Then $\quad r^2 = a^2 + b^2 \qquad r = \sqrt{a^2 + b^2}$

and $\quad \tan\theta = \dfrac{b}{a} \qquad \theta = \tan^{-1}\dfrac{b}{a}$

Also $\quad a = r\cos\theta$ and $b = r\sin\theta$

Since $z = a + ib$, this can be written

$$z = r\cos\theta + i\,r\sin\theta \qquad \text{i.e. } z = r(\cos\theta + i\sin\theta)$$

This is called the *polar form* of the complex number $a + ib$, where

$$r = \sqrt{a^2 + b^2} \text{ and } \theta = \tan^{-1}\frac{b}{a}$$

Let us take a numerical example.

Next frame

47

To express $z = 4 + 3i$ in polar form.

First draw a sketch diagram (that always helps).

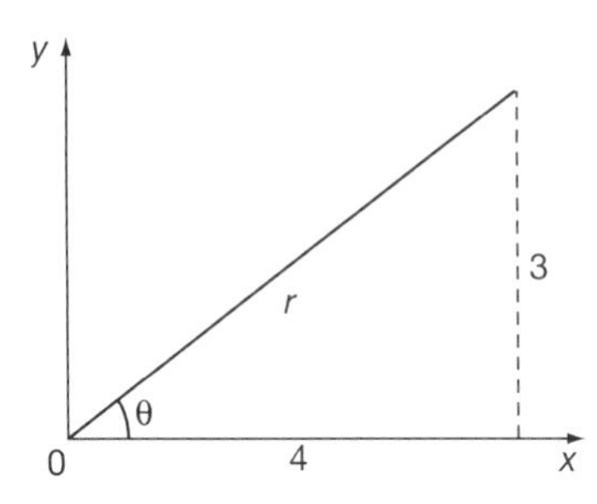

We can see that:

(a) $r^2 = 4^2 + 3^2 = 16 + 9 = 25$

$r = 5$

(b) $\tan\theta = \dfrac{3}{4} = 0.75$

$\theta = 36°52'$

$$z = a + ib = r(\cos\theta + i\sin\theta)$$

So in this case $z = 5(\cos 36°52' + i\sin 36°52')$

Now here is one for you to do.

Find the polar form of the complex number $(2 + 3i)$.

When you have finished it, consult the next frame

48

$z = 3.606(\cos 56°19' + i \sin 56°19')$

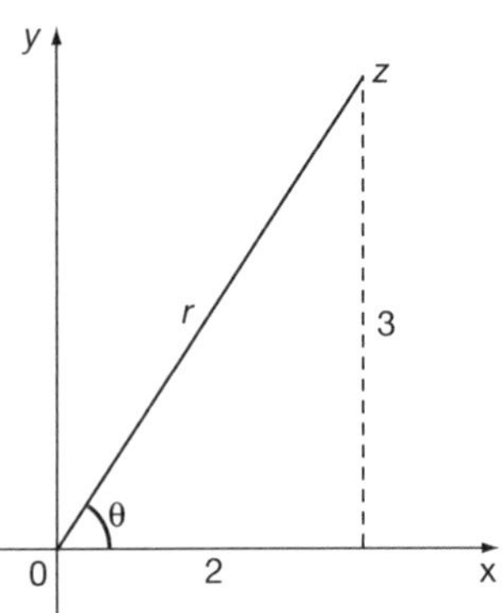

Here is the working:

$z = 2 + 3i = r(\cos\theta + i\sin\theta)$

$r^2 = 4 + 9 = 13 \qquad r = 3.606$ (to 3 dp)

$\tan\theta = \dfrac{3}{2} = 1.5 \qquad \theta = 56°19'$

$z = 3.606(\cos 56°19' + i \sin 56°19')$

We have special names for the values of r and θ:

$z = a + ib = r(\cos\theta + i\sin\theta)$

(a) r is called the *modulus* of the complex number z and is often abbreviated to 'mod z' or indicated by $|z|$.

Thus if $z = 2 + 5i$, then $|z| = \sqrt{2^2 + 5^2} = \sqrt{4 + 25} = \sqrt{29}$

(b) θ is called the *argument* of the complex number and can be abbreviated to 'arg z'.

So if $z = 2 + 5i$, then $\arg z =$ ....

49

$\arg z = 68°12'$

$z = 2 + 5i$. Then $\arg z = \theta = \tan^{-1}\dfrac{5}{2} = 68°12'$

Warning: In finding θ, there are of course two angles between 0° and 360°, the tangent of which has the value $\dfrac{b}{a}$. We must be careful to use the angle in the correct quadrant. *Always* draw a sketch of the complex number to ensure you have the right one.

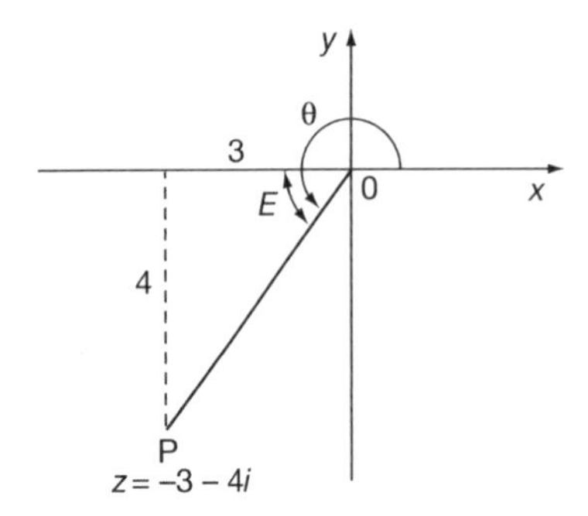

e.g. Find arg z when $z = -3 - 4i$

θ is measured from OX to OP. We first find E, the equivalent acute angle from the triangle shown:

$\tan E = \dfrac{4}{3} = 1.333\ldots \qquad \therefore\ E = 53°8'$

Then in this case:

$\theta = 180° + E = 233°8' \qquad \arg z = 233°8'$

Now you find $\arg(-5 + 2i)$

Move on when you have finished

50

$$\arg z = 158°12'$$

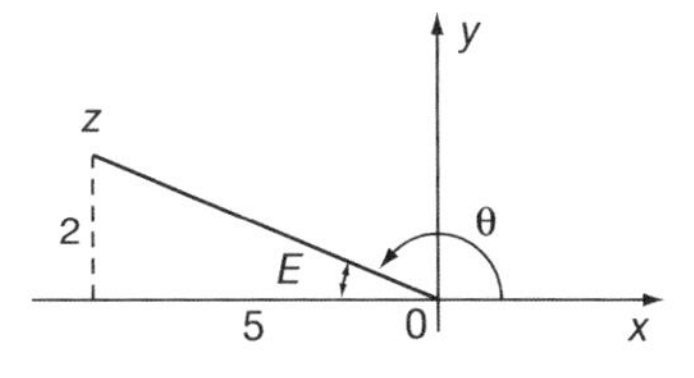

$z = -5 + 2i$

$\tan E = \dfrac{2}{5} = 0.4 \quad \therefore\ E = 21°48'$

In this particular case, $\theta = 180° - E$

$\therefore\ \theta = 158°12'$

Complex numbers in polar form are always of the same shape and differ only in the actual values of r and θ. We often use the shorthand version $r\underline{|\theta}$ to denote the polar form.

e.g. If $z = -5 + 2i$, $r = \sqrt{25+4} = \sqrt{29} = 5.385$ and from above $\theta = 158°12'$

$\therefore$ The full polar form is $z = 5.385(\cos 158°12' + i \sin 158°12')$ and this can be shortened to $z = 5.385\underline{|158°12'}$

Express in shortened form, the polar form of $(4 - 3i)$

Do not forget to draw a sketch diagram first.

51

$$z = 5\underline{|323°8'}$$

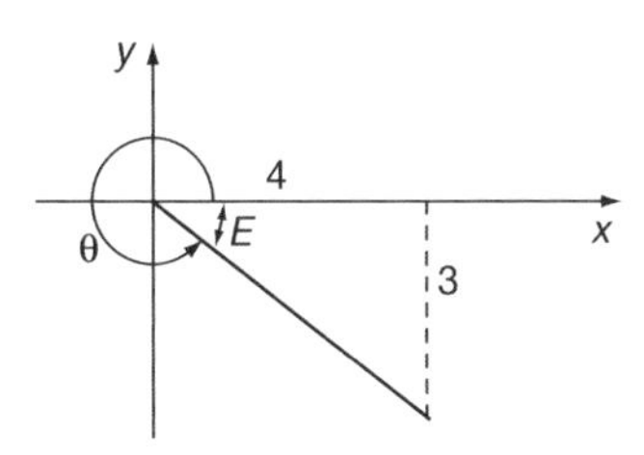

$r = \sqrt{4^2 + 3^2} \qquad r = 5$

$\tan E = 0.75 \qquad \therefore\ E = 36°52'$

$\therefore\ \theta = 360° - E = 323°8'$

$\therefore\ z = 5(\cos 323°8' + i \sin 323°8')$
$= 5\underline{|323°8'}$

Of course, given a complex number in polar form, you can convert it into basic form $a + ib$ simply by evaluating the cosine and the sine and multiplying by the value of r.

e.g. $z = 5(\cos 35° + i \sin 35°) = 5(0.8192 + 0.5736i)$
$z = 4.0960 + 2.8680i$ (to 4 dp)

Now you do this one.

Express in the form $a + ib$, $4(\cos 65° + i \sin 65°)$

52

$$z = 1.6905 + 3.6252i$$

Because

$z = 4(\cos 65° + i \sin 65°) = 4(0.4226 + 0.9063i) = 1.6905 + 3.6252i$ (to 4 dp)

If the argument is greater than 90°, care must be taken in evaluating the cosine and sine to include the appropriate signs.

▶

e.g. If $z = 2(\cos 210^\circ + i \sin 210^\circ)$ the complex number lies in the third quadrant.

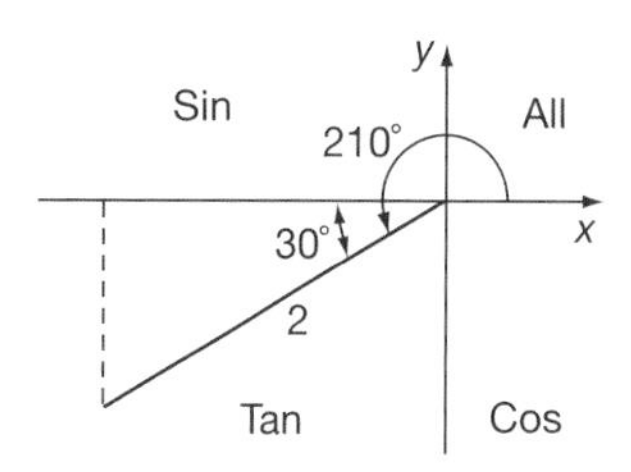

$\cos 210^\circ = -\cos 30^\circ$
$\sin 210^\circ = -\sin 30^\circ$

Then $z = 2(-\cos 30^\circ - i \sin 30^\circ)$
$= 2(-0.8660 - 0.5i)$
$= -1.732 - i$

Here you are. What about this one?

Express $z = 5(\cos 140^\circ + i \sin 140^\circ)$ in the form $a + ib$

What do you make it?

53

$$z = -3.8300 + 3.2140i$$

Here are the details:

$\cos 140^\circ = -\cos 40^\circ$
$\sin 140^\circ = \sin 40^\circ$

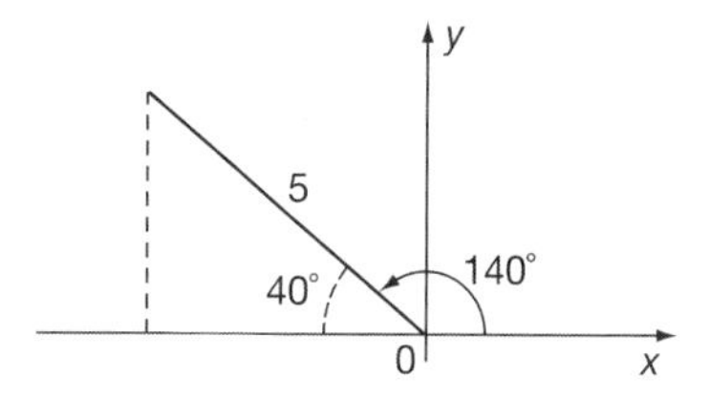

$z = 5(\cos 140^\circ + i \sin 140^\circ)$
$= 5(-\cos 40^\circ + i \sin 40^\circ)$
$= 5(-0.7660 + 0.6428i)$
$= -3.830 + 3.214i$ (to 3 dp)

Fine. Now by way of revision, work out the following:

(a) Express $-5 + 4i$ in polar form
(b) Express $3\underline{|300^\circ}$ in the form $a + ib$

When you have finished both of them, check your results with those in Frame 52

54

Here is the working:

(a) $r^2 = 4^2 + 5^2 = 16 + 25 = 41$

$\therefore\ r = 6.403$ (to 3 dp)

$\tan E = 0.8 \quad \therefore\ E = 38^\circ 40'$

$\therefore\ \theta = 141^\circ 20'$

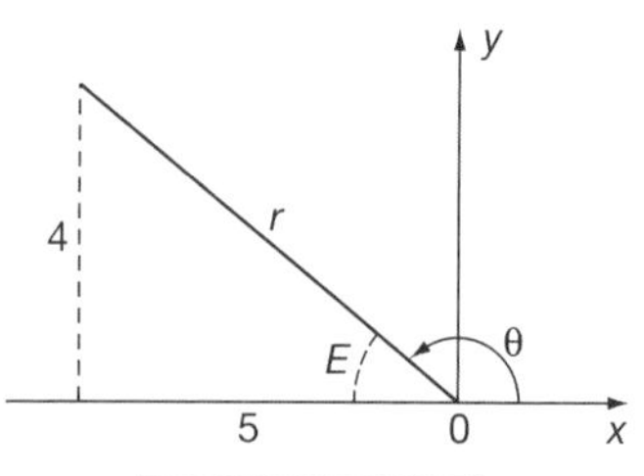

$$-5 + 4i = 6.403(\cos 141^\circ 20' + i \sin 141^\circ 20') = \boxed{6.403\underline{|141^\circ 20'}}$$

▶

(b) $3\underline{|300^\circ} = 3(\cos 300^\circ + i \sin 300^\circ)$

$\cos 300^\circ = \cos 60^\circ$

$\sin 300^\circ = -\sin 60^\circ$

$3\underline{|300^\circ} = 3(\cos 60^\circ - i \sin 60^\circ)$

$= 3(0.500 - i0.866)$ (to 3 dp)

$= \boxed{1.500 - i2.598}$

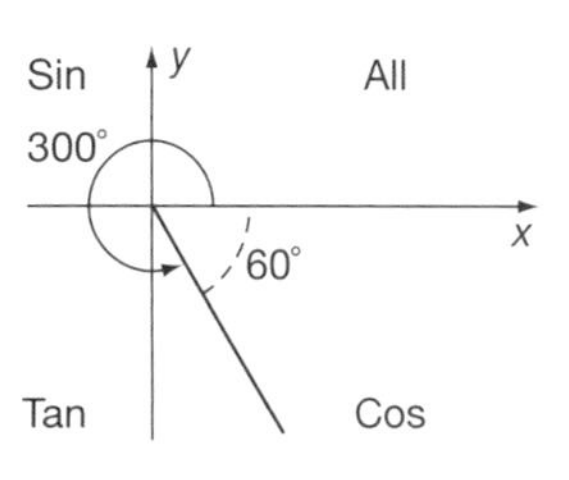

Move to Frame 55

55

We see then that there are two ways of expressing a complex number:

(a) in standard form $z = a + ib$

(b) in polar form $z = r(\cos\theta + i\sin\theta)$

where $r = \sqrt{a^2 + b^2}$

and $\theta = \tan^{-1}\dfrac{b}{a}$

If we remember the simple diagram, we can easily convert from one system to the other:

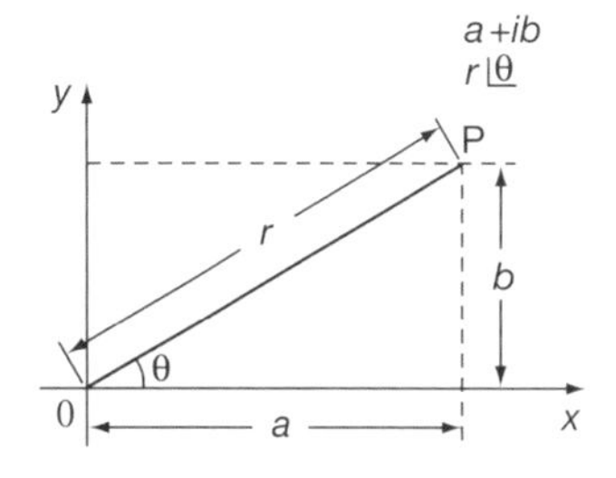

So on now to Frame 56

Exponential form of a complex number

56

There is still another way of expressing a complex number which we must deal with, for it too has its uses. We shall arrive at it this way:

Many functions can be expressed as series. for example,

$$e^x = 1 + x + \frac{x^2}{2!} + \frac{x^3}{3!} + \frac{x^4}{4!} + \frac{x^5}{5!} + \dots\dots\dots$$

$$\sin x = x - \frac{x^3}{3!} + \frac{x^5}{5!} - \frac{x^7}{7!} + \frac{x^9}{9!} + \dots\dots\dots$$

$$\cos x = 1 - \frac{x^2}{2!} + \frac{x^4}{4!} - \frac{x^6}{6!} + \dots\dots\dots$$

You no doubt have hazy recollections of these series. You had better make a note of them since they have turned up here.

57

If we now take the series for e^x and write $i\theta$ in place of x, we get:

$$e^{i\theta} = 1 + i\theta + \frac{(i\theta)^2}{2!} + \frac{(i\theta)^3}{3!} + \frac{(i\theta)^4}{4!} + \dots \dots \dots$$
$$= 1 + i\theta + \frac{i^2\theta^2}{2!} + \frac{i^3\theta^3}{3!} + \frac{i^4\theta^4}{4!} + \dots \dots \dots$$
$$= 1 + i\theta - \frac{\theta^2}{2!} - \frac{i\theta^3}{3!} + \frac{\theta^4}{4!} + \dots \dots \dots$$
$$= \left(1 - \frac{\theta^2}{2!} + \frac{\theta^4}{4!} - \dots \dots \dots\right) + i\left(\theta - \frac{\theta^3}{3!} + \frac{\theta^5}{5!} - \dots \dots \dots\right)$$
$$= \cos\theta + i\sin\theta$$

Therefore, $r(\cos\theta + i\sin\theta)$ can now be written as $re^{i\theta}$. This is called the *exponential form* of the complex number. It can be obtained from the polar quite easily since the r value is the same and the angle θ is the same in both. It is important to note, however, that in the exponential form, the angle must be in *radians*.

Move on to the next frame

58

The three ways of expressing a complex number are therefore:

(a) $z = a + ib$
(b) $z = r(\cos\theta + i\sin\theta)$ Polar form
(c) $z = r \cdot e^{i\theta}$ Exponential form

Remember that the exponential form is obtained from the polar form:

(a) the r value is the same in each case

(b) the angle is also the same in each case, but in the exponential form the angle must be in radians.

So, knowing that, change the polar form $5(\cos 60^\circ + i\sin 60^\circ)$ into the exponential form.

Then turn to Frame 59

59

$$\boxed{5e^{i\frac{\pi}{3}}}$$

Because we have $5(\cos 60^\circ + i\sin 60^\circ)$ $r = 5$

$\theta = 60^\circ = \frac{\pi}{3}$ radians

$\therefore$ Exponential form is $5e^{i\frac{\pi}{3}}$

And now a word about negative angles:

We know $e^{i\theta} = \cos\theta + i\sin\theta$

If we replace θ by $-\theta$ in this result, we get

$$e^{-i\theta} = \cos(-\theta) + i\sin(-\theta)$$
$$= \cos\theta - i\sin\theta$$

▶

So we have

$$\left.\begin{aligned} e^{i\theta} &= \cos\theta + i\sin\theta \\ e^{-i\theta} &= \cos\theta - i\sin\theta \end{aligned}\right\}$$

Make a note of these

Logarithm of a complex number

60

There is one operation that we have been unable to carry out with complex numbers before this. That is to find the logarithm of a complex number. The exponential form now makes this possible, since the exponential form consists only of products and powers.

For, if we have:

$$z = re^{i\theta}$$

then we can say:

$$\ln z = \ln r + i\theta$$

e.g. If $z = 6.42e^{1.57i}$ then

$$\begin{aligned} \ln z &= \ln 6.42 + 1.57i \\ &= 1.86 + 1.57i \text{ (to 2 dp)} \end{aligned}$$

and the result is once again a complex number.

And if $z = 3.8e^{-0.236i}$, then $\ln z =$ .... to 3 dp

61

$$\ln z = \ln 3.8 - 0.236i = \boxed{1.335 - 0.236i}$$

Finally, here is an example of a rather different kind. Once you have seen it done, you will be able to deal with others of this kind. Here it is.

Express $e^{1-i\pi/4}$ in the form $a + ib$

Well now, we can write:

$$\begin{aligned} e^{1-i\pi/4} \text{ as } e^1e^{-i\pi/4} & \\ &= e(\cos\pi/4 - i\sin\pi/4) \\ &= e\left\{\frac{1}{\sqrt{2}} - i\frac{1}{\sqrt{2}}\right\} \\ &= \frac{e}{\sqrt{2}}(1 - i) \end{aligned}$$

This brings us to the end of this Program, except for the **Can You?** checklist and the **Test exercise**. Before you do them, read down the **Summary** that follows in the next frame and review any points on which you are not completely sure.

Move to Frame 62

Summary

62

1 *Powers of i*

$$i = \sqrt{-1}, \quad i^2 = -1, \quad i^3 = -i, \quad i^4 = 1.$$
$$i^{-1} = -i, \quad i^{-2} = -1, \quad i^{-3} = i, \quad i^{-4} = 1$$

A factor i turns a number through 90° in the positive direction.

2 *Complex numbers*

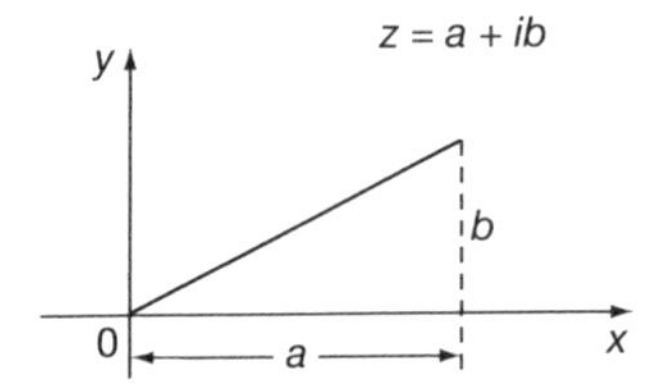

a = real part
b = imaginary part

3 *Conjugate complex numbers* $(a + ib)$ and $(a - ib)$
The product of two conjugate numbers is always real:

$$(a + ib)(a - ib) = a^2 + b^2$$

4 *Equal complex numbers*

If $a + ib = c + id$, then $a = c$ and $b = d$.

5 *Polar form of a complex number*

$$\begin{aligned} z &= a + ib \\ &= r(\cos\theta + i\sin\theta) \\ &= r\underline{|\theta} \end{aligned}$$

$$r = \sqrt{a^2 + b^2}; \quad \theta = \tan^{-1}\left\{\frac{b}{a}\right\}$$

also $\quad a = r\cos\theta; \; b = r\sin\theta$

$r =$ the modulus of z written 'mod z' or $|z|$

$\theta =$ the argument of z, written 'arg z'

6 *Exponential form of a complex number*

$$\left.\begin{aligned} z = r(\cos\theta + i\sin\theta) &= re^{i\theta} \\ \text{and} \quad r(\cos\theta - i\sin\theta) &= re^{-i\theta} \end{aligned}\right\} \theta \text{ in radians}$$

7 *Logarithm of a complex number*

$$z = re^{i\theta} \quad \therefore \; \ln z = \ln r + i\theta$$

or if $\quad z = re^{-i\theta} \quad \therefore \; \ln z = \ln r - i\theta$

Can You?

Checklist 2

63

Check this list before and after you try the end of Program test.

On a scale of 1 to 5 how confident are you that you can:

	Frames
• Recognize i as standing for $\sqrt{-1}$ and be able to reduce the powers of i to $\pm i$ or ± 1? *Yes* ☐ ☐ ☐ ☐ ☐ *No*	1 to 13
• Recognize that all complex numbers are in the form (real part) $+ i$(imaginary part)? *Yes* ☐ ☐ ☐ ☐ ☐ *No*	14 to 15
• Add, subtract and multiply complex numbers? *Yes* ☐ ☐ ☐ ☐ ☐ *No*	16 to 22
• Find the complex conjugate of a complex number? *Yes* ☐ ☐ ☐ ☐ ☐ *No*	23 to 27
• Divide complex numbers? *Yes* ☐ ☐ ☐ ☐ ☐ *No*	30 to 34
• State the conditions for the equality of two complex numbers? *Yes* ☐ ☐ ☐ ☐ ☐ *No*	35 to 39
• Draw complex numbers and recognize the parallel law of addition? *Yes* ☐ ☐ ☐ ☐ ☐ *No*	40 to 45
• Convert a complex number from Cartesian to polar form and vice versa? *Yes* ☐ ☐ ☐ ☐ ☐ *No*	46 to 55
• Write a complex number in its exponential form? *Yes* ☐ ☐ ☐ ☐ ☐ *No*	56 to 59
• Obtain the logarithm of a complex number? *Yes* ☐ ☐ ☐ ☐ ☐ *No*	60

Test exercise 2

64 You will find the questions quite straightforward and easy.

1 Simplify: (a) i^3 (b) i^5 (c) i^{12} (d) i^{14}.

2 Express in the form $a + ib$:

(a) $(4 - 7i)(2 + 3i)$ (b) $(-1 + i)^2$

(c) $(5 + 2i)(4 - 5i)(2 + 3i)$ (d) $\dfrac{4 + 3i}{2 - i}$

3 Find the values of x and y that satisfy the equation:
$(x + y) + i(x - y) = 14.8 + 6.2i$

4 Express in polar form:
(a) $3 + 5i$ (b) $-6 + 3i$ (c) $-4 - 5i$

5 Express in the form $a + ib$:
(a) $5(\cos 225^\circ + i \sin 225^\circ)$ (b) $4\underline{|330^\circ}$

6 Express in exponential form:
(a) $z_1 = 10\underline{|37^\circ 15'}$ and (b) $z_2 = 10\underline{|322^\circ 45'}$
Hence find $\ln z_1$ and $\ln z_2$.

7 Express $z = e^{1+i\pi/2}$ in the form $a + ib$.

Now are you ready to start Part 2 of the work on complex numbers

Further problems 2

65 **1** Simplify:

(a) $(5 + 4i)(3 + 7i)(2 - 3i)$ (b) $\dfrac{(2 - 3i)(3 + 2i)}{(4 - 3i)}$ (c) $\dfrac{\cos 3x + i \sin 3x}{\cos x + i \sin x}$

2 Express $\dfrac{2 + 3i}{i(4 - 5i)} + \dfrac{2}{i}$ in the form $a + ib$.

3 If $z = \dfrac{1}{2 + 3i} + \dfrac{1}{1 - 2i}$, express z in the form $a + ib$.

4 If $z = \dfrac{2 + i}{1 - i}$, find the real and imaginary parts of the complex number $z + \dfrac{1}{z}$.

5 Simplify $(2 + 5i)^2 + \dfrac{5(7 + 2i)}{3 - 4i} - i(4 - 6i)$, expressing the result in the form $a + ib$.

▶

6 If $z_1 = 2 + i$, $z_2 = -2 + 4i$ and $\frac{1}{z_3} = \frac{1}{z_1} + \frac{1}{z_2}$, evaluate z_3 in the form $a + ib$.
If z_1, z_2, z_3 are represented on an Argand diagram by the points P, Q, R, respectively, prove that R is the foot of the perpendicular from the origin on to the line PQ.

7 Points A, B, C, D, on an Argand diagram, represent the complex numbers $9 + i$, $4 + 13i$, $-8 + 8i$, $-3 - 4i$ respectively. Prove that ABCD is a square.

8 If $(2 + 3i)(3 - 4i) = x + iy$, evaluate x and y.

9 If $(a + b) + i(a - b) = (2 + 5i)^2 + i(2 - 3i)$, find the values of a and b.

10 If x and y are real, solve the equation:

$$\frac{ix}{1 + iy} = \frac{3x + 4i}{x + 3y}$$

11 If $z = \frac{a + ib}{c + id}$, where a, b, c and d are real quantities, show that (a) if z is real then $\frac{a}{b} = \frac{c}{d}$ and (b) if z is entirely imaginary then $\frac{a}{b} = -\frac{d}{c}$.

12 Given that $(a + b) + i(a - b) = (1 + i)^2 + i(2 + i)$, obtain the values of a and b.

13 Express $(-1 + i)$ in the form $re^{i\theta}$ where r is positive and $-\pi < \theta < \pi$.

14 Find the modulus of $z = (2 - i)(5 + 12i)/(1 + 2i)^3$.

15 If x is real, show that $(2 + i)e^{(1+3i)x} + (2 - i)e^{(1-3i)x}$ is also real.

16 Given that $z_1 = R_1 + R + i\omega L$; $z_2 = R_2$; $z_3 = \frac{1}{i\omega C_3}$; and $z_4 = R_4 + \frac{1}{i\omega C_4}$; and also that $z_1 z_3 = z_2 z_4$, express R and L in terms of the real constants R_1, R_2, R_4, C_3 and C_4.

17 If $z = x + iy$, where x and y are real, and if the real part of $(z + 1)/(z + i)$ is equal to 1, show that the point z lies on a straight line in the Argand diagram.

18 When $z_1 = 2 + 3i$, $z_2 = 3 - 4i$, $z_3 = -5 + 12i$, then $z = z_1 + \frac{z_2 z_3}{z_2 + z_3}$. If $E = Iz$, E when $I = 5 + 6i$.

19 If $\frac{R_1 + i\omega L}{R_3} = \frac{R_2}{R_4 - i\frac{1}{\omega C}}$, where R_1, R_2, R_3, R_4, ω, L and C are real, show that

$$L = \frac{CR_2R_3}{\omega^2 C^2 R_4^2 + 1}$$

20 If z and $\bar{z}$ are conjugate complex numbers, find two complex numbers, $z = z_1$ and $z = z_2$, that satisfy the equation:

$$3z\bar{z} + 2(z - \bar{z}) = 39 + 12i$$

On an Argand diagram, these two numbers are represented by the points P and Q. If R represents the number $1i$, show that the angle PRQ is a right angle.

The polar form

Learning outcomes

When you have completed this Program you will be able to:

- Use the shorthand form for a complex number in polar form
- Write complex numbers in polar form using negative angles
- Multiply and divide complex numbers in polar form
- Use DeMoivre's theorem
- Find the roots of a complex number
- Demonstrate trigonometric identities of multiple angles using complex numbers
- Solve loci problems using complex numbers

Introduction

1

In the previous program on complex numbers, we discovered how to manipulate them in adding, subtracting, multiplying and dividing. We also finished by seeing that a complex number $a + ib$ can also be expressed in polar form, which is always of the form $r(\cos\theta + i\sin\theta)$.

You will remember that values of r and θ can easily be found from the diagram of the given complex number:

$$r^2 = a^2 + b^2 \quad \therefore \quad r = \sqrt{a^2 + b^2}$$

$$\text{and } \tan\theta = \frac{b}{a} \quad \therefore \quad \theta = \tan^{-1}\frac{b}{a}$$

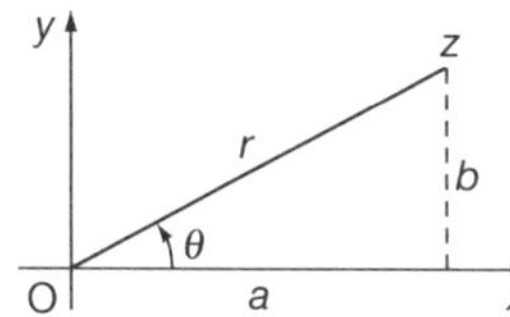

To be sure that you have taken the correct value of θ, always *draw a sketch diagram* to see which quadrant the complex number is in.

Remember that θ is always measured from ....

2

$\boxed{Ox}$ i.e. the positive x-axis.

Right. As a warming-up exercise, do the following:

Express $z = 12 - 5i$ in polar form

Do not forget the sketch diagram. It ensures that you get the correct value for θ.

When you have finished, and not before, move on to Frame 3 to check your result

3

$$\boxed{13(\cos 337°23' + i\sin 337°23')}$$

Here it is, worked out in full:

$$r^2 = 12^2 + 5^2 = 144 + 25 = 169$$

$$\therefore \ r = 13$$

$$\tan E = \frac{5}{12} = 0.4167 \quad \therefore \quad E = 22°37'$$

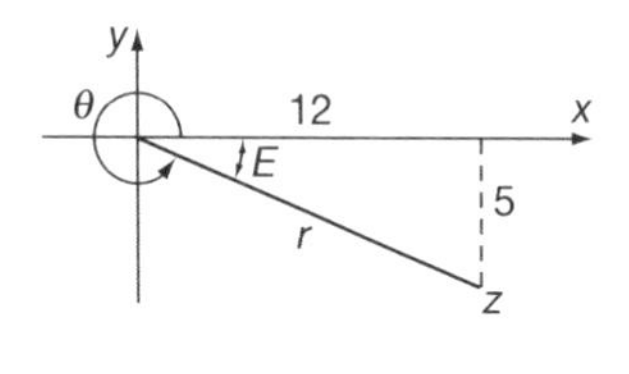

In this case, $\theta = 360° - E = 360° - 22°37' \quad \therefore \quad \theta = 337°23'$

$$z = r(\cos\theta + i\sin\theta) = 13(\cos 337°23' + i\sin 337°23')$$

Did you get that right? Here is one more, done in just the same way:

Express $-5 - 4i$ in polar form.

Diagram first of all! Then you cannot go wrong.

When you have the result, on to Frame 4

4

$$z = 6.403(\cos 218^\circ 40' + i\sin 218^\circ 40')$$

Here is the working: check yours.

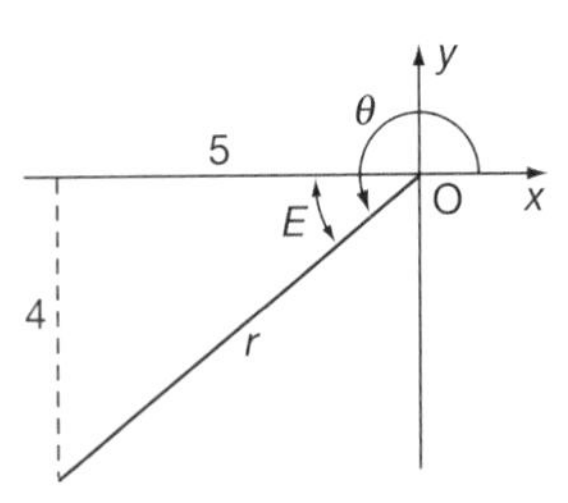

$r^2 = 5^2 + 4^2 = 25 + 16 = 41$

$\therefore\ r = \sqrt{41} = 6.403$ (to 3 dp)

$\tan E = \dfrac{4}{5} = 0.8 \ \therefore\ E = 38^\circ 40'$

In this case, $\theta = 180^\circ + E = 218^\circ 40'$

So $z = -5 - 4i = 6.403(\cos 218^\circ 40' + i\sin 218^\circ 40')$

Shorthand notation of polar form

Since every complex number in polar form is of the same structure, i.e. $r(\cos\theta + i\sin\theta)$ and differs from another complex number simply by the values of r and θ, we have a shorthand method of quoting the result in polar form. Do you remember what it is? The shorthand way of writing the result above, i.e. $6.403(\cos 218^\circ 40' + i\sin 218^\circ 40')$ is ....

5

$$6.403\underline{\lvert 218^\circ 40'}$$

Correct. Likewise:

$5.72(\cos 322^\circ 15' + i\sin 322^\circ 15')$ is written $5.72\underline{\lvert 322^\circ 15'}$

$5(\cos 105^\circ + i\sin 105^\circ)$ is written $5\underline{\lvert 105^\circ}$

$3.4\left(\cos\dfrac{\pi}{6} + i\sin\dfrac{\pi}{6}\right)$ is written $3.4\underline{\left\lvert \dfrac{\pi}{6}\right.}$

They are all complex numbers in polar form. They are all the same structure and differ one from another simply by the values of

.... and ....

6

r and θ

Now let us consider the following example.

Express $z = 4 - 3i$ in polar form.

First the diagram:

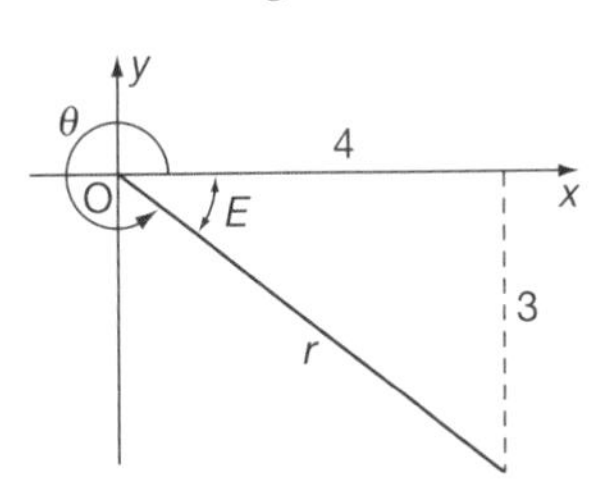

From this:

$r = 5$

$\tan E = \dfrac{3}{4} = 0.75 \ \therefore\ E = 36^\circ 52'$

$\theta = 360^\circ - 36^\circ 52' = 323^\circ 8'$

$z = 4 - 3i = 5(\cos 323^\circ 8' + i\sin 323^\circ 8')$

or in shortened form, $z =$ ....

7

$$z = 5\underline{|323°8'}$$

Negative angles

In the previous example, we have:

$z = 5(\cos 323°8' + i \sin 323°8')$

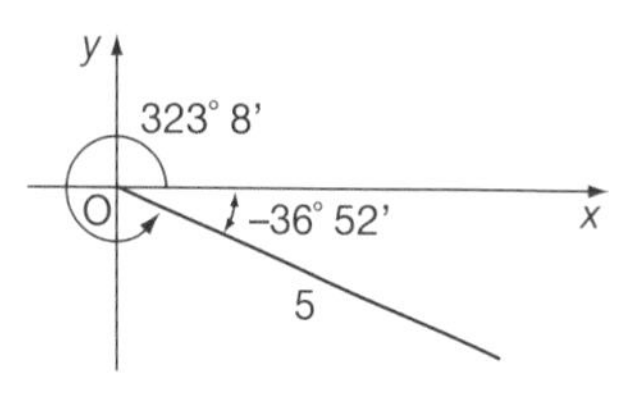

But the direction of the complex number, measured from Ox, could be given as $-36°52'$, the minus sign showing that we are measuring the angle in the opposite sense from the usual positive direction.

We could write $z = 5(\cos[-36°52'] + i \sin[-36°52'])$. But you already know that $\cos[-\theta] = \cos\theta$ and $\sin[-\theta] = -\sin\theta$.

$z = 5(\cos 36°52' - i \sin 36°52')$

i.e. very much like the polar form but with a minus sign in the middle. This comes about whenever we use negative angles. In the same way:

$$\begin{aligned} z &= 4(\cos 250° + i \sin 250°) \\ &= 4(\cos[-110°] + i \sin[-110°]) \\ &= 4(\ldots\ldots\ldots\ldots) \end{aligned}$$

8

$$z = 4(\cos 110° - i \sin 110°)$$

since $\cos(-110°) = \cos 110°$
and $\sin(-110°) = -\sin 110°$

It is sometimes convenient to use this form when the value of θ is greater than 180°, i.e. in the 3rd and 4th quadrants.

Here are some examples:

$$\begin{aligned} z &= 3(\cos 230° + i \sin 230°) \\ &= 3(\cos 130° - i \sin 130°) \end{aligned}$$

Similarly: $z = 3(\cos 300° + i \sin 300°) = 3(\cos 60° - i \sin 60°)$
$z = 4(\cos 290° + i \sin 290°) = 4(\cos 70° - i \sin 70°)$
$z = 2(\cos 215° + i \sin 215°) = 2(\cos 145° - i \sin 145°)$

and $z = 6(\cos 310° + i \sin 310°) = \ldots\ldots\ldots\ldots$

9

$$\boxed{z = 6(\cos 50^\circ - i \sin 50^\circ)}$$

Because $\cos 310^\circ = \cos 50^\circ$
and $\sin 310^\circ = -\sin 50^\circ$

A moment ago we agreed that the minus sign comes about by the use of negative angles. To convert a complex number given in this way back into proper polar form, i.e. with a '+' in the middle, we simply work back the way we came. A complex number with a negative sign in the middle is equivalent to the same complex number with a positive sign, but with the angles made negative.

e.g. $z = 4(\cos 30^\circ - i \sin 30^\circ)$
$= 4(\cos[-30^\circ] + i \sin[-30^\circ])$
$= 4(\cos 330^\circ + i \sin 330^\circ)$ and we are back in proper polar form.

You do this one: Convert $z = 5(\cos 40^\circ - i \sin 40^\circ)$ into proper polar form.

Then on to Frame 10

10

$$\boxed{z = 5(\cos 320^\circ + i \sin 320^\circ)}$$

Because

$$z = 5(\cos 40^\circ - i \sin 40^\circ) = 5(\cos[-40^\circ] + i \sin[-40^\circ])$$
$$= 5(\cos 320^\circ + i \sin 320^\circ)$$

Here is another for you to do.

Express $z = 4(\cos 100^\circ - i \sin 100^\circ)$ in proper polar form.

Do not forget, it all depends on the use of negative angles.

11

$$\boxed{z = 4(\cos 260^\circ + i \sin 260^\circ)}$$

Because

$$z = 4(\cos 100^\circ - i \sin 100^\circ) = 4(\cos[-100^\circ] + i \sin[-100^\circ])$$
$$= 4(\cos 260^\circ + i \sin 260^\circ)$$

We ought to see how this modified polar form affects our shorthand notation.

▶

Remember, $5(\cos 60^\circ + i \sin 60^\circ)$ is written $5\underline{|60^\circ}$

How then shall we write $5(\cos 60^\circ - i \sin 60^\circ)$?

We know that this really stands for $5(\cos[-60^\circ] + i \sin[-60^\circ])$ so we could write $5\underline{|-60^\circ}$. But instead of using the negative angle we use a different symbol, i.e. $5\underline{|-60^\circ}$ becomes $5\overline{|60^\circ}$.

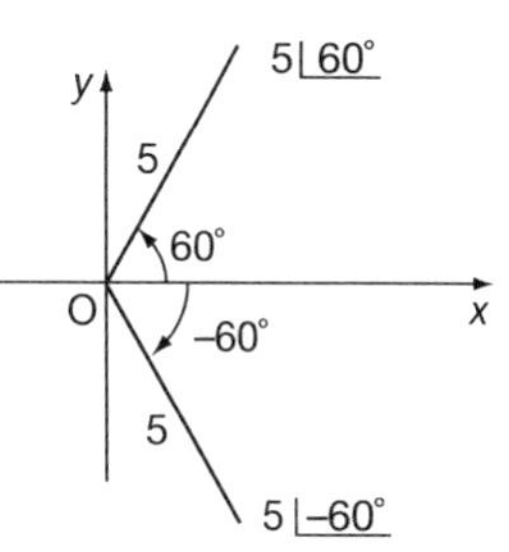

Similarly, $3(\cos 45^\circ - i \sin 45^\circ) = 3\underline{|-45^\circ} = \dots\dots\dots\dots$

12

$$3\overline{|45^\circ}$$

This is easy to remember,

for the sign $\underline{|\;\;}$ resembles the first quadrant and indicates

measuring angles $\circlearrowleft$ i.e. in the positive direction,

while the sign $\overline{|\;\;}$ resembles the fourth quadrant and indicates

measuring angles $\circlearrowright$ i.e. in the negative direction.

e.g. $(\cos 15^\circ + i \sin 15^\circ)$ is written $\underline{|15^\circ}$

but $(\cos 15^\circ - i \sin 15^\circ)$, which is really $(\cos[-15^\circ] + i \sin[-15^\circ])$

is written $\overline{|15^\circ}$

So how do we write (a) $(\cos 120^\circ + i \sin 120^\circ)$
and (b) $(\cos 135^\circ - i \sin 135^\circ)$
in the shorthand way?

13

(a) $\underline{|120^\circ}$ (b) $\overline{|135^\circ}$

The polar form at first sight seems to be a complicated way of representing a complex number. However it is very useful, as we shall see. Suppose we multiply together two complex numbers in this form:

Let $z_1 = r_1(\cos\theta_1 + i \sin\theta_1)$ and $z_2 = r_2(\cos\theta_2 + i \sin\theta_2)$

Then $z_1 z_2 = r_1(\cos\theta_1 + i \sin\theta_1) r_2(\cos\theta_2 + i \sin\theta_2)$

$$= r_1 r_2(\cos\theta_1 \cos\theta_2 + i \sin\theta_1 \cos\theta_2 + i \cos\theta_1 \sin\theta_2 + i^2 \sin\theta_1 \sin\theta_2)$$

Rearranging the terms and remembering that $i^2 = -1$, we get

$$z_1 z_2 = r_1 r_2[(\cos\theta_1 \cos\theta_2 - \sin\theta_1 \sin\theta_2) + i(\sin\theta_1 \cos\theta_2 + \cos\theta_1 \sin\theta_2)]$$

Now the brackets $(\cos\theta_1 \cos\theta_2 - \sin\theta_1 \sin\theta_2)$
and $(\sin\theta_1 \cos\theta_2 + \cos\theta_1 \sin\theta_2)$
ought to ring a bell. What are they?

14

$$\boxed{\begin{aligned}\cos\theta_1\cos\theta_2 - \sin\theta_1\sin\theta_2 &= \cos(\theta_1+\theta_2)\\ \sin\theta_1\cos\theta_2 + \cos\theta_1\sin\theta_2 &= \sin(\theta_1+\theta_2)\end{aligned}}$$

In that case, $z_1z_2 = r_1r_2[\cos(\theta_1+\theta_2) + i\sin(\theta_1+\theta_2)]$.
Note this important result. We have just shown that

$$r_1(\cos\theta_1 + i\sin\theta_1)\cdot r_2(\cos\theta_2 + i\sin\theta_2) = r_1r_2[\cos(\theta_1+\theta_2) + i\sin(\theta_1+\theta_2)]$$

i.e. To multiply together two complex numbers in polar form,

(a) multiply the r's together,

(b) add the angles, θ, together.

It is just as easy as that!

$$\begin{aligned}\text{e.g. } & 2(\cos 30^\circ + i\sin 30^\circ)\times 3(\cos 40^\circ + i\sin 40^\circ)\\ &= 2\times 3(\cos[30^\circ+40^\circ] + i\sin[30^\circ+40^\circ])\\ &= 6(\cos 70^\circ + i\sin 70^\circ)\end{aligned}$$

So if we multiply together $5(\cos 50^\circ + i\sin 50^\circ)$ and $2(\cos 65^\circ + i\sin 65^\circ)$ we get ....

15

$$\boxed{10(\cos 115^\circ + i\sin 115^\circ)}$$

Remember, multiply the r's; add the θ's.

Here you are then; all done the same way:

(a) $2(\cos 120^\circ + i\sin 120^\circ)\times 4(\cos 20^\circ + i\sin 20^\circ)$
$\quad = 8(\cos 140^\circ + i\sin 140^\circ)$

(b) $a(\cos\theta + i\sin\theta)\times b(\cos\phi + i\sin\phi)$
$\quad = ab(\cos[\theta+\phi] + i\sin[\theta+\phi])$

(c) $6(\cos 210^\circ + i\sin 210^\circ)\times 3(\cos 80^\circ + i\sin 80^\circ)$
$\quad = 18(\cos 290^\circ + i\sin 290^\circ)$

(d) $5(\cos 50^\circ + i\sin 50^\circ)\times 3(\cos[-20^\circ] + i\sin[-20^\circ])$
$\quad = 15(\cos 30^\circ + i\sin 30^\circ)$

Have you got it? No matter what the angles are, all we do is:

(a) multiply the moduli, (b) add the arguments.

So therefore, $4(\cos 35^\circ + i\sin 35^\circ)\times 3(\cos 20^\circ + i\sin 20^\circ) =$ ....

16

$$\boxed{12(\cos 55^\circ + i\sin 55^\circ)}$$

Now let us see if we can discover a similar set of rules for division.

We already know that to simplify $\dfrac{5+6i}{3+4i}$ we first obtain a denominator that is entirely real by multiplying top and bottom by ....

17

the conjugate of the denominator, i.e. $3 - 4i$

Right. Then let us do the same with $\dfrac{r_1(\cos\theta_1 + i\sin\theta_1)}{r_2(\cos\theta_2 + i\sin\theta_2)}$

$$\frac{r_1(\cos\theta_1 + i\sin\theta_1)}{r_2(\cos\theta_2 + i\sin\theta_2)} = \frac{r_1(\cos\theta_1 + i\sin\theta_1)(\cos\theta_2 - i\sin\theta_2)}{r_2(\cos\theta_2 + i\sin\theta_2)(\cos\theta_2 - i\sin\theta_2)}$$
$$= \frac{r_1}{r_2}\frac{(\cos\theta_1\cos\theta_2 + i\sin\theta_1\cos\theta_2 - i\cos\theta_1\sin\theta_2 + \sin\theta_1\sin\theta_2)}{(\cos^2\theta_2 + \sin^2\theta_2)}$$
$$= \frac{r_1}{r_2}\frac{[(\cos\theta_1\cos\theta_2 + \sin\theta_1\sin\theta_2) + i(\sin\theta_1\cos\theta_2 - \cos\theta_1\sin\theta_2)]}{1}$$
$$= \frac{r_1}{r_2}[\cos(\theta_1 - \theta_2) + i\sin(\theta_1 - \theta_2)]$$

So, for division, the rule is ....

18

divide the r's and subtract the angle

That is correct.

$$\text{e.g. } \frac{6(\cos 72^\circ + i\sin 72^\circ)}{2(\cos 41^\circ + i\sin 41^\circ)} = 3(\cos 31^\circ + i\sin 31^\circ)$$

So we now have two important rules:

If $z_1 = r_1(\cos\theta_1 + i\sin\theta_1)$ and $z_2 = r_2(\cos\theta_2 + i\sin\theta_2)$

then (a) $z_1z_2 = r_1r_2[\cos(\theta_1 + \theta_2) + i\sin(\theta_1 + \theta_2)]$

and (b) $\dfrac{z_1}{z_2} = \dfrac{r_1}{r_2}[\cos(\theta_1 - \theta_2) + i\sin(\theta_1 - \theta_2)]$

The results are still, of course, in polar form.

Now here is one for you to think about.

If $z_1 = 8(\cos 65^\circ + i\sin 65^\circ)$ and $z_2 = 4(\cos 23^\circ + i\sin 23^\circ)$

then (a) $z_1z_2 =$ .... and (b) $\dfrac{z_1}{z_2} =$ ....

19

(a) $z_1z_2 = 32(\cos 88^\circ + i\sin 88^\circ)$

(b) $\dfrac{z_1}{z_2} = 2(\cos 42^\circ + i\sin 42^\circ)$

▶

Of course, we can combine the rules in a single example:

$$\text{e.g.}\quad \frac{5(\cos 60^\circ + i\sin 60^\circ) \times 4(\cos 30^\circ + i\sin 30^\circ)}{2(\cos 50^\circ + i\sin 50^\circ)}$$
$$= \frac{20(\cos 90^\circ + i\sin 90^\circ)}{2(\cos 50^\circ + i\sin 50^\circ)}$$
$$= 10(\cos 40^\circ + i\sin 40^\circ)$$

What does the following product become?

$$4(\cos 20^\circ + i\sin 20^\circ) \times 3(\cos 30^\circ + i\sin 30^\circ) \times 2(\cos 40^\circ + i\sin 40^\circ)$$

Result in next frame

20

$$\boxed{24(\cos 90^\circ + i\sin 90^\circ)}$$

$$\text{i.e.}\quad (4 \times 3 \times 2)[\cos(20^\circ + 30^\circ + 40^\circ) + i\sin(20^\circ + 30^\circ + 40^\circ)]$$
$$= 24(\cos 90^\circ + i\sin 90^\circ)$$

Now what about a few review examples on the work we have done so far?

Move to the next frame

Review exercise

21

Work all these questions and then turn on to Frame 22 and check your results.

1 Express in polar form, $z = -4 + 2i$.

2 Express in true polar form, $z = 5(\cos 55^\circ - i\sin 55^\circ)$.

3 Simplify the following, giving the results in polar form:

(a) $3(\cos 143^\circ + i\sin 143^\circ) \times 4(\cos 57^\circ + i\sin 57^\circ)$

(b) $\dfrac{10(\cos 126^\circ + i\sin 126^\circ)}{2(\cos 72^\circ + i\sin 72^\circ)}$

4 Express in the form $a + ib$:

(a) $2(\cos 30^\circ + i\sin 30^\circ)$

(b) $5(\cos 57^\circ - i\sin 57^\circ)$

Solutions are in Frame 22. Move on and see how you have fared

22

1

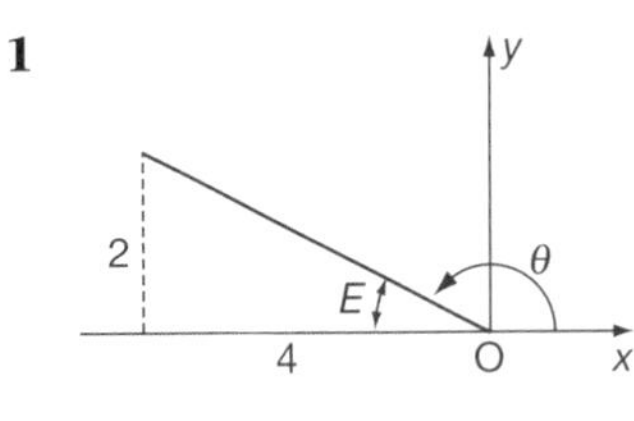

$r^2 = 2^2 + 4^2 = 4 + 16 = 20$

$\therefore\ r = 4.472$

$\tan E = 0.5\ \therefore\ E = 26^\circ 34'$

$\therefore\ \theta = 153^\circ 26'$

$z = -4 + 2i = 4.472(\cos 153^\circ 26' + i\sin 153^\circ 26')$

▶

2 $z = 5(\cos 55° - i\sin 55°) = 5[\cos(-55°) + i\sin(-55°)]$
$= 5(\cos 305° + i\sin 305°)$

3 (a) $3(\cos 143° + i\sin 143°) \times 4(\cos 57° + i\sin 57°)$
$= 3 \times 4[\cos(143° + 57°) + i\sin(143° + 57°)]$
$= 12(\cos 200° + i\sin 200°)$

(b) $\dfrac{10(\cos 126° + i\sin 126°)}{2(\cos 72° + i\sin 72°)}$
$= \dfrac{10}{2}[\cos(126° - 72°) + i\sin(126° - 72°)]$
$= 5(\cos 54° + i\sin 54°)$

4 (a) $2(\cos 30° + i\sin 30°)$
$= 2(0.866 + 0.5i) = 1.732 + i$

(b) $5(\cos 57° - i\sin 57°)$
$= 5(0.5446 - 0.8387i)$
$= 2.723 - 4.193i$

Now continue the Program in Frame 23

Multiplication and division of the polar form

23

Now we are ready to go on to a very important section which follows from our work on multiplication of complex numbers in polar form.

We have already established that:

if $z_1 = r_1(\cos\theta_1 + i\sin\theta_1)$ and $z_2 = r_2(\cos\theta_2 + i\sin\theta_2)$

then $z_1z_2 = r_1r_2[\cos(\theta_1 + \theta_2) + i\sin(\theta_1 + \theta_2)]$

So if $z_3 = r_3(\cos\theta_3 + i\sin\theta_3)$ then we have

$z_1z_2z_3 = r_1r_2[\cos(\theta_1 + \theta_2) + i\sin(\theta_1 + \theta_2)]r_3(\cos\theta_3 + i\sin\theta_3)$

$= \ldots\ldots\ldots\ldots$

24

$$z_1z_2z_3 = r_1r_2r_3[\cos(\theta_1 + \theta_2 + \theta_3) + i\sin(\theta_1 + \theta_2 + \theta_3)]$$

because in multiplication, we multiply the moduli and add the arguments

Now suppose that z_1, z_2, z_3 are all alike and that each is equal to $z = r(\cos\theta + i\sin\theta)$. Then the result above becomes:

$$z_1z_2z_3 = z^3 = r \cdot r \cdot r[\cos(\theta + \theta + \theta) + i\sin(\theta + \theta + \theta)] = r^3(\cos 3\theta + i\sin 3\theta)$$

or $$z^3 = [r(\cos\theta + i\sin\theta)]^3 = r^3(\cos\theta + i\sin\theta)^3 = r^3(\cos 3\theta + i\sin 3\theta)$$

That is, if we wish to cube a complex number in polar form, we just cube the modulus (r value) and multiply the argument (θ) by 3.

Similarly, to square a complex number in polar form, we square the modulus (r value) and multiply the argument (θ) by $\ldots\ldots\ldots\ldots$

25

$\boxed{2}$ i.e. $[r(\cos\theta + i\sin\theta)]^2 = r^2(\cos 2\theta + i\sin 2\theta)$

Let us take another look at these results:

$$[r(\cos\theta + i\sin\theta)]^2 = r^2(\cos 2\theta + i\sin 2\theta)$$
$$[r(\cos\theta + i\sin\theta)]^3 = r^3(\cos 3\theta + i\sin 3\theta)$$

Similarly:

$$[r(\cos\theta + i\sin\theta)]^4 = r^4(\cos 4\theta + i\sin 4\theta)$$
$$[r(\cos\theta + i\sin\theta)]^5 = r^5(\cos 5\theta + i\sin 5\theta) \quad \text{and so on}$$

In general, then, we can say:

$[r(\cos\theta + i\sin\theta)]^n = \ldots\ldots\ldots\ldots$

26

$$[r(\cos\theta + i\sin\theta)]^n = \boxed{r^n(\cos n\theta + i\sin n\theta)}$$

De Moivre's theorem

This general result is very important and is called *DeMoivre's theorem*. It says that to raise a complex number in polar form to any power n, we raise the r to the power n and multiply the angle by n:

e.g. $[4(\cos 50° + i\sin 50°]^2 = 4^2[\cos(2 \times 50°) + i\sin(2 \times 50°)]$
$= 16(\cos 100° + i\sin 100°)$

and $[3(\cos 110° + i\sin 110°)]^3 = 27(\cos 330° + i\sin 330°)$

and in the same way:

$[2(\cos 37° + i\sin 37°)]^4 = \ldots\ldots\ldots\ldots$

27

$$\boxed{16(\cos 148° + i\sin 148°)}$$

Roots of a complex number

This is where the polar form really comes into its own! For DeMoivre's theorem also applies when we are raising the complex number to a fractional power, i.e. when we are finding the roots of a complex number.

e.g. To find the square root of $z = 4(\cos 70° + i\sin 70°)$

We have $\sqrt{z} = z^{\frac{1}{2}} = [4(\cos 70° + i\sin 70°)]^{\frac{1}{2}}$ i.e. $n = \dfrac{1}{2}$

$$= 4^{\frac{1}{2}}\left(\cos\frac{70°}{2} + i\sin\frac{70°}{2}\right)$$
$$= 2(\cos 35° + i\sin 35°)$$

▶

It works every time, no matter whether the power is positive, negative, whole number or fraction. In fact, DeMoivre's theorem is so important, let us write it down again. Here goes:

If $z = r(\cos\theta + i\sin\theta)$, then $z^n = \dots\dots\dots\dots$

28

$z = r(\cos\theta + i\sin\theta)$, then $\boxed{z^n = r^n(\cos n\theta + i\sin n\theta)}$

for any value of n.

Look again at finding a root of a complex number. Let us find the cube root of $z = 8(\cos 120° + i\sin 120°)$. Here is the given complex number shown on an Argand diagram:

$z = 8\underline{|120°}$

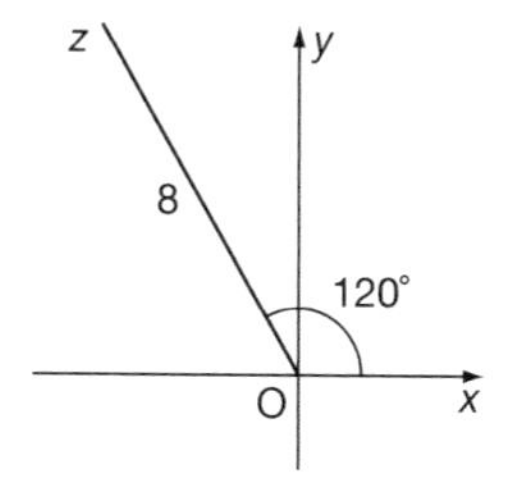

Of course, we could say that θ was '1 revolution + 120°': the vector would still be in the same position, or, for that matter (2 revs + 120°), (3 revs + 120°) etc. i.e. $z = 8\underline{|120°}$ or $8\underline{|480°}$ or $8\underline{|840°}$ or $8\underline{|1200°}$ etc. and if we now apply DeMoivre's theorem to each of these, we get:

$$z^{\frac{1}{3}} = 8^{\frac{1}{3}}\underline{\left|\frac{120°}{3}\right.} \text{ or } 8^{\frac{1}{3}}\underline{\left|\frac{480°}{3}\right.} \text{ or } \dots\dots\dots\dots \text{ or } \dots\dots\dots\dots \text{ etc.}$$

29

$$\boxed{z^{\frac{1}{3}} = 8^{\frac{1}{3}}\underline{\left|\frac{120°}{3}\right.} \text{ or } 8^{\frac{1}{3}}\underline{\left|\frac{480°}{3}\right.} \text{ or } 8^{\frac{1}{3}}\underline{\left|\frac{840°}{3}\right.} \text{ or } 8^{\frac{1}{3}}\underline{\left|\frac{1200°}{3}\right.}}$$

If we simplify these, we get:

$z^{\frac{1}{3}} = 2\underline{|40°}$ or $2\underline{|160°}$ or $2\underline{|280°}$ or $2\underline{|400°}$etc.

If we put each of these on an Argand diagram, as follows:

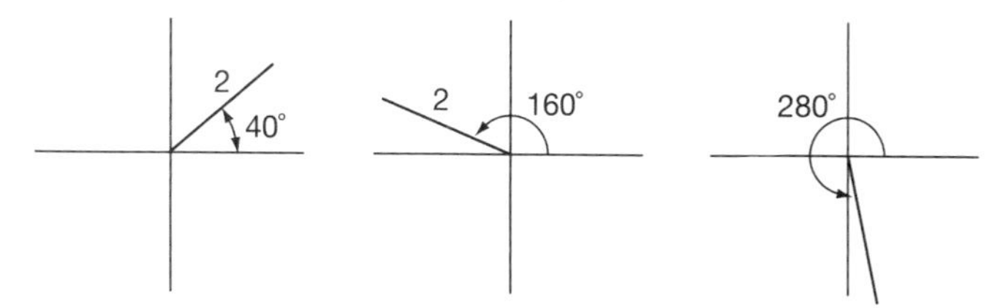

we see we have three quite different results for the cube root of z and also that the fourth diagram would be a repetition of the first. Any subsequent calculations merely repeat these three positions.

Make a sketch of the first three vectors on a single Argand diagram

30

Here they are. The cube roots of $z = 8(\cos 120^\circ + i\sin 120^\circ)$:

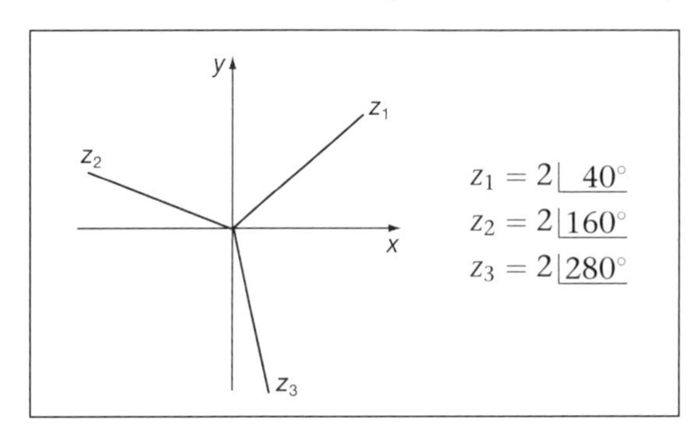

We see, therefore, that there are 3 cube roots of a complex number. Also, if you consider the angles, you see that the 3 roots are equally spaced round the diagram, any two adjacent complex numbers being separated by .... degrees.

31

$$\boxed{120^\circ}$$

That is right. Therefore all we need to do in practice is to find the first of the roots and simply add 120° on to get the next – and so on.

Notice that the three cube roots of a complex number are equal in modulus (or length) and equally spaced at intervals of $\dfrac{360^\circ}{3}$ i.e. 120°.

Now let us take another example. On to the next frame

32

To find the three cube roots of $z = 5(\cos 225^\circ + i\sin 225^\circ)$

The first root is given by

$$z_1 = z^{\frac{1}{3}} = 5^{\frac{1}{3}}\left(\cos\frac{225^\circ}{3} + i\sin\frac{225^\circ}{3}\right)$$
$$= 1.71(\cos 75^\circ + i\sin 75^\circ)$$
$$z_1 = 1.71\lfloor 75^\circ$$

We know that the other cube roots are the same length (modulus), i.e. 1.71, and separated at intervals of $\dfrac{360^\circ}{3}$, i.e. 120°.

So the three cube roots are:

$$z_1 = 1.71\lfloor 75^\circ$$
$$z_2 = 1.71\lfloor 195^\circ$$
$$z_3 = 1.71\lfloor 315^\circ$$

It helps to see them on an Argand diagram, so sketch them on a combined diagram.

33

Here they are:

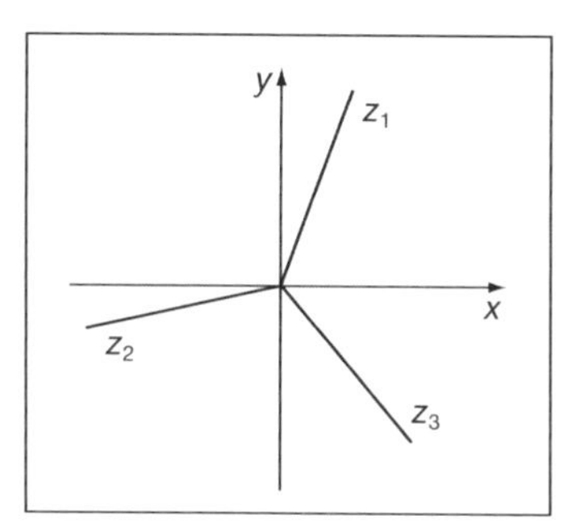

We find any roots of a complex number in the same way:

(a) Apply DeMoivre's theorem to find the first of the n roots.
(b) The other roots will then be distributed round the diagram at regular intervals of $\dfrac{360^\circ}{n}$.

A complex number, therefore, has:

2 square roots, separated by $\dfrac{360^\circ}{2}$ i.e. 180°

3 cube roots, separated by $\dfrac{360^\circ}{3}$ i.e. 120°

4 fourth roots, separated by $\dfrac{360^\circ}{4}$ i.e. 90°

5 fifth roots, separated by ....etc.

34

$$\boxed{\frac{360^\circ}{5} \text{ i.e. } 72^\circ}$$

And now: To find the 5 fifth roots of $12\underline{|300^\circ}$

$$z = 12\underline{|300^\circ} \quad \therefore \quad z_1 = 12^{\frac{1}{5}}\underline{\left|\frac{300^\circ}{5}\right.} = 12^{\frac{1}{5}}\underline{|60^\circ}$$

Now, $12^{\frac{1}{5}} = 1.644$ to 3 dp and so the first of the 5 fifth roots is therefore $z_1 = 1.644\underline{|60^\circ}$

The others will be of the same modulus, i.e. 1.644, and equally separated at intervals of $\dfrac{360^\circ}{5}$ i.e. 72°.

So the required 5 fifth roots of $12\underline{|300^\circ}$ are:

$z_1 = 1.644\underline{|60^\circ}$ $\quad z_2 = 1.644\underline{|132^\circ}$ $\quad z_3 = 1.644\underline{|204^\circ}$

$z_4 = 1.644\underline{|276^\circ}$ $\quad z_5 = 1.644\underline{|348^\circ}$

Sketch them on an Argand diagram, as before.

35

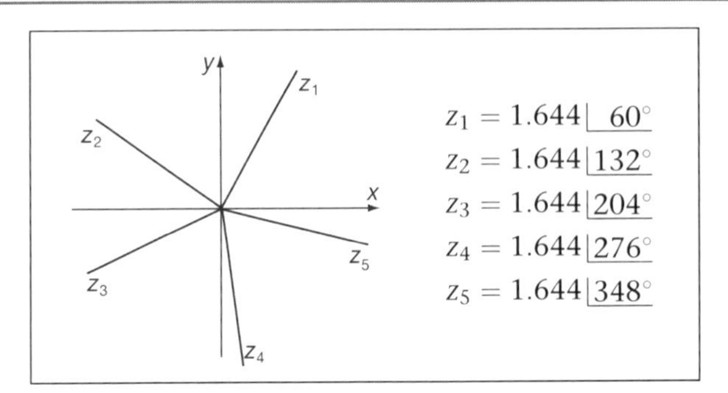

Although there are 5 fifth roots of a complex number, we are sometimes asked to find the *principal root*. This is always the root which is nearest to the positive x-axis.

In some cases, it may be the first root. In others, it may be the last root. The only test is to see which root is nearest to the positive x-axis. If the first and last root are equidistant from the positive x-axis, the principal root is taken to be the first root.

In the example above, the *principal root* is therefore ....

36

$$z_5 = 1.644\underline{|348^\circ}$$

Good. Now here is another example worked in detail. Follow it.

We have to find the 4 fourth roots of $z = 7(\cos 80^\circ + i \sin 80^\circ)$

The first root $z_1 = 7^{\frac{1}{4}}\underline{\left|\dfrac{80^\circ}{4}\right.} = 7^{\frac{1}{4}}\underline{|20^\circ}$

Since $7^{\frac{1}{4}} = 1.627$ to 3 dp

$z_1 = 1.627\underline{|20^\circ}$

The other roots will be separated by intervals of $\dfrac{360^\circ}{4} = 90^\circ$

Therefore the 4 fourth roots are:

$z_1 = 1.627\underline{|20^\circ}$ $\quad z_2 = 1.627\underline{|110^\circ}$

$z_3 = 1.627\underline{|200^\circ}$ $\quad z_4 = 1.627\underline{|290^\circ}$

And once again, draw an Argand diagram to illustrate these roots.

37

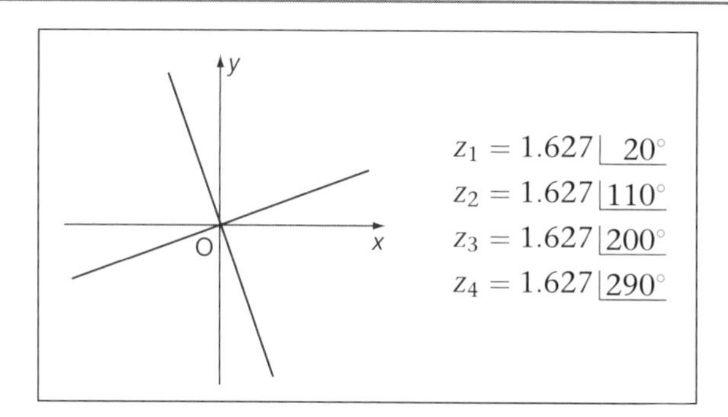

And in this example the principal fourth root is . . . .

38

$$\boxed{z_1 = 1.627\underline{|20^\circ}}$$

since it is the root nearest to the positive x-axis.

Now you can do one entirely on your own. Here it is.

Find the three cube roots of $6(\cos 240^\circ + i \sin 240^\circ)$. Represent them on an Argand diagram and indicate which is the principal cube root.

When you have finished it, move on to Frame 39 and check your results

39

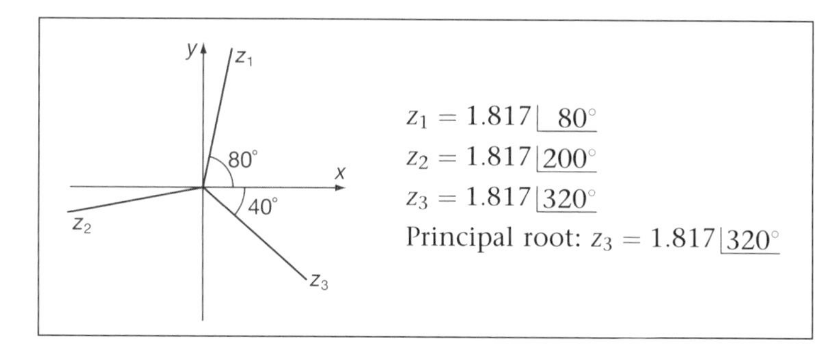

Here is the working:

$$z = 6\underline{|240^\circ} \qquad z_1 = 6^{\frac{1}{3}}\underline{\left|\frac{240^\circ}{3}\right.} = 1.817\underline{|80^\circ}$$

$$\text{Interval between roots} = \frac{360^\circ}{3} = 120^\circ$$

Therefore the roots are:

$$z_1 = 1.817\underline{|80^\circ} \qquad z_2 = 1.817\underline{|200^\circ} \qquad z_3 = 1.817\underline{|320^\circ}$$

The principal root is the root nearest to the positive x-axis. In this case, then, the principal root is $z_3 = 1.817\underline{|320^\circ}$

On to the next frame

40

Expansions of $\sin n\theta$ and $\cos n\theta$, where n is a positive integer

By DeMoivre's theorem, we know that:

$$\cos n\theta + i\sin n\theta = (\cos\theta + i\sin\theta)^n$$

The method is simply to expand the right-hand side as a binomial series, after which we can equate real and imaginary parts.

An example will soon show you how it is done:

To find expansions for $\cos 3\theta$ and $\sin 3\theta$.

We have:

$$\cos 3\theta + i\sin 3\theta = (\cos\theta + i\sin\theta)^3$$

$$= (c + is)^3 \qquad \text{where } c \equiv \cos\theta,\ s \equiv \sin\theta$$

Now expand this by the binomial series – like $(a+b)^3$ so that

$$\cos 3\theta + i\sin 3\theta = \ldots\ldots\ldots\ldots$$

41

$$\boxed{c^3 + i3c^2s - 3cs^2 - is^3}$$

Because

$$\begin{aligned}\cos 3\theta + i\sin 3\theta &= c^3 + 3c^2(is) + 3c(is)^2 + (is)^3 \\ &= c^3 + i3c^2s - 3cs^2 - is^3 \qquad \text{since } i^2 = -1,\ i^3 = -i \\ &= (c^3 - 3cs^2) + i(3c^2s - s^3)\end{aligned}$$

Now, equating real parts and imaginary parts, we get

$$\cos 3\theta = \ldots\ldots\ldots\ldots$$

and $$\sin 3\theta = \ldots\ldots\ldots\ldots$$

42

$$\boxed{\begin{aligned}\cos 3\theta &= \cos^3\theta - 3\cos\theta\sin^2\theta \\ \sin 3\theta &= 3\cos^2\theta\sin\theta - \sin^3\theta\end{aligned}}$$

If we wish, we can replace $\sin^2\theta$ by $(1 - \cos^2\theta)$
and $\cos^2\theta$ by $(1 - \sin^2\theta)$

so that we could write the results above as:

$$\cos 3\theta = \ldots\ldots\ldots \text{(all in terms of } \cos\theta)$$

$$\sin 3\theta = \ldots\ldots\ldots \text{(all in terms of } \sin\theta)$$

43

$$\boxed{\begin{aligned}\cos 3\theta &= 4\cos^3\theta - 3\cos\theta \\ \sin 3\theta &= 3\sin\theta - 4\sin^3\theta\end{aligned}}$$

Because

$$\begin{aligned}\cos 3\theta &= \cos^3\theta - 3\cos\theta(1 - \cos^2\theta) \\ &= \cos^3\theta - 3\cos\theta + 3\cos^3\theta \\ &= 4\cos^3\theta - 3\cos\theta\end{aligned}$$

and

$$\begin{aligned}\sin 3\theta &= 3(1 - \sin^2\theta)\sin\theta - \sin^3\theta \\ &= 3\sin\theta - 3\sin^3\theta - \sin^3\theta \\ &= 3\sin\theta - 4\sin^3\theta\end{aligned}$$

While these results are useful, it is really the method that counts.

So now do this one in just the same way:

Obtain an expression for $\cos 4\theta$ in terms of $\cos\theta$.

When you have finished, check your result with the next frame

44

$$\boxed{\cos 4\theta = 8\cos^4\theta - 8\cos^2\theta + 1}$$

Working: $\cos 4\theta + i\sin 4\theta = (\cos\theta + i\sin\theta)^4$

$$\begin{aligned}&= (c + is)^4 \\ &= c^4 + 4c^3(is) + 6c^2(is)^2 + 4c(is)^3 + (is)^4 \\ &= c^4 + i4c^3s - 6c^2s^2 - i4cs^3 + s^4 \\ &= (c^4 - 6c^2s^2 + s^4) + i(4c^3s - 4cs^3)\end{aligned}$$

Equating real parts: $\cos 4\theta = c^4 - 6c^2s^2 + s^4$

$$\begin{aligned}&= c^4 - 6c^2(1 - c^2) + (1 - c^2)^2 \\ &= c^4 - 6c^2 + 6c^4 + 1 - 2c^2 + c^4 \\ &= 8c^4 - 8c^2 + 1 \\ &= 8\cos^4\theta - 8\cos^2\theta + 1\end{aligned}$$

Now for a different problem.

On to the next frame

45

Expansions for $\cos^n\theta$ and $\sin^n\theta$, in terms of sines and cosines of multiples of θ

Let $$z = \cos\theta + i\sin\theta$$

then $$\frac{1}{z} = z^{-1} = \cos\theta - i\sin\theta$$

$$\therefore \; z + \frac{1}{z} = 2\cos\theta \text{ and } z - \frac{1}{z} = 2i\sin\theta$$

Also, by DeMoivre's theorem:

$$z^n = \cos n\theta + i\sin n\theta$$

and $$\frac{1}{z^n} = z^{-n} = \cos n\theta - i\sin n\theta$$

$$\therefore \; z^n + \frac{1}{z^n} = 2\cos n\theta \text{ and } z^n - \frac{1}{z^n} = 2i\sin n\theta$$

Let us collect these four results together: $z = \cos\theta + i\sin\theta$

$z + \frac{1}{z} = 2\cos\theta$	$z - \frac{1}{z} = 2i\sin\theta$
$z^n + \frac{1}{z^n} = 2\cos n\theta$	$z^n - \frac{1}{z^n} = 2i\sin n\theta$

Make a note of these results in your record book. Then move on and we will see how we use them

46

We shall expand $\cos^3\theta$ as an example.

From our results: $z + \frac{1}{z} = 2\cos\theta$

$$\therefore \; (2\cos\theta)^3 = \left(z + \frac{1}{z}\right)^3$$

$$= z^3 + 3z^2\left(\frac{1}{z}\right) + 3z\left(\frac{1}{z^2}\right) + \frac{1}{z^3}$$

$$= z^3 + 3z + 3\frac{1}{z} + \frac{1}{z^3}$$

Now here is the trick: we rewrite this, collecting the terms up in pairs from the two extreme ends, thus:

$$(2\cos\theta)^3 = \left(z^3 + \frac{1}{z^3}\right) + 3\left(z + \frac{1}{z}\right)$$

And, from the four results that we noted:

$$z + \frac{1}{z} = \dots\dots\dots\dots$$

and $$z^3 + \frac{1}{z^3} = \dots\dots\dots\dots$$

47

$$z + \frac{1}{z} = 2\cos\theta; \quad z^3 + \frac{1}{z^3} = 2\cos 3\theta$$

$$\therefore \quad (2\cos\theta)^3 = 2\cos 3\theta + 3 \times 2\cos\theta$$
$$8\cos^3\theta = 2\cos 3\theta + 6\cos\theta$$
$$4\cos^3\theta = \cos 3\theta + 3\cos\theta$$
$$\cos^3\theta = \frac{1}{4}(\cos 3\theta + 3\cos\theta)$$

Now one for you:

Find an expression for $\sin^4\theta$.

Work in the same way, but, this time, remember that

$z - \frac{1}{z} = 2i\sin\theta$ and $z^n - \frac{1}{z^n} = 2i\sin n\theta$.

When you have obtained a result, check it with the next frame

48

$$\sin^4\theta = \frac{1}{8}[\cos 4\theta - 4\cos 2\theta + 3]$$

Because we have:

$$z - \frac{1}{z} = 2i\sin\theta; \quad z^n - \frac{1}{z^n} = 2i\sin n\theta$$

$$\therefore \quad (2i\sin\theta)^4 = \left(z - \frac{1}{z}\right)^4$$
$$= z^4 - 4z^3\left(\frac{1}{z}\right) + 6z^2\left(\frac{1}{z^2}\right) - 4z\left(\frac{1}{z^3}\right) + \frac{1}{z^4}$$
$$= \left(z^4 + \frac{1}{z^4}\right) - 4\left(z^2 + \frac{1}{z^2}\right) + 6$$

Now: $\quad z^n + \frac{1}{z^n} = 2\cos n\theta$

$$\therefore \quad 16\sin^4\theta = 2\cos 4\theta - 4 \times 2\cos 2\theta + 6$$
$$\therefore \quad \sin^4\theta = \frac{1}{8}[\cos 4\theta - 4\cos 2\theta + 3]$$

They are all done in the same way: once you know the trick, the rest is easy.

Now let us move on to something new

Loci problems

49

We are sometimes required to find the locus of a point which moves in the Argand diagram according to some stated condition. Before we work through one or two examples of this kind, let us just review a couple of useful points.

You will remember that when we were representing a complex number in polar form, i.e. $z = a + ib = r(\cos\theta + i\sin\theta)$, we said that:

(a) r is called the *modulus* of z and is written 'mod z' or $|z|$ and
(b) θ is called the *argument* of z and is written 'arg z'.

Also $r = \sqrt{a^2 + b^2}$ and $\theta = \tan^{-1}\left\{\frac{b}{a}\right\}$

so that $|z| = \sqrt{a^2 + b^2}$ and $\arg z = \tan^{-1}\left\{\frac{a}{b}\right\}$

Similarly, if $z = x + iy$ then $|z| = \ldots\ldots\ldots\ldots$

and $\arg z = \ldots\ldots\ldots\ldots$

50

$$|z| = \sqrt{x^2 + y^2} \text{ and } \arg z = \tan^{-1}\left\{\frac{y}{x}\right\}$$

Keep those in mind and we are now ready to tackle some examples.

Example 1

If $z = x + iy$, find the locus defined as $|z| = 5$.

Now we know that in this case, $|z| = \sqrt{x^2 + y^2}$

The locus is defined as $\sqrt{x^2 + y^2} = 5$

$$\therefore\; x^2 + y^2 = 25$$

This is a circle, with center at the origin and with radius 5.

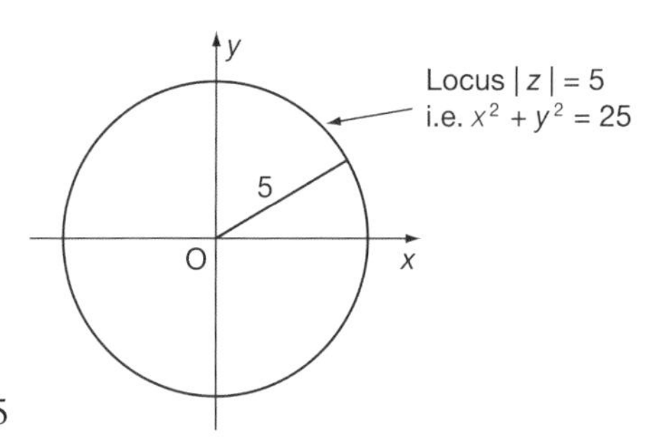

That was easy enough. Move on for Example 2

51

Example 2

If $z = x + iy$, find the locus defined as $\arg z = \frac{\pi}{4}$.

In this case: $\arg z = \tan^{-1}\left\{\frac{y}{x}\right\}$ $\therefore$ $\tan^{-1}\left\{\frac{y}{x}\right\} = \frac{\pi}{4}$

$$\therefore \frac{y}{x} = \tan\frac{\pi}{4} = \tan 45^\circ = 1 \quad \therefore \frac{y}{x} = 1 \therefore y = x$$

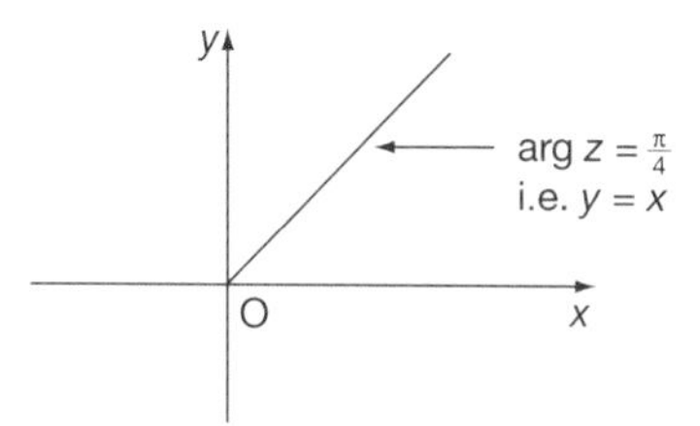

So the locus $\arg z = \frac{\pi}{4}$ is therefore the straight line $y = x$ and $y > 0$.

All locus problems at this stage are fundamentally of one of these kinds. Of course, the given condition may look a trifle more involved, but the approach is always the same.

Let us look at a more complicated one. Next frame

52

Example 3

If $z = x + iy$, find the equation of the locus $\left|\frac{z+1}{z-1}\right| = 2$.

Since $z = x + iy$:

$$z + 1 = x + iy + 1 = (x + 1) + iy = r_1\underline{|\theta_1} = z_1$$

$$z - 1 = x + iy - 1 = (x - 1) + iy = r_2\underline{|\theta_2} = z_2$$

$$\therefore \frac{z+1}{z-1} = \frac{r_1\underline{|\theta_1}}{r_2\underline{|\theta_2}} = \frac{r_1}{r_2}\underline{|\theta_1 - \theta_2}$$

$$\therefore \left|\frac{z+1}{z-1}\right| = \frac{r_1}{r_2} = \frac{|z_1|}{|z_2|} = \frac{\sqrt{(x+1)^2 + y^2}}{\sqrt{(x-1)^2 + y^2}}$$

$$\therefore \frac{\sqrt{(x+1)^2 + y^2}}{\sqrt{(x-1)^2 + y^2}} = 2$$

$$\therefore \frac{(x+1)^2 + y^2}{(x-1)^2 + y^2} = 4$$

All that now remains is to multiply across by the denominator and tidy up the result. So finish it off in its simplest form.

The answer is in the next frame

53

$$\boxed{3x^2 - 10x + 3 + 3y^2 = 0}$$

We had $\dfrac{(x+1)^2 + y^2}{(x-1)^2 + y^2} = 4$

So therefore $\quad (x+1)^2 + y^2 = 4\left\{(x-1)^2 + y^2\right\}$

$$x^2 + 2x + 1 + y^2 = 4(x^2 - 2x + 1 + y^2)$$
$$= 4x^2 - 8x + 4 + 4y^2$$
$$\therefore\ 3x^2 - 10x + 3 + 3y^2 = 0$$

This is the equation of the given locus.

Although this takes longer to write out than either of the first two examples, the basic principle is the same. The given condition must be a function of either the modulus or the argument.

Move on now to Frame 54 for Example 4

54

Example 4

If $z = x + iy$, find the equation of the locus $\arg(z^2) = -\dfrac{\pi}{4}$.

$z = x + iy = r\underline{|\theta} \quad \therefore\ \arg z = \theta = \tan^{-1}\left\{\dfrac{y}{x}\right\}$

$\therefore\ \tan\theta = \dfrac{y}{x}$

$\therefore$ By DeMoivre's theorem, $z^2 = r^2\underline{|2\theta}$

$\therefore\ \arg(z^2) = 2\theta = -\dfrac{\pi}{4}$

$\therefore\ \tan 2\theta = \tan\left(-\dfrac{\pi}{4}\right) = -1$

$\therefore\ \dfrac{2\tan\theta}{1 - \tan^2\theta} = -1$

$\therefore\ 2\tan\theta = \tan^2\theta - 1$

But $\quad \tan\theta = \dfrac{y}{x} \quad \therefore\ \dfrac{2y}{x} = \dfrac{y^2}{x^2} - 1$

$$2xy = y^2 - x^2 \quad \therefore\ y^2 = x^2 + 2xy$$

In that example, the given condition was a function of the argument. Here is one for you to do:

If $z = x + iy$, find the equation of the locus $\arg(z + 1) = \dfrac{\pi}{3}$.

Do it carefully; then check with the next frame

55

$$y = \sqrt{3}(x+1) \text{ for } y > 0$$

Here is the solution set out in detail.
If $z = x + iy$, find the locus $\arg(z+1) = \dfrac{\pi}{3}$.

$$z = x + iy \quad \therefore \; z + 1 = x + iy + 1 = (x+1) + iy$$

$$\arg(z+1) = \tan^{-1}\left\{\frac{y}{x+1}\right\} = \frac{\pi}{3} \quad \therefore \; \frac{y}{x+1} = \tan\frac{\pi}{3} = \sqrt{3}$$

$$y = \sqrt{3}(x+1) \text{ for } y > 0$$

And that is all there is to that.

Now do this one. You will have no trouble with it.

If $z = x + iy$, find the equation of the locus $|z - 1| = 5$

When you have finished it, move on to Frame 56

56

$$x^2 - 2x + y^2 = 24$$

Here it is: $z = x + iy$, given locus $|z - 1| = 5$

$$z - 1 = x + iy - 1 = (x-1) + iy$$

$$\therefore \; |z-1| = \sqrt{(x-1)^2 + y^2} = 5 \qquad \therefore \; (x-1)^2 + y^2 = 25$$

$$\therefore \; x^2 - 2x + 1 + y^2 = 25 \qquad \therefore \; x^2 - 2x + y^2 = 24$$

Every one is very much the same.

This brings us to the end of this Program, except for the final **Can You?** checklist and **Test exercise**. Before you work through them, read down the **Summary** (Frame 57), just to refresh your memory of what we have covered in this Program.

So on now to Frame 57

Summary

57

1

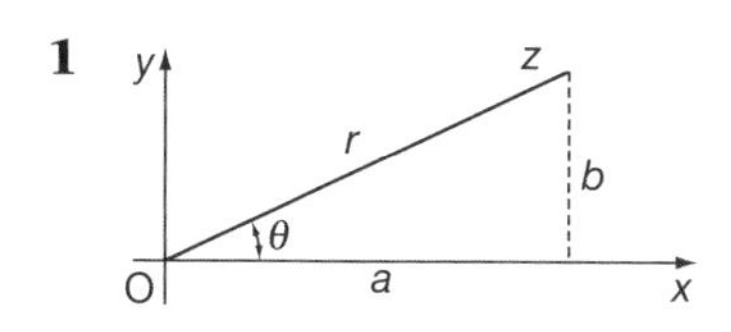

Polar form of a complex number

$$z = a + ib = r(\cos\theta + i\sin\theta) = r\underline{|\theta}$$

$$r = \text{mod } z = |z| = \sqrt{a^2 + b^2}$$

$$\theta = \arg z = \tan^{-1}\left\{\frac{a}{b}\right\}$$

2

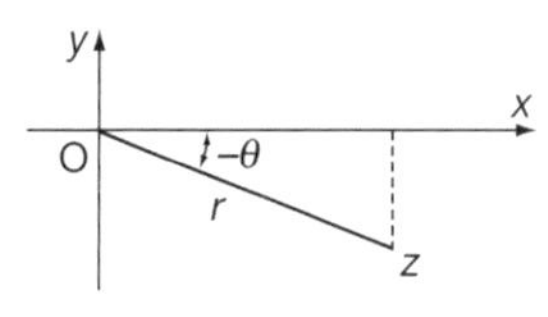

Negative angles

$$z = r(\cos[-\theta] + i\sin[-\theta])$$
$$\cos[-\theta] = \cos\theta$$
$$\sin[-\theta] = -\sin\theta$$
$$\therefore\; z = r(\cos\theta - i\sin\theta) = r\overline{|\theta}$$

3 *Multiplication and division in polar form*

If $\quad z_1 = r_1\underline{|\theta_1}; \quad z_2 = r_2\underline{|\theta_2}$

then $\quad z_1 z_2 = r_1 r_2\underline{|\theta_1 + \theta_2}$

$$\frac{z_1}{z_2} = \frac{r_1}{r_2}\underline{|\theta_1 - \theta_2}$$

4 *DeMoivre's theorem*

If $z = r(\cos\theta + i\sin\theta)$, then $z^n = r^n(\cos n\theta + i\sin n\theta)$

5 *Loci problems*

If $\quad z = x + iy, \quad |z| = \sqrt{x^2 + y^2}$

$$\arg z = \tan^{-1}\left\{\frac{y}{x}\right\}$$

That's it! Now you are ready for the ***Can You?*** *checklist in Frame 58 and the* ***Test exercise*** *in Frame 59*

Can You?

58 Checklist 3

Check this list before and after you try the end of Program test.

On a scale of 1 to 5 how confident are you that you can: **Frames**

- Use the shorthand form for a complex number in polar form? — 1 to 6
  Yes ☐ ☐ ☐ ☐ ☐ *No*
- Write complex numbers in polar form using negative angles? — 7 to 12
  Yes ☐ ☐ ☐ ☐ ☐ *No*
- Multiply and divide complex numbers in polar form? — 13 to 20
  Yes ☐ ☐ ☐ ☐ ☐ *No*
- Use DeMoivre's theorem? — 23 to 26
  Yes ☐ ☐ ☐ ☐ ☐ *No*
- Find the roots of a complex number? — 27 to 39

Frames

- Demonstrate trigonometric identities of multiple angles using complex numbers?
  Yes ☐ ☐ ☐ ☐ ☐ *No* — 40 to 48
- Solve loci problems using complex numbers?
  Yes ☐ ☐ ☐ ☐ ☐ *No* — 49 to 56

Test exercise 3

59

1. Express in polar form, $z = -5 - 3i$.
2. Express in the form $a + ib$:
   (a) $2\underline{|156^\circ}$
   (b) $5\overline{|37^\circ}$
3. If $z_1 = 12(\cos 125^\circ + i\sin 125^\circ)$ and $z_2 = 3(\cos 72^\circ + i\sin 72^\circ)$, find (a) $z_1 z_2$ and (b) $\dfrac{z_1}{z_2}$ giving the results in polar form.
4. If $z = 2(\cos 25^\circ + i\sin 25^\circ)$, find z^3 in polar form.
5. Find the three cube roots of $8(\cos 264^\circ + i\sin 264^\circ)$ and state which of them is the principal cube root. Show all three roots on an Argand diagram.
6. Expand $\sin 4\theta$ in powers of $\sin\theta$ and $\cos\theta$.
7. Express $\cos^4\theta$ in terms of cosines of multiples of θ.
8. If $z = x + iy$, find the equations of the two loci defined by:
   (a) $|z - 4| = 3$ (b) $\arg(z + 2) = \dfrac{\pi}{6}$

Further problems 3

60

1. If $z = x + iy$, where x and y are real, find the values of x and y when $\dfrac{3z}{1-i} + \dfrac{3z}{i} = \dfrac{4}{3-i}$.
2. In the Argand diagram, the origin is the center of an equilateral triangle and one vertex of the triangle is the point $3 + \sqrt{3}i$. Find the complex numbers representing the other vertices.
3. Express $2 + 3i$ and $1 - 2i$ in polar form and apply DeMoivre's theorem to evaluate $\dfrac{(2+3i)^4}{1-2i}$. Express the result in the form $a + ib$ and in exponential form.

▶

4 Find the fifth roots of $-3+3i$ in polar form and in exponential form.

5 Express $5+12i$ in polar form and hence evaluate the principal value of $\sqrt[3]{(5+12i)}$, giving the results in the form $a+ib$ and in the form $re^{i\theta}$.

6 Determine the fourth roots of -16, giving the results in the form $a+ib$.

7 Find the fifth roots of -1, giving the results in polar form. Express the principal root in the form $re^{i\theta}$.

8 Determine the roots of the equation $x^3+64=0$ in the form $a+ib$, where a and b are real.

9 Determine the three cube roots of $\dfrac{2-i}{2+i}$ giving the result in modulus/argument form. Express the principal root in the form $a+ib$.

10 Show that the equation $z^3=1$ has one real root and two other roots which are not real, and that, if one of the non-real roots is denoted by ω, the other is then ω^2. Mark on the Argand diagram the points which represent the three roots and show that they are the vertices of an equilateral triangle.

11 Determine the fifth roots of $(2-5i)$, giving the results in modulus/argument form. Express the principal root in the form $a+ib$ and in the form $re^{i\theta}$.

12 Solve the equation $z^2+2(1+i)z+2=0$, giving each result in the form $a+ib$, with a and b correct to 2 places of decimals.

13 Express $e^{1-i\pi/2}$ in the form $a+ib$.

14 Obtain the expansion of $\sin 7\theta$ in powers of $\sin\theta$.

15 Express $\sin^6 x$ as a series of terms which are cosines of angles that are multiples of x.

16 If $z=x+iy$, where x and y are real, show that the locus $\left|\dfrac{z-2}{z+2}\right|=2$ is a circle and determine its center and radius.

17 If $z=x+iy$, show that the locus $\arg\left\{\dfrac{z-1}{z-i}\right\}=\dfrac{\pi}{6}$ is a circle. Find its center and radius.

18 If $z=x+iy$, determine the Cartesian equation of the locus of the point z which moves in the Argand diagram so that

$$|z+2i|^2+|z-2i|^2=40.$$

19 If $z=x+iy$, determine the equations of the two loci:

(a) $\left|\dfrac{z+2}{z}\right|=3$ and (b) $\arg\left\{\dfrac{z+2}{z}\right\}=\dfrac{\pi}{4}$

20 If $z = x + iy$, determine the equations of the loci in the Argand diagram, defined by:

(a) $\left|\dfrac{z+2}{z-1}\right| = 2$ and (b) $\arg\left\{\dfrac{z-1}{z+2}\right\} = \dfrac{\pi}{2}$

21 Prove that:

(a) if $|z_1 + z_2| = |z_1 - z_2|$, the difference of the arguments of z_1 and z_2 is $\dfrac{\pi}{2}$

(b) if $\arg\left\{\dfrac{z_1 + z_2}{z_1 - z_2}\right\} = \dfrac{\pi}{2}$, then $|z_1| = |z_2|$

22 If $z = x + iy$, determine the loci in the Argand diagram, defined by:

(a) $|z + 2i|^2 - |z - 2i|^2 = 24$

(b) $|z + ik|^2 + |z - ik|^2 = 10k^2 \quad (k > 0)$

Hyperbolic functions

Learning outcomes

When you have completed this Program you will be able to:

- Define the hyperbolic functions in terms of the exponential function
- Express the hyperbolic functions as power series
- Recognize the graphs of the hyperbolic functions
- Evaluate hyperbolic functions and their inverses
- Determine the logarithmic form of the inverse hyperbolic functions
- Prove hyperbolic trigonometric identities
- Understand the relationship between the circular and the hyperbolic trigonometric functions

Introduction

1

The cosine of an angle was first defined as the ratio of two sides of a right-angled triangle – adjacent over hypotenuse. In Program 1 you learnt how to extend the definition of a cosine to *any* angle, positive or negative. You might just check that out to refresh your memory by re-reading Frame 29 of Program 1.

Now, in Frames 46 to 59 of Program 2 you learnt how a complex number of unit length could be written in either polar or exponential form, giving rise to the equations:

$$\cos\theta + i\sin\theta = e^{i\theta}$$

$$\cos\theta - i\sin\theta = e^{-i\theta}$$

If these two equations are added you find that:

$$2\cos\theta = e^{i\theta} + e^{-i\theta} \text{ so that } \cos\theta = \frac{e^{i\theta} + e^{-i\theta}}{2}$$

If θ is replaced by ix in this last equation you find that:

$$\cos ix = \frac{e^{iix} + e^{-iix}}{2} = \frac{e^{-x} + e^{x}}{2}$$

where the right-hand side is *entirely real*. In fact, you have seen this before in Frame 73 of Program 1pgoto 115

, it is the even part of the exponential function which is called the *hyperbolic cosine*:

$$\cosh x = \frac{e^{x} + e^{-x}}{2} \text{ so that } \cos ix = \cosh x$$

The graph of $y = \cosh x$ is called a *catenary* from the Latin word *catena* meaning chain because the shape of the graph is the shape of a hanging chain.

Move on to Frame 2 and start the Program

2

You may remember that of the many functions that can be expressed as a series of powers of x, a common one is e^{x}:

$$e^{x} = 1 + x + \frac{x^2}{2!} + \frac{x^3}{3!} + \frac{x^4}{4!} + \dots$$

If we replace x by $-x$, we get:

$$e^{-x} = 1 - x + \frac{x^2}{2!} - \frac{x^3}{3!} + \frac{x^4}{4!} - \dots$$

and these two functions e^{x} and e^{-x} are the foundations of the definitions we are going to use.

▶

(a) If we take the value of e^x, subtract e^{-x}, and divide by 2, we form what is defined as the hyperbolic sine of x:

$$\frac{e^x - e^{-x}}{2} = \textit{hyperbolic sine of } x$$

This is a lot to write every time we wish to refer to it, so we shorten it to sinh x, the h indicating its connection with the hyperbola. We pronounce it 'shine x'.

$$\frac{e^x - e^{-x}}{2} = \sinh x$$

So, in the same way, $\dfrac{e^y - e^{-y}}{2}$ would be written as ....

3

$$\boxed{\sinh y}$$

In much the same way, we have two other definitions:

(b) $\dfrac{e^x + e^{-x}}{2} = \textit{hyperbolic cosine}$ of x

$= \cosh x$ [pronounced 'cosh x']

(c) $\dfrac{e^x - e^{-x}}{e^x + e^{-x}} = \textit{hyperbolic tangent}$ of x

$= \tanh x$ [pronounced 'than x']

We must start off by learning these definitions, for all the subsequent developments depend on them.

So now then; what was the definition of sinh x?

$\sinh x =$ ....

4

$$\boxed{\sinh x = \frac{e^x - e^{-x}}{2}}$$

Here they are together so that you can compare them:

$$\sinh x = \frac{e^x - e^{-x}}{2}$$
$$\cosh x = \frac{e^x + e^{-x}}{2}$$
$$\tanh x = \frac{e^x - e^{-x}}{e^x + e^{-x}}$$

Make a copy of these in your record book for future reference when necessary.

5

$$\sinh x = \frac{e^x - e^{-x}}{2}; \quad \cosh x = \frac{e^x + e^{-x}}{2}; \quad \tanh x = \frac{e^x - e^{-x}}{e^x + e^{-x}}$$

We started the program by referring to e^x and e^{-x} as series of powers of x. It should not be difficult therefore to find series at least for $\sinh x$ and for $\cosh x$. Let us try.

(a) *Series for sinh x*

$$e^x = 1 + x + \frac{x^2}{2!} + \frac{x^3}{3!} + \frac{x^4}{4!} + \dots$$

$$e^{-x} = 1 - x + \frac{x^2}{2!} - \frac{x^3}{3!} + \frac{x^4}{4!} - \dots$$

If we subtract, we get:

$$e^x - e^{-x} = 2x + \frac{2x^3}{3!} + \frac{2x^5}{5!} \dots$$

Divide by 2:

$$\frac{e^x - e^{-x}}{2} = \sinh x = x + \frac{x^3}{3!} + \frac{x^5}{5!} + \dots$$

(b) If we add the series for e^x and e^{-x}, we get a similar result. What is it?

When you have decided, move on to Frame 6

6

$$\cosh x = 1 + \frac{x^2}{2!} + \frac{x^4}{4!} + \frac{x^6}{6!} + \dots$$

Because we have:

$$e^x = 1 + x + \frac{x^2}{2!} + \frac{x^3}{3!} + \frac{x^4}{4!} + \dots$$

$$e^{-x} = 1 - x + \frac{x^2}{2!} - \frac{x^3}{3!} + \frac{x^4}{4!} - \dots$$

$$\therefore \; e^x + e^{-x} = 2 + \frac{2x^2}{2!} + \frac{2x^4}{4!} + \dots$$

$$\therefore \; \frac{e^x + e^{-x}}{2} = \cosh x = 1 + \frac{x^2}{2!} + \frac{x^4}{4!} + \dots$$

Move on to Frame 7

7

So we have:

$$\sinh x = x + \frac{x^3}{3!} + \frac{x^5}{5!} + \frac{x^7}{7!} + \dots$$

$$\cosh x = 1 + \frac{x^2}{2!} + \frac{x^4}{4!} + \frac{x^6}{6!} + \dots$$

Note: All terms positive: $\sinh x$ has all the odd powers
$\cosh x$ has all the even powers

▶

We cannot easily get a series for $\tanh x$ by this process, so we will leave that one to some other time.

Make a note of these two series in your record book. Then, cover up what you have done so far and see if you can write down the definitions of:

(a) $\sinh x = \dots\dots\dots\dots$

(b) $\cosh x = \dots\dots\dots\dots$

(c) $\tanh x = \dots\dots\dots\dots$ No looking!

8

$$\sinh x = \frac{e^x - e^{-x}}{2}; \quad \cosh x = \frac{e^x + e^{-x}}{2}; \quad \tanh x = \frac{e^x - e^{-x}}{e^x + e^{-x}}$$

All correct? Right.

Graphs of hyperbolic functions

9

We shall get to know quite a lot about these hyperbolic functions if we sketch the graphs of these functions. Since they depend on the values of e^x and e^{-x}, we had better just refresh our memories of what these graphs look like.

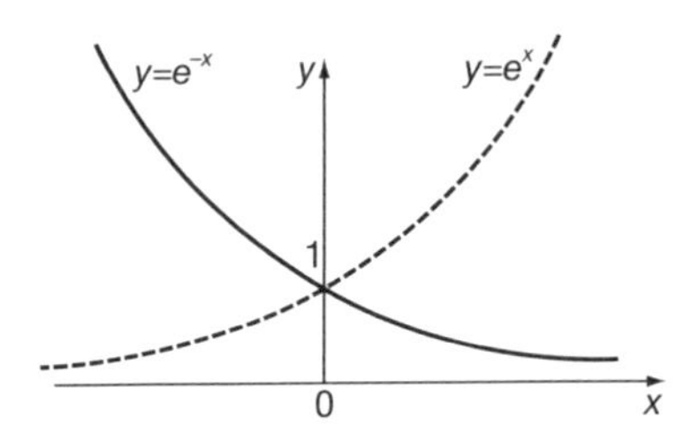

$y = e^x$ and $y = e^{-x}$ cross the y-axis at the point $y = 1$ $(e^0 = 1)$. Each graph then approaches the x-axis as an asymptote, getting nearer and nearer to it as it goes away to infinity in each direction, without actually crossing it,.

So, for what range of values of x are e^x and e^{-x} positive?

10

e^x and e^{-x} are positive for all values of x

Correct, since the graphs are always above the x-axis.

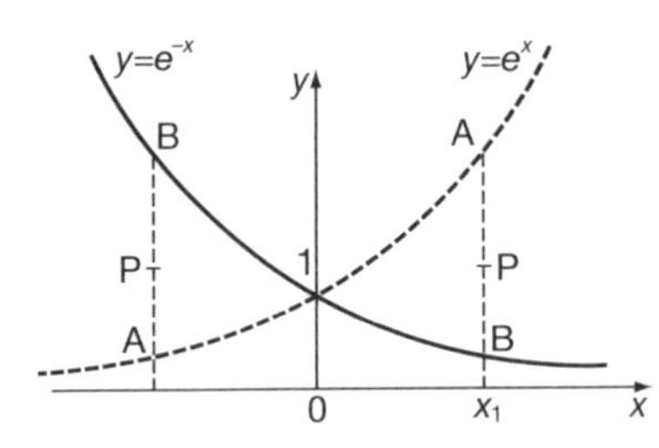

At any value of x, e.g. $x = x_1$, $\cosh x = \dfrac{e^x + e^{-x}}{2}$, i.e. the value of $\cosh x$ is the average of the values of e^x and e^{-x} at that value of x. This is given by P, the mid-point of AB.

If we can imagine a number of ordinates (or verticals) like AB and we plot their mid-points, we shall obtain the graph of $y = \cosh x$.

Can you sketch in what the graph will look like?

11

Here it is:

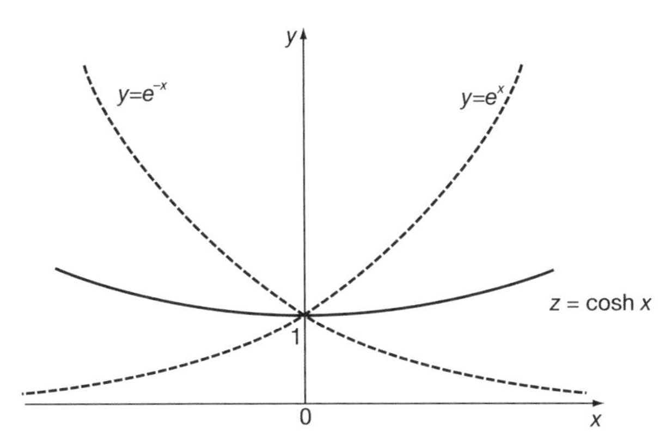

We see from the graph of $y = \cosh x$ that:

(a) $\cosh 0 = 1$

(b) the value of $\cosh x$ is never less than 1

(c) the curve is symmetrical about the y-axis, i.e.

$$\cosh(-x) = \cosh x$$

(d) for any given value of $\cosh x$, there are two values of x, equally spaced about the origin, i.e. $x = \pm a$.

Now let us see about the graph of $y = \sinh x$ in the same sort of way.

12

$$\sinh x = \frac{e^x - e^{-x}}{2}$$

On the diagram:

$$CA = e^x$$
$$CB = e^{-x}$$
$$BA = e^x - e^{-x}$$
$$BP = \frac{e^x - e^{-x}}{2}$$

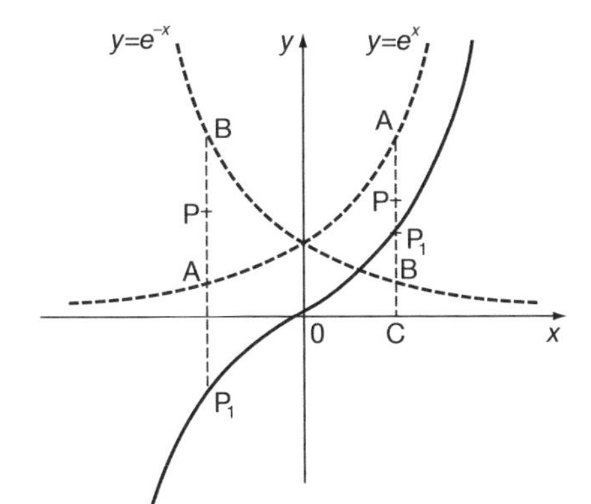

The corresponding point on the graph of $y = \sinh x$ is thus obtained by standing the ordinate BP on the x-axis at C, i.e. P_1.

Note that on the left of the origin, BP is negative and is therefore placed below the x-axis.

So what can we say about $y = \sinh x$?

13

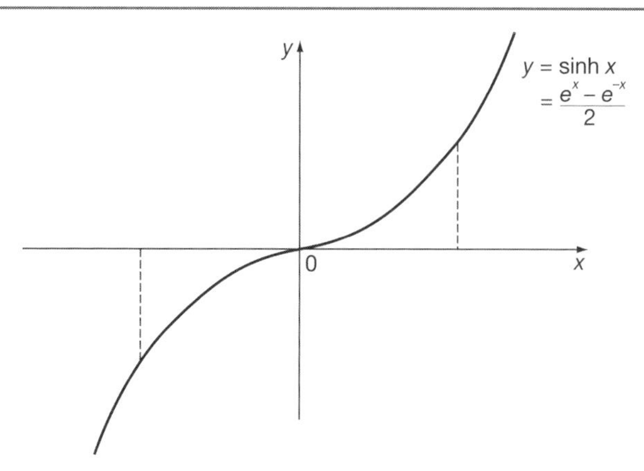

From the graph of $y = \sinh x$, we see:

(a) $\sinh 0 = 0$

(b) $\sinh x$ can have all values from $-\infty$ to $+\infty$

(c) the curve is symmetrical about the origin, i.e.

$$\sinh(-x) = -\sinh x$$

(d) for a given value of $\sinh x$, there is only one real value of x.

If we draw $y = \sinh x$ and $y = \cosh x$ on the same graph, what do we get?

14

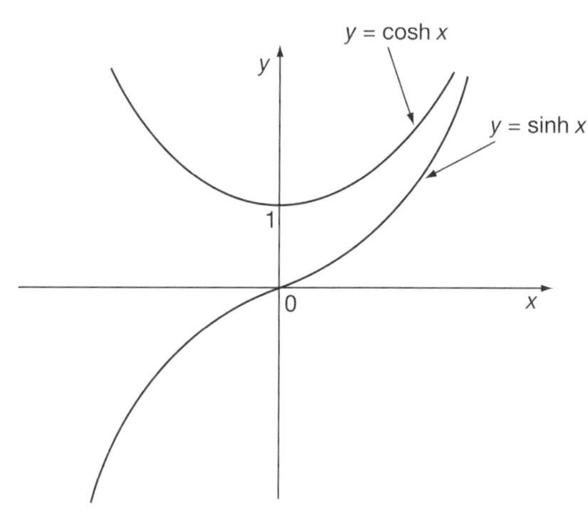

Note that $y = \sinh x$ is always outside $y = \cosh x$, but gets nearer to it as x increases:

i.e. as $x \to \infty$, $\sinh x \to \cosh x$

And now let us consider the graph of $y = \tanh x$.

Move on

15

It is not easy to build $y = \tanh x$ directly from the graphs of $y = e^x$ and $y = e^{-x}$. If, however, we take values of e^x and e^{-x} and then calculate $y = \dfrac{e^x - e^{-x}}{e^x + e^{-x}}$ and plot points, we get a graph as shown:

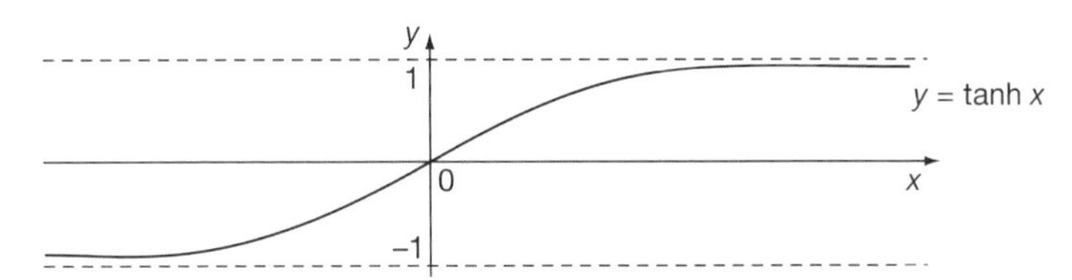

We see:

(a) $\tanh 0 = 0$

(b) $\tanh x$ always lies between $y = -1$ and $y = 1$

(c) $\tanh(-x) = -\tanh x$

(d) as $x \to \infty$, $\tanh x \to 1$
as $x \to -\infty$, $\tanh x \to -1$

Finally, let us now sketch all three graphs on one diagram so that we can compare them and distinguish between them.

16

Here they are:

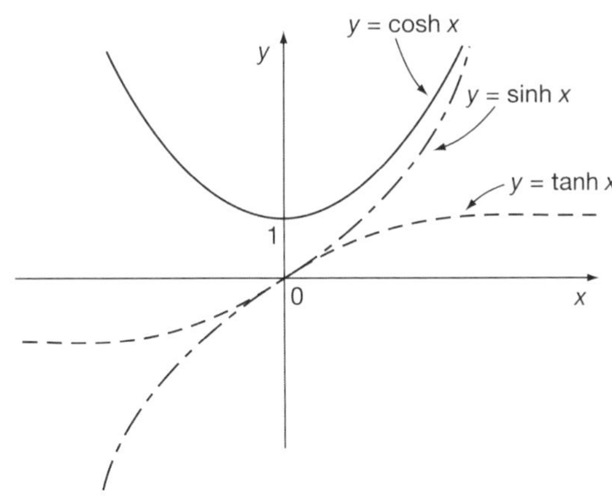

One further point to note:

At the origin, $y = \sinh x$ and $y = \tanh x$ have the same slope. The two graphs therefore slide into each other and out again.

It is worth while to remember this combined diagram: sketch it in your record book for reference.

Review exercise

17

Fill in the following:

(a) $\dfrac{e^x + e^{-x}}{2} = \dots\dots\dots\dots$

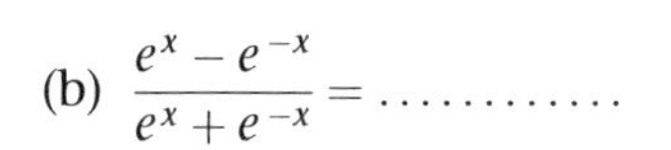

(b) $\dfrac{e^x - e^{-x}}{e^x + e^{-x}} = \dots\dots\dots\dots$

(c) $\dfrac{e^x - e^{-x}}{2} = \dots\dots\dots\dots$

(d)

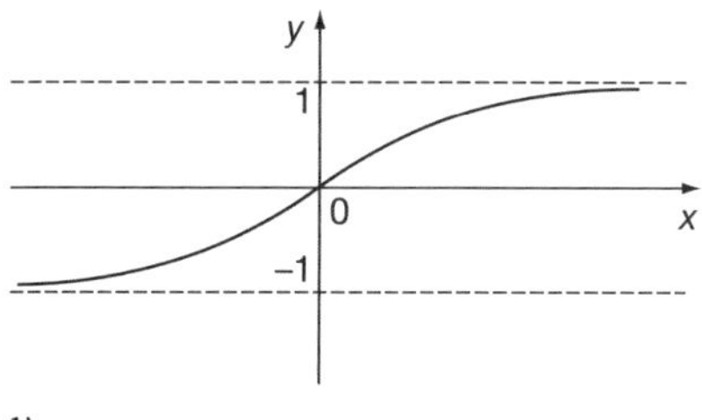

$y = \dots\dots\dots\dots$

(e)

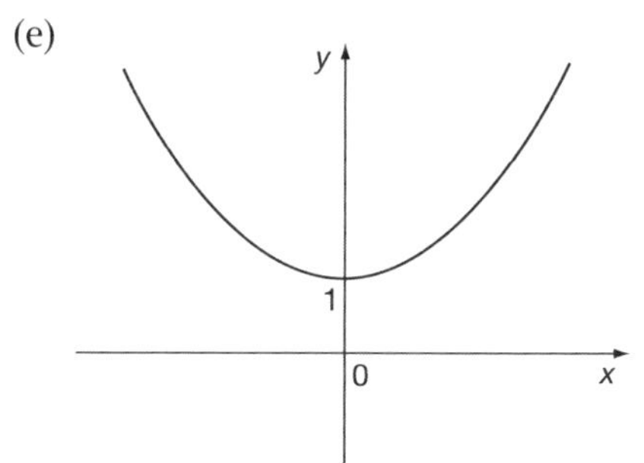

$y = \dots\dots\dots\dots$

(f)

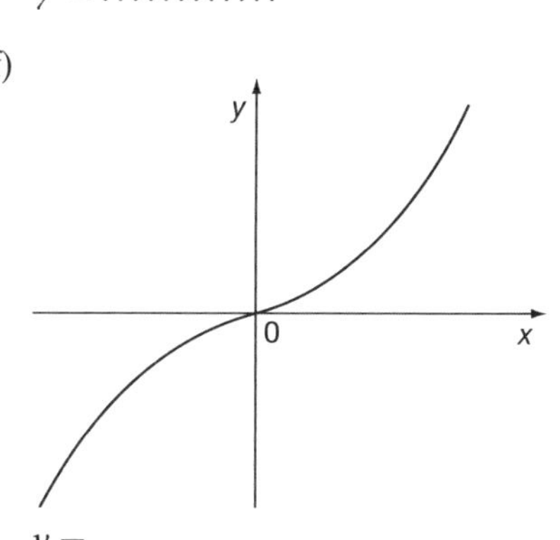

$y = \dots\dots\dots\dots$

Results in the next frame. Check your answers carefully

18

Here are the results: check yours.

(a) $\dfrac{e^x + e^{-x}}{2} = \cosh x$

(b) $\dfrac{e^x - e^{-x}}{e^x + e^{-x}} = \tanh x$

(c) $\dfrac{e^x - e^{-x}}{2} = \sinh x$

(d)

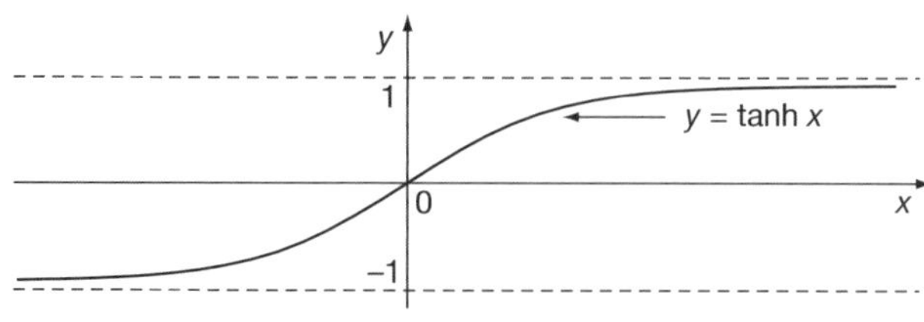

$y = \tanh x$

▶

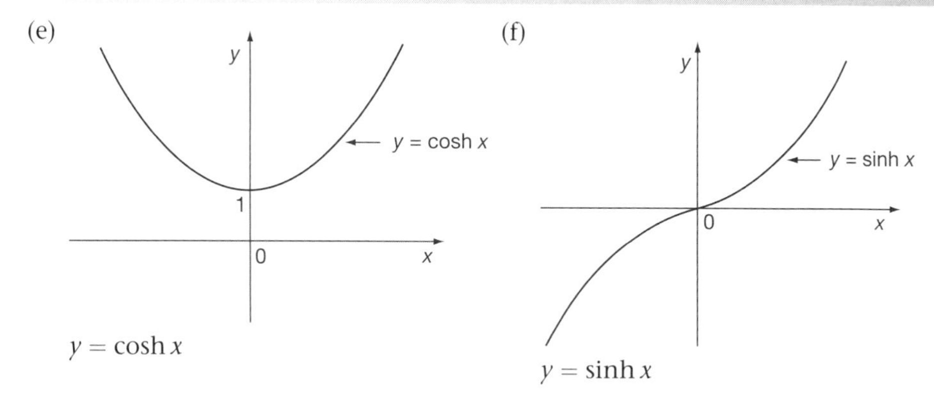

Now we can continue with the next piece of work

Evaluation of hyperbolic functions

19

The values of $\sinh x$, $\cosh x$ and $\tanh x$ can be found using a calculator in just the same manner as the values of the circular trigonometric expressions were found. However, if your calculator does not possess the facility to work out hyperbolic expressions then their values can still be found by using the exponential key instead.

Example 1

To evaluate sinh 1.275

$$\text{Now } \sinh x = \frac{1}{2}(e^x - e^{-x}) \quad \therefore \quad \sinh 1.275 = \frac{1}{2}(e^{1.275} - e^{-1.275}).$$

We now have to evaluate $e^{1.275}$ and $e^{-1.275}$.

Using your calculator, you will find that:

$$e^{1.275} = 3.579 \text{ and } e^{-1.275} = \frac{1}{3.579} = 0.2794$$

$$\therefore \quad \sinh 1.275 = \frac{1}{2}(3.579 - 0.279)$$

$$= \frac{1}{2}(3.300) = 1.65$$

$$\therefore \quad \sinh 1.275 = 1.65$$

In the same way, you now find the value of $\cosh 2.156$.

When you have finished, move on to Frame 20

20

$$\boxed{\cosh 2.156 = 4.377}$$

Here is the working:

Example 2

$$\cosh 2.156 = \frac{1}{2}(e^{2.156} + e^{-2.156})$$

$$\therefore \quad \cosh 2.156 = \frac{1}{2}(8.637 + 0.116)$$

$$= \frac{1}{2}(8.753) = 4.377$$

$$\therefore \quad \cosh 2.156 = 4.377$$

Right, one more. Find the value of $\tanh 1.27$.

When you have finished, move on to Frame 21

21

$$\boxed{\tanh 1.27 = 0.8538}$$

Here is the working.

Example 3

$$\tanh 1.27 = \frac{e^{1.27} - e^{-1.27}}{e^{1.27} + e^{-1.27}}$$

$$\therefore \quad \tanh 1.27 = \frac{3.561 - 0.281}{3.561 + 0.281} = \frac{3.280}{3.842}$$

$$\tanh 1.27 = 0.8538$$

So, evaluating sinh, cosh and tanh is easy enough and depends mainly on being able to evaluate e^k, where k is a given number.

And now let us look at the reverse process. So on to Frame 22

Inverse hyperbolic functions

22

Example 1

To find $\sinh^{-1} 1.475$, i.e. to find the value of x such that $\sinh x = 1.475$.

Here it is: $\sinh x = 1.475 \quad \therefore \quad \frac{1}{2}(e^x - e^{-x}) = 1.475$

$$\therefore \; e^x - \frac{1}{e^x} = 2.950$$

Multiplying both sides by e^x : $(e^x)^2 - 1 = 2.95(e^x)$

$$(e^x)^2 - 2.95(e^x) - 1 = 0$$

This is a quadratic equation and can be solved as usual, giving:

$$e^x = \frac{2.95 \pm \sqrt{2.95^2 + 4}}{2} = \frac{2.95 \pm \sqrt{8.703 + 4}}{2}$$

$$= \frac{2.95 \pm \sqrt{12.703}}{2} = \frac{2.95 \pm 3.564}{2}$$

$$= \frac{6.514}{2} \text{ or } -\frac{0.614}{2} = 3.257 \text{ or } -0.307$$

But e^x is always positive for real values of x. Therefore the only real solution is given by $e^x = 3.257$.

$$\therefore \; x = \ln 3.257 = 1.1808$$

$$\therefore \; x = 1.1808$$

Example 2

Now you find $\cosh^{-1} 2.364$ in the same way.

23

$$\boxed{\cosh^{-1} 2.364 = \pm 1.5054 \text{ to 4 dp}}$$

Because

To evaluate $\cosh^{-1} 2.364$, let $x = \cosh^{-1} 2.364$

$$\therefore \; \cosh x = 2.364 \quad \therefore \; \frac{e^x + e^{-x}}{2} = 2.364 \quad \therefore \; e^x + \frac{1}{e^x} = 4.728$$

$$(e^x)^2 - 4.728(e^x) + 1 = 0$$

$$e^x = \frac{4.728 \pm \sqrt{4.728^2 - 4}}{2}$$

$$= \frac{1}{2}(4.728 \pm 4.284\ldots)$$

$$= 4.5060\ldots \text{ or } 0.2219\ldots$$

Therefore:

$$x = \ln 4.5060\ldots \text{ or } \ln 0.2219\ldots$$

$$= \pm 1.5054 \text{ to 4 dp}$$

Before we do the next one, do you remember the exponential definition of $\tanh x$? Well, what is it?

24

$$\tanh x = \frac{e^x - e^{-x}}{e^x + e^{-x}}$$

That being so, we can now evaluate $\tanh^{-1} 0.623$.

Let $x = \tanh^{-1} 0.623 \quad \therefore \quad \tanh x = 0.623$

$$\therefore \quad \frac{e^x - e^{-x}}{e^x + e^{-x}} = 0.623$$

$$\therefore \quad e^x - e^{-x} = 0.623(e^x + e^{-x})$$

$$\therefore \quad (1 - 0.623)e^x = (1 + 0.623)e^{-x}$$

$$0.377e^x = 1.623e^{-x}$$

$$= \frac{1.623}{e^x}$$

$$\therefore \quad (e^x)^2 = \frac{1.623}{0.377}$$

$$\therefore \quad e^x = 2.075$$

$$\therefore \quad x = \ln 2.075 = 0.7299$$

$$\therefore \quad \tanh^{-1} 0.623 = 0.730$$

Now one for you to do on your own. Evaluate $\sinh^{-1} 0.5$.

25

$$\sinh^{-1} 0.5 = 0.4812$$

Check your working.

Let $x = \sinh^{-1} 0.5 \quad \therefore \quad \sinh x = 0.5$

$$\therefore \quad \frac{e^x - e^{-x}}{2} = 0.5 \quad \therefore \quad e^x - \frac{1}{e^x} = 1$$

$$\therefore \quad (e^x)^2 - 1 = e^x$$

$$\therefore \quad (e^x)^2 - (e^x) - 1 = 0$$

$$e^x = \frac{1 \pm \sqrt{1+4}}{2} = \frac{1 \pm \sqrt{5}}{2}$$

$$= \frac{3.2361}{2} \text{ or } \frac{-1.2361}{2}$$

$$= 1.6180 \text{ or } -0.6180$$

$$\therefore \quad x = \ln 1.6180 = 0.4812$$

$$\sinh^{-1} 0.5 = 0.4812$$

$e^x = -0.6180$ gives no real value of x

And just one more! Evaluate $\tanh^{-1} 0.75$.

26

$$\boxed{\tanh^{-1} 0.75 = 0.9730}$$

Let $x = \tanh^{-1} 0.75 \quad \therefore \tanh x = 0.75$

$$\therefore \frac{e^x - e^{-x}}{e^x + e^{-x}} = 0.75$$

$$e^x - e^{-x} = 0.75(e^x + e^{-x})$$

$$(1 - 0.75)e^x = (1 + 0.75)e^{-x}$$

$$0.25e^x = 1.75e^{-x}$$

$$(e^x)^2 = \frac{1.75}{0.25} = 7$$

$$e^x = \pm\sqrt{7} = \pm 2.6458$$

But remember that e^x cannot be negative for real values of x.
Therefore $e^x = 2.6458$ is the only real solution.

$$\therefore x = \ln 2.6458 = 0.9730$$

$$\tanh^{-1} 0.75 = 0.9730$$

Log form of the inverse hyperbolic functions

27

Let us do the same thing in a general way.

To find $\tanh^{-1} x$ in log form.
As usual, we start off with: Let $y = \tanh^{-1} x \quad \therefore x = \tanh y$

$$\therefore \frac{e^y - e^{-y}}{e^y + e^{-y}} = x \quad \therefore e^y - e^{-y} = x(e^y + e^{-y})$$

$$e^y(1 - x) = e^{-y}(1 + x) = \frac{1}{e^y}(1 + x)$$

$$e^{2y} = \frac{1 + x}{1 - x}$$

$$\therefore 2y = \ln\left\{\frac{1 + x}{1 - x}\right\}$$

$$\therefore y = \tanh^{-1} x = \frac{1}{2}\ln\left\{\frac{1 + x}{1 - x}\right\}$$

So that: $\tanh^{-1} 0.5 = \frac{1}{2}\ln\left\{\frac{1.5}{0.5}\right\}$

$$= \frac{1}{2}\ln 3 = \frac{1}{2}(1.0986) = 0.5493$$

And similarly: $\tanh^{-1}(-0.6) = \dots\dots\dots\dots$

28

$$\boxed{\tanh^{-1}(-0.6) = -0.6931}$$

Because

$$\tanh^{-1} x = \frac{1}{2}\ln\left\{\frac{1+x}{1-x}\right\}$$

$$\therefore\ \tanh^{-1}(-0.6) = \frac{1}{2}\ln\left\{\frac{1-0.6}{1+0.6}\right\} = \frac{1}{2}\ln\left\{\frac{0.4}{1.6}\right\}$$

$$= \frac{1}{2}\ln 0.25$$

$$= \frac{1}{2}(-1.3863)$$

$$= -0.6931$$

Now, in the same way, find an expression for $\sinh^{-1} x$.
Start off by saying: Let $y = \sinh^{-1} x \quad \therefore\ x = \sinh y$

$$\therefore\ \frac{e^y - e^{-y}}{2} = x \quad \therefore\ e^y - e^{-y} = 2x \quad \therefore\ e^y - \frac{1}{e^y} = 2x$$

$$(e^y)^2 - 2x(e^y) - 1 = 0$$

Now finish it off – result in Frame 29

29

$$\boxed{\sinh^{-1} x = \ln\left\{x + \sqrt{x^2+1}\right\}}$$

Because

$$(e^y)^2 - 2x(e^y) - 1 = 0$$

$$e^y = \frac{2x \pm \sqrt{4x^2+4}}{2} = \frac{2x \pm 2\sqrt{x^2+1}}{2}$$

$$= x \pm \sqrt{x^2+1}$$

$$e^y = x + \sqrt{x^2+1} \text{ or } e^y = x - \sqrt{x^2+1}$$

At first sight, there appear to be two results, but notice this:

In the second result $\quad \sqrt{x^2+1} > x$
$\therefore\ e^y = x -$ (something $> x$), i.e. negative

Therefore we can discard the second result as far as we are concerned since powers of e are always positive. (Remember the graph of e^x.)
The only real solution then is given by $e^y = x + \sqrt{x^2+1}$

$$y = \sinh^{-1} x = \ln\left\{x + \sqrt{x^2+1}\right\}$$

30

Finally, let us find the general expression for $\cosh^{-1} x$.

$$\text{Let} \quad y = \cosh^{-1} x \qquad \therefore\ x = \cosh y = \frac{e^y + e^{-y}}{2}$$

$$\therefore\ e^y + \frac{1}{e^y} = 2x \qquad \therefore\ (e^y)^2 - 2x(e^y) + 1 = 0$$

$$\therefore\ e^y = \frac{2x \pm \sqrt{4x^2 - 4}}{2} = x \pm \sqrt{x^2 - 1}$$

$$\therefore\ e^y = x + \sqrt{x^2 - 1} \quad \text{and} \quad e^y = x - \sqrt{x^2 - 1}$$

Both these results are positive, since $\sqrt{x^2 - 1} < x$.

$$\text{However} \quad \frac{1}{x + \sqrt{x^2 - 1}} = \frac{1}{x + \sqrt{x^2 - 1}} \cdot \frac{x - \sqrt{x^2 - 1}}{x - \sqrt{x^2 - 1}}$$

$$= \frac{x - \sqrt{x^2 - 1}}{x^2 - (x^2 - 1)} = x - \sqrt{x^2 - 1}$$

So our results can be written:

$$e^y = x + \sqrt{x^2 - 1} \quad \text{and} \quad e^y = \frac{1}{x + \sqrt{x^2 - 1}}$$

$$e^y = x + \sqrt{x^2 - 1} \text{ or } \left\{x + \sqrt{x^2 - 1}\right\}^{-1}$$

$$\therefore\ y = \ln\left\{x + \sqrt{x^2 - 1}\right\} \text{ or } -\ln\left\{x + \sqrt{x^2 - 1}\right\}$$

$$\therefore \qquad \cosh^{-1} x = \pm \ln\left\{x + \sqrt{x^2 - 1}\right\}$$

Notice that the plus and minus signs give two results which are symmetrical about the y-axis (agreeing with the graph of $y = \cosh x$).

31

Here are the three general results collected together:

$$\sinh^{-1} x = \ln\left\{x + \sqrt{x^2 + 1}\right\}$$

$$\cosh^{-1} x = \pm \ln\left\{x + \sqrt{x^2 - 1}\right\}$$

$$\tanh^{-1} x = \frac{1}{2}\ln\left\{\frac{1 + x}{1 - x}\right\}$$

Add these to your list in your record book. They will be useful. Compare the first two carefully, for they are very nearly alike. Note also that:

(a) $\sinh^{-1} x$ has only one value

(b) $\cosh^{-1} x$ has two values.

So what comes next? We shall see in Frame 32

Hyperbolic identities

32

There is no need to recoil in horror. You will see before long that we have an easy way of doing these. First of all, let us consider one or two relationships based on the basic definitions.

(1) The first set are really definitions themselves. Like the trig ratios, we have reciprocal hyperbolic functions:

(a) $\coth x$ (i.e. hyperbolic cotangent) $= \dfrac{1}{\tanh x}$

(b) $\operatorname{sech} x$ (i.e. hyperbolic secant) $= \dfrac{1}{\cosh x}$

(c) $\operatorname{cosech} x$ (i.e. hyperbolic cosecant) $= \dfrac{1}{\sinh x}$

These, by the way, are pronounced (a) coth, (b) sheck and (c) co-sheck respectively.

These remind us, once again, how like trig functions these hyperbolic functions are.

Make a list of these three definitions: then move on to Frame 33

33

(2) Let us consider
$$\frac{\sinh x}{\cosh x} = \frac{e^x - e^{-x}}{2} \div \frac{e^x + e^{-x}}{2}$$
$$= \frac{e^x - e^{-x}}{e^x + e^{-x}} = \tanh x$$

$$\therefore \tanh x = \frac{\sinh x}{\cosh x} \qquad \left\{ \begin{matrix} \text{Very much like} \\ \tan\theta = \dfrac{\sin\theta}{\cos\theta} \end{matrix} \right\}$$

(3) $\cosh x = \frac{1}{2}(e^x + e^{-x})$; $\qquad \sinh x = \frac{1}{2}(e^x - e^{-x})$

Add these results: $\qquad \cosh x + \sinh x = e^x$

Subtract: $\qquad \cosh x - \sinh x = e^{-x}$

Multiply these two expressions together:

$(\cosh x + \sinh x)(\cosh x - \sinh x) = e^x \cdot e^{-x}$

$\therefore \cosh^2 x - \sinh^2 x = 1$

$\left\{ \begin{matrix} \text{In trig, we have } \cos^2\theta + \sin^2\theta - 1, \\ \text{so there is a difference in sign here.} \end{matrix} \right\}$

On to Frame 34

34

(4) We have just established that $\cosh^2 x - \sinh^2 x = 1$.

Divide by $\cosh^2 x$: $$1 - \frac{\sinh^2 x}{\cosh^2 x} = \frac{1}{\cosh^2 x}$$

$$\therefore\ 1 - \tanh^2 x = \operatorname{sech}^2 x$$

$$\therefore\ \operatorname{sech}^2 x = 1 - \tanh^2 x$$

{Something like $\sec^2\theta = 1 + \tan^2\theta$, isn't it?}

(5) If we start again with $\cosh^2 x - \sinh^2 x = 1$ and divide this time by $\sinh^2 x$, we get:

$$\frac{\cosh^2 x}{\sinh^2 x} - 1 = \frac{1}{\sinh^2 x}$$

$$\therefore\ \coth^2 x - 1 = \operatorname{cosech}^2 x$$

$$\therefore\ \operatorname{cosech}^2 x = \coth^2 x - 1$$

$\left\{\begin{array}{l}\text{In trig, we have } \operatorname{cosec}^2\theta = 1 + \cot^2\theta, \\ \text{so there is a sign difference here too.}\end{array}\right\}$

Move on to Frame 35

35

(6) We have already used the fact that

$\cosh x + \sinh x = e^x$ and $\cosh x - \sinh x = e^{-x}$

If we square each of these statements, we obtain

(a) ....

(b) ....

36

(a) $\cosh^2 x + 2\sinh x\cosh x + \sinh^2 x = e^{2x}$

(b) $\cosh^2 x - 2\sinh x\cosh x + \sinh^2 x = e^{-2x}$

So if we subtract as they stand, we get:

$$4\sinh x\cosh x = e^{2x} - e^{-2x}$$

$$\therefore\ 2\sinh x\cosh x = \frac{e^{2x} - e^{-2x}}{2} = \sinh 2x$$

$$\therefore\ \sinh 2x = 2\sinh x\cosh x$$

If however we add the two lines together, we get ....

37

$$2(\cosh^2 x + \sinh^2 x) = e^{2x} + e^{-2x}$$

$$\therefore \quad \cosh^2 x + \sinh^2 x = \frac{e^{2x} + e^{-2x}}{2} = \cosh 2x$$

$$\therefore \quad \cosh 2x = \cosh^2 x + \sinh^2 x$$

We already know that $\cosh^2 x - \sinh^2 x = 1$

$$\therefore \quad \cosh^2 x = 1 + \sinh^2 x$$

Substituting this in our previous result, we have:

$$\cosh 2x = 1 + \sinh^2 x + \sinh^2 x$$

$$\therefore \quad \cosh 2x = 1 + 2\sinh^2 x$$

Or we could say $\cosh^2 x - 1 = \sinh^2 x$

$$\therefore \quad \cosh 2x = \cosh^2 x + (\cosh^2 x - 1)$$

$$\therefore \quad \cosh 2x = 2\cosh^2 x - 1$$

Now we will collect all these hyperbolic identities together and compare them with the corresponding trig identities.

These are all listed in the next frame, so move on

38

	Trig identities	*Hyperbolic identities*
(1)	$\cot x = 1/\tan x$	$\coth x = 1/\tanh x$
	$\sec x = 1/\cos x$	$\operatorname{sech} x = 1/\cosh x$
	$\operatorname{cosec} x = 1/\sin x$	$\operatorname{cosech} x = 1/\sinh x$
(2)	$\cos^2 x + \sin^2 x = 1$	$\cosh^2 x - \sinh^2 x = 1$
	$\sec^2 x = 1 + \tan^2 x$	$\operatorname{sech}^2 x = 1 - \tanh^2 x$
	$\operatorname{cosec}^2 x = 1 + \cot^2 x$	$\operatorname{cosech}^2 x = \coth^2 x - 1$
(3)	$\sin 2x = 2\sin x \cos x$	$\sinh 2x = 2\sinh x \cosh x$
	$\cos 2x = \cos^2 x - \sin^2 x$	$\cosh 2x = \cosh^2 x + \sinh^2 x$
	$= 1 - 2\sin^2 x$	$= 1 + 2\sinh^2 x$
	$= 2\cos^2 x - 1$	$= 2\cosh^2 x - 1$

If we look at these results, we find that some of the hyperbolic identities follow exactly the trig identities: others have a difference in sign. This change of sign occurs whenever $\sin^2 x$ in the trig results is being converted into $\sinh^2 x$ to form the corresponding hyperbolic identities. This sign change also occurs when $\sin^2 x$ is involved without actually being written as such. For example, $\tan^2 x$ involves $\sin^2 x$ since $\tan^2 x$ could be written as $\dfrac{\sin^2 x}{\cos^2 x}$. The change of sign therefore occurs with

$\tan^2 x$ when it is being converted into $\tanh^2 x$
$\cot^2 x$ when it is being converted into $\coth^2 x$
$\operatorname{cosec}^2 x$ when it is being converted into $\operatorname{cosech}^2 x$

The sign change also occurs when we have a product of two sinh terms, e.g. the trig identity $\cos(A+B) = \cos A \cos B - \sin A \sin B$ gives the hyperbolic identity $\cosh(A+B) = \cosh A \cosh B + \sinh A \sinh B$.

Apart from this one change, the hyperbolic identities can be written down from the trig identities which you already know.
For example:

$$\tan 2x = \frac{2\tan x}{1-\tan^2 x} \text{ becomes } \tanh 2x = \frac{2\tanh x}{1+\tanh^2 x}$$

So provided you know your trig identities, you can apply the rule to form the corresponding hyperbolic identities.

Relationship between trigonometric and hyperbolic functions

39

From our previous work on complex numbers, we know that:

$$e^{i\theta} = \cos\theta + i\sin\theta$$

and $$e^{-i\theta} = \cos\theta - i\sin\theta$$

Adding these two results together, we have:

$e^{i\theta} + e^{-i\theta} = \ldots\ldots\ldots\ldots$

40

$$\boxed{2\cos\theta}$$

So that: $\cos\theta = \dfrac{e^{i\theta}+e^{-i\theta}}{2}$

which is of the form $\dfrac{e^x+e^{-x}}{2}$, with x replaced by $(i\theta)$

$\therefore \quad \cos\theta = \ldots\ldots\ldots\ldots$

41

$$\boxed{\cosh i\theta}$$

Here, then, is our first relationship:

$\cos\theta = \cosh i\theta$

Make a note of that for the moment: then on to Frame 42

42

If we return to our two original statements:

$$e^{i\theta} = \cos\theta + i\sin\theta$$
$$e^{-i\theta} = \cos\theta - i\sin\theta$$

and this time subtract, we get a similar kind of result

$$e^{i\theta} - e^{-i\theta} = \ldots\ldots\ldots\ldots$$

43

$$\boxed{2i\sin\theta}$$

So that: $i\sin\theta = \dfrac{e^{i\theta} - e^{-i\theta}}{2}$

$= \ldots\ldots\ldots\ldots$

44

$$\boxed{\sinh i\theta}$$

So: $\sinh i\theta = i\sin\theta$

Make a note of that also

45

So far, we have two important results:

(a) $\cosh i\theta = \cos\theta$

(b) $\sinh i\theta = i\sin\theta$

Now if we substitute $\theta = ix$ in the first of these results, we have:

$$\cos ix = \cosh(i^2x) = \cosh(-x) \quad \therefore \quad \cos ix = \cosh x \quad [\text{since } \cosh(-x) = \cosh x]$$

Writing this in reverse order, gives:

$\cosh x = \cos ix$ Another result to note

Now do exactly the same with the second result above, i.e. put $\theta = ix$ in the relationship $i\sin\theta = \sinh i\theta$ and simplify the result. What do you get?

46

$$\boxed{i\sinh x = \sin ix}$$

Because we have:

$$\begin{aligned} i\sin\theta &= \sinh i\theta \\ i\sin ix &= \sinh(i^2x) \\ &= \sinh(-x) \\ &= -\sinh x \qquad [\text{since } \sinh(-x) = -\sinh x] \end{aligned}$$

Finally, divide both sides by i, and we have

$\sin ix = i\sinh x$

On to the next frame

47

Now let us collect together the results we have established. They are so nearly alike, that we must distinguish between them:

$\sin ix = i \sinh x$	$\sinh ix = i \sin x$
$\cos ix = \cosh x$	$\cosh ix = \cos x$

and, by division, we can also obtain:

$\tan ix = i \tanh x$	$\tanh ix = i \tan x$

Copy the complete table into your record book for future use.

48

Here is an example of an application of these results:

Find an expansion for $\sin(x + iy)$.
Now we know that:

$$\sin(A + B) = \sin A \cos B + \cos A \sin B$$
$$\therefore \sin(x + iy) = \sin x \cos iy + \cos x \sin iy$$

so using the results we have listed, we can replace

$\cos iy$ by ....

and $\sin iy$ by ....

49

$\cos iy = \cosh y$	$\sin iy = i \sinh y$

So that:

$$\sin(x + iy) = \sin x \cos iy + \cos x \sin iy$$
becomes $$\sin(x + iy) = \sin x \cosh y + i \cos x \sinh y$$

Note: $\sin(x + iy)$ is a function of the angle $(x + iy)$, which is, of course, a complex quantity.

Meanwhile, here is just one example for you to work through:
Find an expansion for $\cos(x - iy)$.

Then check with Frame 50

50

$$\boxed{\cos(x - iy) = \cos x \cosh y + i \sin x \sinh y}$$

Here is the working:

$$\cos(A - B) = \cos A \cos B + \sin A \sin B$$
$$\therefore \cos(x - iy) = \cos x \cos iy + \sin x \sin iy$$
But $\cos iy = \cosh y$
and $\sin iy = i \sinh y$
$$\therefore \cos(x - iy) = \cos x \cosh y + i \sin x \sinh y$$

51

All that now remains is the **Can You?** checklist and the **Test exercise**, but before working through them, look through your notes, or review any parts of the Program on which you are not perfectly clear.

When you are ready, move on to the next frame

Can You?

52

Checklist 4

Check this list before and after you try the end of Program test.

On a scale of 1 to 5 how confident are you that you can: **Frames**

- Define the hyperbolic functions in terms of the exponential function? 1 to 4
  Yes ☐ ☐ ☐ ☐ ☐ *No*
- Express the hyperbolic functions as power series? 5 to 8
  Yes ☐ ☐ ☐ ☐ ☐ *No*
- Recognize the graphs of the hyperbolic functions? 9 to 16
  Yes ☐ ☐ ☐ ☐ ☐ *No*
- Evaluate hyperbolic functions and their inverses? 19 to 26
  Yes ☐ ☐ ☐ ☐ ☐ *No*
- Determine the logarithmic form of the inverse hyperbolic functions? 27 to 31
  Yes ☐ ☐ ☐ ☐ ☐ *No*
- Prove hyperbolic trigonometric identities? 32 to 38
  Yes ☐ ☐ ☐ ☐ ☐ *No*
- Understand the relationship between the circular and the hyperbolic trigonometric functions? 39 to 50
  Yes ☐ ☐ ☐ ☐ ☐ *No*

Test exercise 4

53

1. On the same axes, draw sketch graphs of (a) $y = \sinh x$, (b) $y = \cosh x$, (c) $y = \tanh x$.
2. If $L = 2C \sinh \dfrac{H}{2C}$, find L when $H = 63$ and $C = 50$.
3. If $v^2 = 1.8L \tanh \dfrac{6.3d}{L}$, find v when $d = 40$ and $L = 315$.
4. Calculate from first principles, the value of:
   (a) $\sinh^{-1} 1.532$ (b) $\cosh^{-1} 1.25$
5. If $\tanh x = \dfrac{1}{3}$, find e^{2x} and hence evaluate x.
6. The curve assumed by a heavy chain or cable is:
   $y = C \cosh \dfrac{x}{C}$
   If $C = 50$, calculate : (a) the value of y when $x = 109$
   (b) the value of x when $y = 75$.
7. Simplify $\dfrac{1 + \sinh 2A + \cosh 2A}{1 - \sinh 2A - \cosh 2A}$.
8. Obtain the expansion of $\sin(x - iy)$ in terms of the trigonometric and hyperbolic functions of x and y.

Further problems 4

54

1. Prove that $\cosh 2x = 1 + 2\sinh^2 x$.
2. Express $\cosh 2x$ and $\sinh 2x$ in exponential form and hence solve for real values of x, the equation:
   $2\cosh 2x - \sinh 2x = 2$
3. If $\sinh x = \tan y$, show that $x = \ln(\tan y \pm \sec y)$.
4. If $a = c \cosh x$ and $b = c \sinh x$, prove that
   $(a + b)^2 e^{-2x} = a^2 - b^2$
5. Evaluate: (a) $\tanh^{-1} 0.75$ and (b) $\cosh^{-1} 2$.
6. Prove that $\tanh^{-1} \left\{\dfrac{x^2 - 1}{x^2 + 1}\right\} = \ln x$.

▶

7 Express (a) $\cosh\dfrac{1+i}{2}$ and (b) $\sinh\dfrac{1+i}{2}$ in the form $a+ib$, giving a and b to 4 significant figures.

8 Prove that:

(a) $\sinh(x+y) = \sinh x \cosh y + \cosh x \sinh y$

(b) $\cosh(x+y) = \cosh x \cosh y + \sinh x \sinh y$

Hence prove that:

$$\tanh(x+y) = \frac{\tanh x + \tanh y}{1 + \tanh x \tanh y}$$

9 Show that the coordinates of any point on the hyperbola $\dfrac{x^2}{a^2} - \dfrac{y^2}{b^2} = 1$ can be represented in the form $x = a\cosh u$, $y = b\sinh u$.

10 Solve for real values of x:

$$3\cosh 2x = 3 + \sinh 2x$$

11 Prove that: $\dfrac{1+\tanh x}{1-\tanh x} = e^{2x}$.

12 If $t = \tanh\dfrac{x}{2}$, prove that $\sinh x = \dfrac{2t}{1-t^2}$ and $\cosh x = \dfrac{1+t^2}{1-t^2}$.

Hence solve the equation:

$$7\sinh x + 20\cosh x = 24$$

13 If $x = \ln\tan\left\{\dfrac{\pi}{4} + \dfrac{\theta}{2}\right\}$, find e^x and e^{-x}, and hence show that $\sinh x = \tan\theta$.

14 Given that $\sinh^{-1} x = \ln\left\{x + \sqrt{x^2+1}\right\}$, determine $\sinh^{-1}(2+i)$ in the form $a+ib$.

15 If $\tan\dfrac{x}{2} = \tan A \tanh B$, prove that: $\tan x = \dfrac{\sin 2A \sinh 2B}{1 + \cos 2A \cosh 2B}$

16 Prove that $\sinh 3\theta = 3\sinh\theta + 4\sinh^3\theta$.

17 If $\lambda = \dfrac{at}{2}\left\{\dfrac{\sinh at + \sin at}{\cosh at - \cos at}\right\}$, calculate λ when $a = 0.215$ and $t = 5$.

18 Prove that $\tanh^{-1}\left\{\dfrac{x^2-a^2}{x^2+a^2}\right\} = \ln\dfrac{x}{a}$.

Complex mappings

Learning outcomes

When you have completed this Program you will be able to:

- Recognize the transformation equation in the form $w = f(z) = u(x, y) + iv(x, y)$
- Illustrate the image of a point in the complex z-plane under a complex mapping onto the w-plane
- Map a straight line in the z-plane onto the w-plane under the transformation $w = f(z)$
- Identify complex mappings that form translations, magnifications, rotations and their combinations
- Deal with the non-linear transformations $w = z^2$, $w = 1/z$, $w = 1/(z - a)$ and $w = (az + b)/(cz + d)$

Functions of a complex variable

1

For a function of a single real variable $f(x)$ we can construct the graph of the function by plotting points against two mutually perpendicular Cartesian axes, the x-axis and the $f(x)$-axis. For a function of a single complex variable $w = f(z) = u(x, y) + iv(x, y)$ we have four real variables, x, y, u and v. For example if $z = x + iy$ and $f(z) = z^2$ then

$$\begin{aligned} f(z) &= (x + iy)^2 \\ &= x^2 + 2ixy + (iy)^2 \\ &= x^2 - y^2 + 2ixy \end{aligned}$$

and so

$$\begin{aligned} u(x, y) &= x^2 - y^2 \\ \text{and} \quad v(x, y) &= 2xy \end{aligned}$$

We cannot plot the graph of the function $f(z)$ against a single set of axes because to do so we would be required to draw four mutually perpendicular axes, which is not possible. Instead, we resort to plotting z-values against x- and y-axes in the complex z-plane and to plotting the corresponding values of $w = f(z)$ against u- and v-axes in the complex w-plane. Accordingly, values of z are plotted on one plane and the corresponding values of $f(z)$ are plotted on another plane. So in our example above for a particular value of z, for example, $z = 4 + 3i$

$u = \ldots\ldots\ldots\ldots$

$v = \ldots\ldots\ldots\ldots$

2

$$u = 7 \quad v = 24$$

Because with $z = 4 + 3i$, $x = 4$ and $y = 3$. Then $u = 16 - 9 = 7$ and $v = 24$.

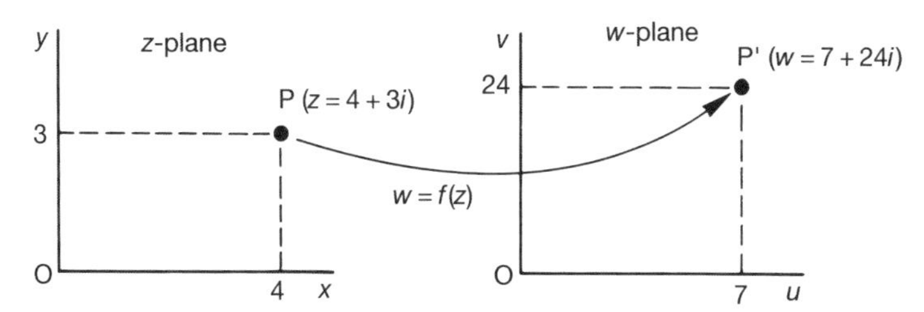

Therefore, z (where $z = x + iy$) and w (where $w = u + iv$) are two complex variables related by the equation $w = f(z)$.

Any other point in the z-plane will similarly be transformed into a corresponding point in the w-plane, the resulting position P′ depending on

(a) the initial position of P

(b) the relationship $w = f(z)$, called the *transformation equation* or *transformation function*.

Complex mapping

The transformation of P in the z-plane onto P′ in the w-plane is said to be a *mapping* of P onto P′ under the transformation $w = f(z)$ and P′ is sometimes referred to as the *image* of P.

Example 1

Determine the image of the point P, $z = 3 + 2i$, on the w-plane under the transformation $w = 3z + 2 - i$.

$$w = u + iv = f(z) = 3z + 2 - i$$
$$= 3(x + iy) + 2 - i$$

so that, for this example,

$u = \ldots\ldots\ldots\ldots$; $v = \ldots\ldots\ldots\ldots$

3

$$u = 3x + 2; \quad v = 3y - 1$$

Then the point P ($z = 3 + 2i$) transforms onto …………

4

$$w = 11 + 5i$$

Because

$z = 3 + 2i \quad \therefore \; x = 3, \; y = 2$

$u = 3x + 2 = 11; \quad v = 3y - 1 = 5; \quad \therefore \; w = 11 + 5i$

We can illustrate the transformation thus:

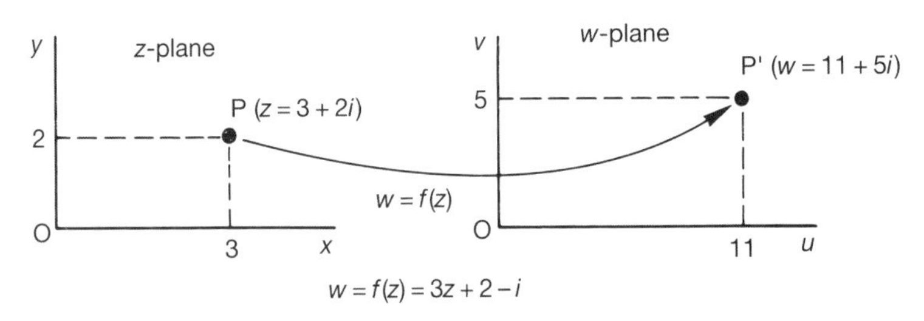

Here is another.

Example 2

Map the points A ($z = -2 + i$) and B ($z = 3 + 4i$) onto the w-plane under the transformation $w = 2zi + 3$ and illustrate the transformation on a diagram.

This is no different from the previous example. Complete the job and check with the next frame.

5

$$A' \; (w = 1 - 4i); \quad B' \; (w = -5 + 6i)$$

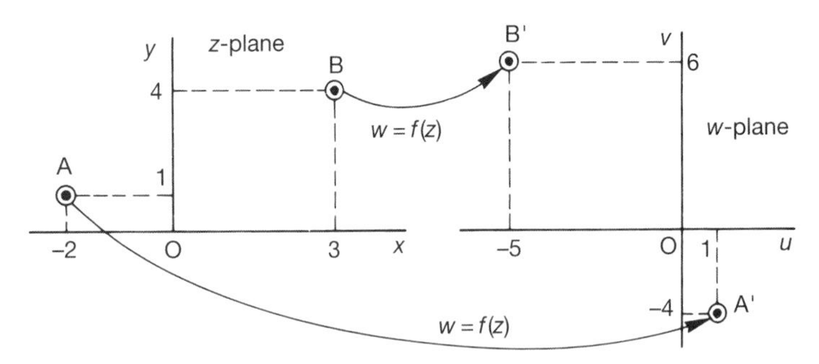

Because

$w = f(z) = 2iz + 3 = 2i(x + iy) + 3 = (3 - 2y) + 2ix$

$w = u + iv \quad \therefore \; u = 3 - 2y; \quad v = 2x$

A: $x = -2, \; y = 1 \quad \therefore \; A'\!: u = 3 - 2 = 1; \; v = -4 \quad \therefore \; A'\!: w = 1 - 4i$

B: $x = 3, \; y = 4 \quad \therefore \; B'\!: u = 3 - 8 = -5; \; v = 6 \quad \therefore \; B'\!: w = -5 + 6i$

There now follows a short practice exercise. Work all four of the items before you check the results. There is no need to illustrate the transformation in each case.

So move on

6

Exercise

Map the following points in the z-plane onto the w-plane under the transformation $w = f(z)$ stated in each case.

1 $z = 4 - 2i$ under $w = 3iz + 2i$

2 $z = -2 - i$ under $w = iz + 3$

3 $z = 3 + 2i$ under $w = (1 + i)z - 2$

4 $z = 2 + i$ under $w = z^2$.

7

1	$w = 6 + 14i$	**2**	$w = 4 - 2i$
3	$w = -1 + 5i$	**4**	$w = 3 + 4i$

That was easy enough. Now let us extend the ideas.

Mapping of a straight line in the *z*-plane onto the *w*-plane under the transformation *w = f (z)*

A typical example will show the method.

Example 1

To map the straight line joining A $(-2 + i)$ and B $(3 + 6i)$ in the z-plane onto the w-plane when $w = 3 + 2iz$.

We first of all map the end points A and B onto the w-plane to obtain A′ and B′ as in the previous cases.

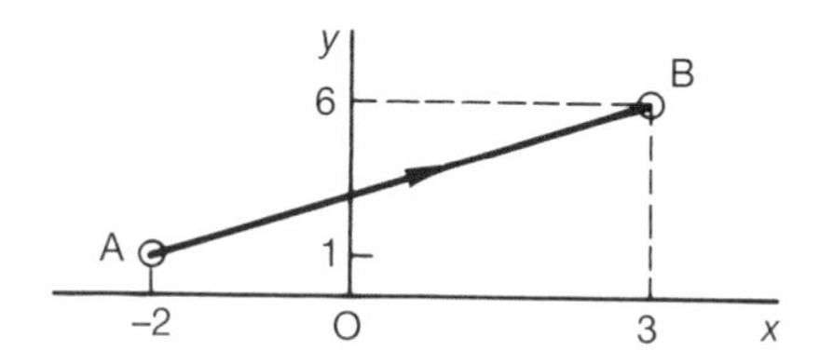

A′: $w =$ . . . .

B′: $w =$ . . . .

8

A′: $w = 1 - 4i$; B′: $w = -9 + 6i$

Because

(1) A: $z = -2 + i$ $\quad w = 3 + 2iz$

$\therefore$ A′: $w = 3 + 2i(-2 + i) = 3 - 4i - 2 = 1 - 4i$

(2) B: $z = 3 + 6i$

$\therefore$ B′: $w = 3 + 2i(3 + 6i) = 3 + 6i - 12 = -9 + 6i$

Then, if we illustrate the transformations on a diagram, as before, we get

. . . .

9

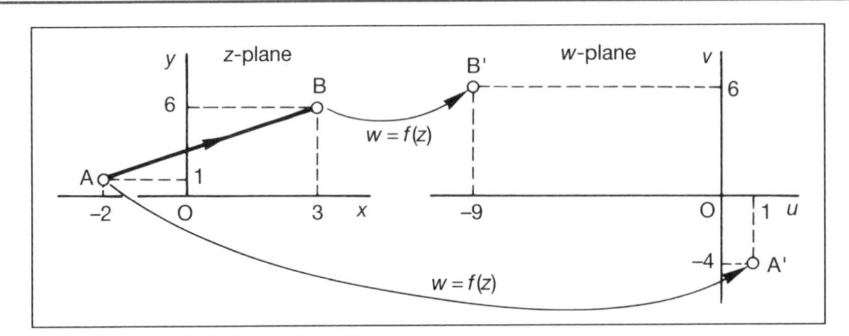

As z moves along the line A to B in the z-plane, we cannot assume that its image in the w-plane travels along a straight line from A′ to B′. As yet, we have no evidence of what the path is. We therefore have to find a general point $w = u + iv$ in the w-plane corresponding to a general point $z = x + iy$ in the z-plane.

$$w = u + iv = f(z) = 3 + 2iz$$
$$= \dots\dots\dots\dots\dots\dots$$

10

$$\boxed{w = u + iv = (3 - 2y) + 2ix}$$

Because

$$w = 3 + 2i(x + iy) = 3 + 2ix - 2y = (3 - 2y) + 2ix$$
$$\therefore\ u = 3 - 2y \quad \text{and} \quad v = 2x$$

Rearranging these results, we also have $y = \dfrac{3-u}{2}$; $\quad x = \dfrac{v}{2}$.

Now the Cartesian equation of AB is $y = x + 3$ and substituting from the previous line, we have $\dfrac{3-u}{2} = \dfrac{v}{2} + 3$ which simplifies to $\dots\dots\dots$

11

$$\boxed{v = -u - 3}$$

which is the equation of a straight line, so, in this case, the path joining A′ and B′ *is* in fact a straight line.

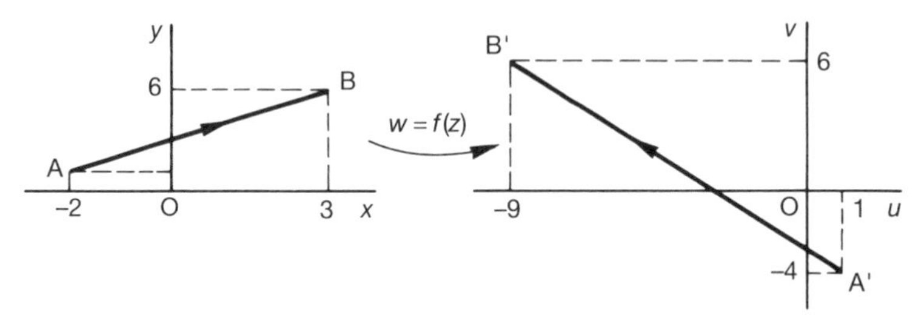

Note that it is useful to attach arrow heads to show the corresponding direction of progression in the transformation.

On to the next

12

Example 2

If $w = z^2$, find the path traced out by w as z moves along the straight line joining A $(2 + 0i)$ and B $(0 + 2i)$.

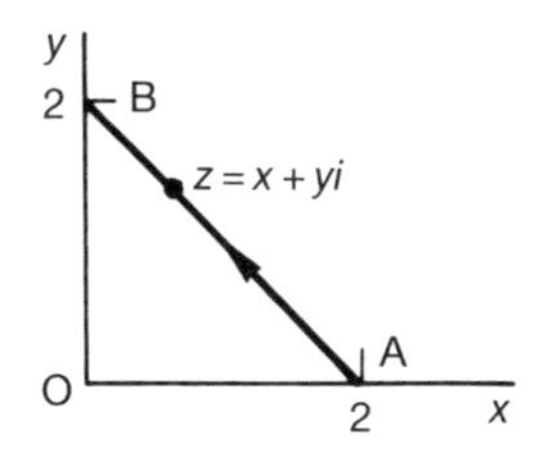

Cartesian equation of AB is

$y = 2 - x$

First we transform the two end points A and B onto A′ and B′ in the w-plane.

A′: ....; B′: ....

13

$$\boxed{A': w = 4 + 0i; \quad B': w = -4 + 0i}$$

Because

$w = z^2$ A: $z = 2$ $\therefore$ A′: $w = 2^2 = 4$

B: $z = 2i$ $\therefore$ B′: $w = (2i)^2 = -4$

So we have

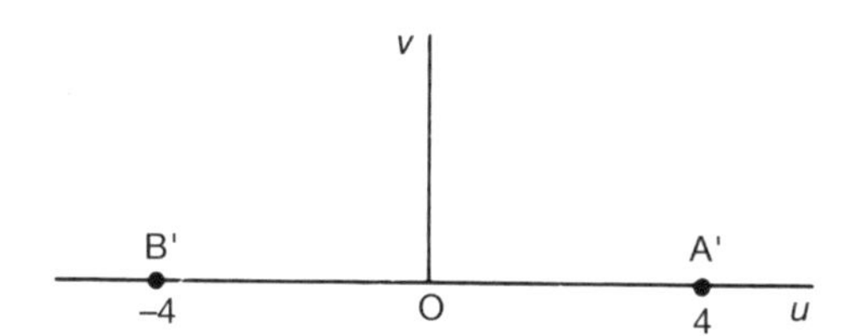

Now we have to find the path from A′ to B′.

The Cartesian equation of AB in the z-plane is $y = 2 - x$.

Also $w = z^2 = (x + iy)^2 = (x^2 - y^2) + 2ixy$

$\therefore$ $u = x^2 - y^2$ and $v = 2xy$

Substituting $y = 2 - x$ in these results we can express u and v in terms of x.

$u =$ ....; $v =$ ....

14

$$u = 4x - 4; \quad v = 4x - 2x^2$$

So, from the first of these $x = \dfrac{u+4}{4}$

Substituting in the second
$$v = 4\left(\frac{u+4}{4}\right) - 2\left(\frac{u+4}{4}\right)^2$$
$$= u + 4 - \frac{1}{8}(u^2 + 8u + 16)$$
$$= -\frac{1}{8}(u^2 - 16)$$

Therefore the path is $v = -\dfrac{1}{8}(u^2 - 16)$ which is a parabola for which at $u = 0$, $v = 2$.

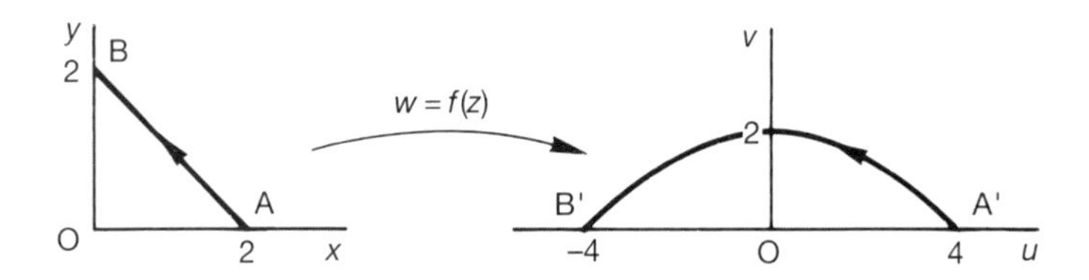

Note that a straight line in the z-plane does not always map onto a straight line in the w-plane. It depends on the particular transformation equation $w = f(z)$.

If the transformation is a *linear equation*, $w = f(z) = az + b$, where a and b may themselves be real or complex, then a straight line in the z-plane maps onto a corresponding straight line in the w-plane.

Example 3

A triangle in the z-plane has vertices at A ($z = 0$), B ($z = 3$) and C ($z = 3 + 2i$). Determine the image of this triangle in the w-plane under the transformation equation $w = (2 + i)z$.

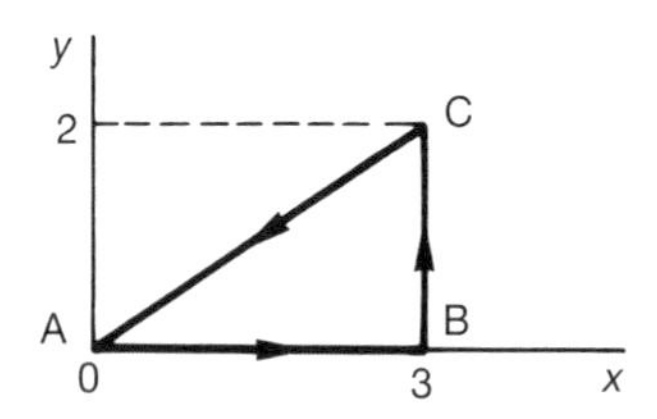

$$w = u + iv = f(z) = (2+i)z = (2+i)(x+iy) = (2x - y) + i(2y + x)$$
$$\therefore \; u = 2x - y; \qquad v = 2y + x$$

We now transform each vertex in turn onto the w-plane to determine A′, B′ and C′.

These are A′: ....; B′: ....; C′: ....

15

$$A': w = 0; \quad B': w = 6 + 3i; \quad C': w = 4 + 7i$$

The transformation is linear (of the form $w = az$) so A′B′, B′C′ and C′A′ are straight lines and the transformation can be illustrated in the diagram

. . . .

16

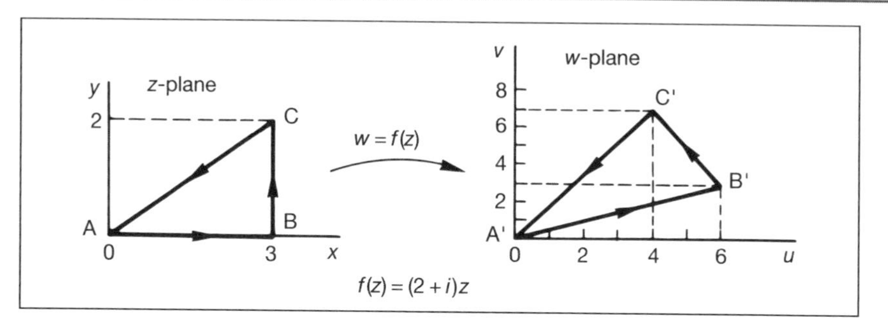

All very straightforward. Let us now take a more detailed look at linear transformations.

Types of transformation of the form $w = az + b$

where the constants a and b may be real or complex.

1 *Translation*

Let $a = 1$ and $b = 2 - i$ i.e. $w = z + (2 - i)$.

If we apply this to the straight line joining A $(0 + i)$ and B $(2 + 3i)$ in the z-plane, then

$$\begin{aligned} w &= x + iy + 2 - i \\ &= (x + 2) + i(y - 1) \end{aligned}$$

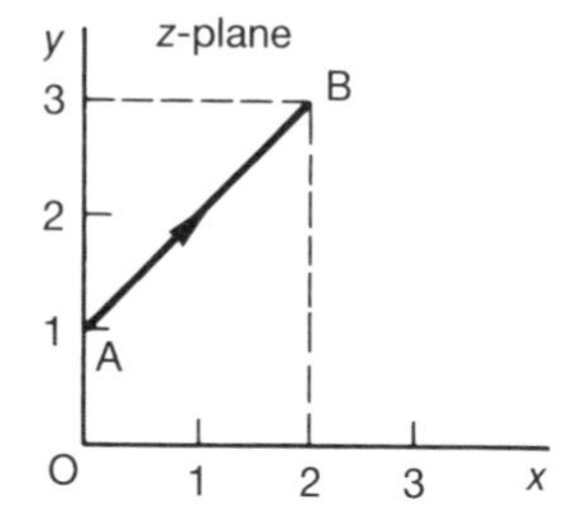

so the corresponding end points A′ and B′ in the w-plane are

A′: . . . .; B′: . . . .

17

$$\text{A}': \ w = 2; \quad \text{B}': \ w = 4 + 2i$$

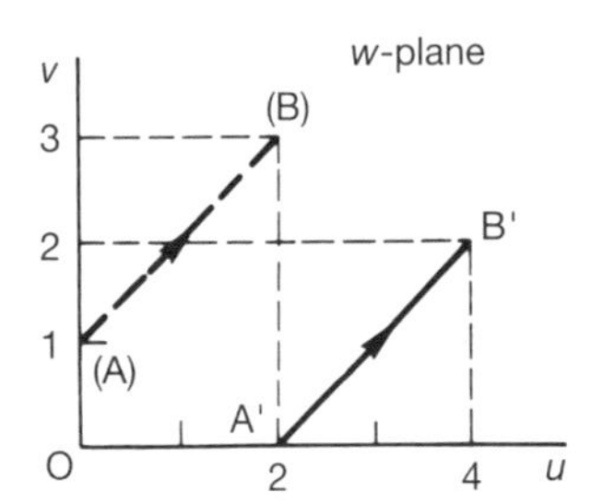

The transformed line A′B′ is then as shown. The broken line (A)(B) indicates the position of the original line AB in the z-plane.

Note that the whole line AB has moved two units to the right and one unit downwards, while retaining its original magnitude (length) and direction.

Such a transformation is called a *translation* and occurs whenever the transformation equation is of the form $w = z + b$. The degree of translation is given by the value of b – in this case $(2 - i)$, i.e. 2 units along the positive real axis and 1 unit in the direction of the negative imaginary axis.

On to the next frame

18

2 *Magnification*

Consider now $w = az + b$ where $b = 0$ and a is real, e.g. $w = 2z$.

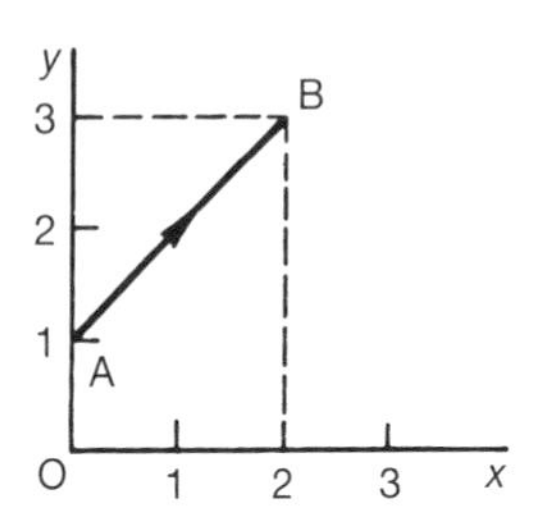

Applying the transformation to the same line AB as before, we have

$$w = u + iv = 2z = 2(x + iy)$$

$$\therefore \ u = 2x \quad \text{and} \quad v = 2y$$

Transforming the end points A $(0 + i)$ and B $(2 + 3i)$ onto A′ and B′ in the w-plane, we have

A′: ....; B′: ....

and the w-plane diagram becomes

....

19

$$\text{A}': \ w = 2i; \quad \text{B}': \ w = 4 + 6i$$

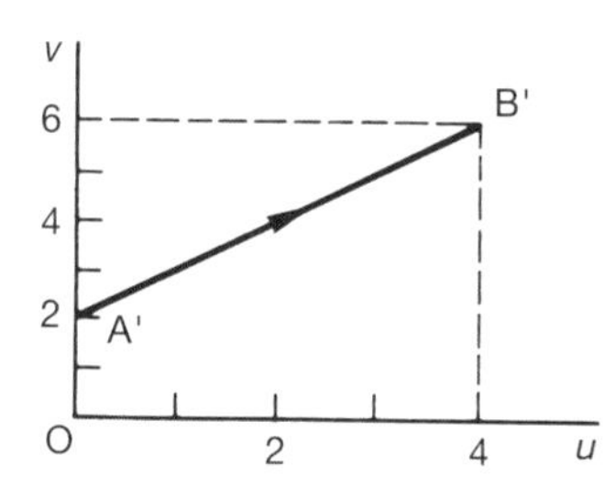

Note that (a) all distances in the z-plane are magnified by a factor 2, and (b) the direction of A′B′ is that of AB unchanged. Any such transformation $w = az$ where a is real, is said to be a *magnification* by the factor a.

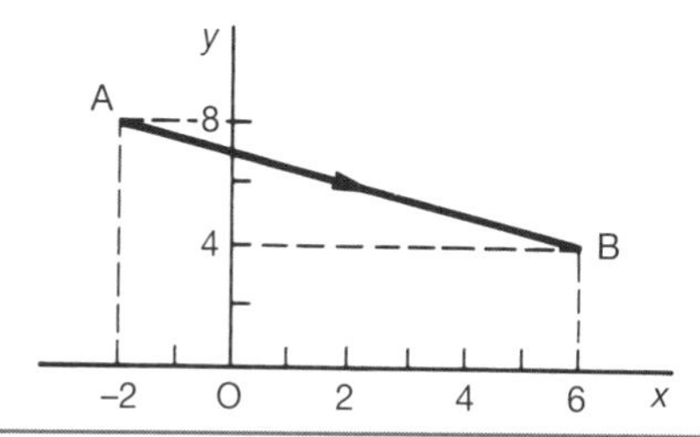

So, if we apply the transformation $w = z/2$ to AB shown here, we can map AB onto A′B′ in the w-plane and obtain

....

Sketch the result

20

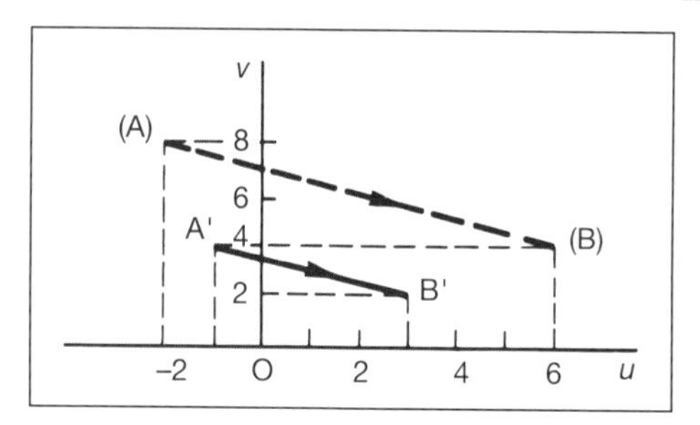

3 *Rotation*

Consider next $w = az + b$ with $b = 0$ and a complex,

e.g. $w = iz$.

$$w = u + iv = iz$$
$$= i(x + iy)$$
$$= -y + ix$$

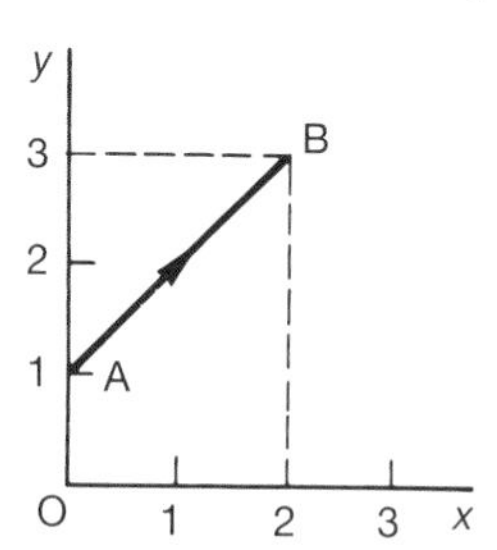

Transforming the end points as usual, we can sketch the original line AB and the mapping A′B′, which gives ....

21

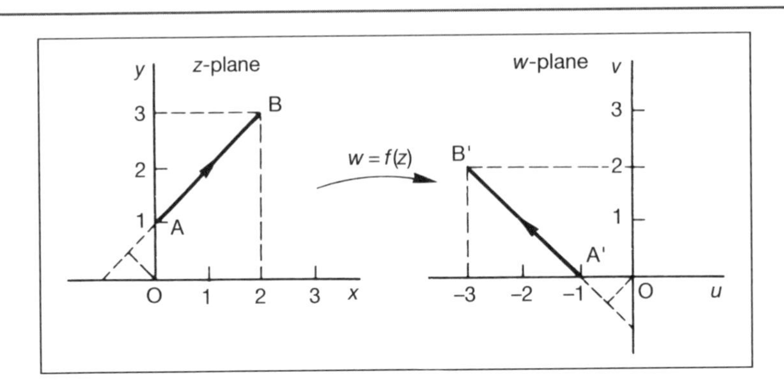

A′ is the point $w = -1 + 0i$; B′ is the point $w = -3 + 2i$

Note $AB = 2\sqrt{2}$ $A'B' = 2\sqrt{2}$

Slope of $AB = m = 1$ Slope of $A'B' = m_1 = -1$

$mm_1 = 1(-1) = -1$

Therefore in transformation by $w = iz$, AB retains its original length but is rotated about the origin, in this case through 90° in a positive (anticlockwise) direction.

Some degree of rotation always occurs when the transformation equation is of the form $w = az + b$ with a complex.

Move on to the next frame

22

4 *Combined magnification and rotation*

If $w = (a + ib)z$, the effect of transformation is

(a) magnification $|a + ib| = \sqrt{a^2 + b^2}$

(b) rotation counter-clockwise through $\arg(a + ib)$, i.e. $\arctan \frac{b}{a}$.

Let us see this with an example.

Example

Map the straight line joining A $(0 + 2i)$ and B $(4 + 6i)$ in the z-plane onto the w-plane under the transformation $w = (3 + 2i)z$.

The working is just as before. Draw the z-plane and w-plane diagrams, which give

. . . .

23

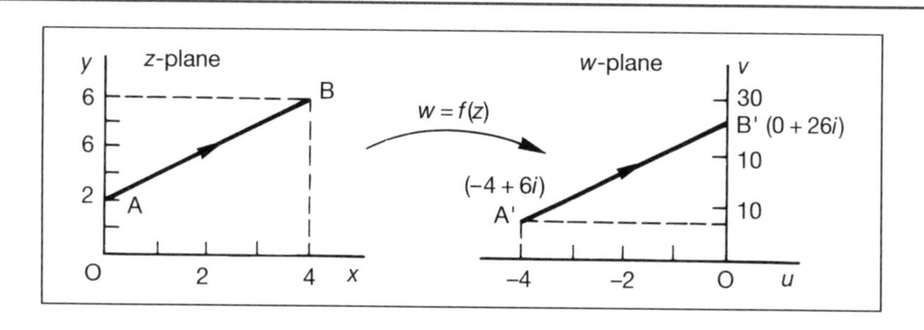

$$w = (3 + 2i)z$$

$$\therefore \quad u + iv = (3 + 2i)(x + iy) = (3x - 2y) + i(2x + 3y)$$

$$\therefore \quad u = 3x - 2y \quad \text{and} \quad v = 2x + 3y$$

A: $z = 0 + 2i$, i.e. $x = 0$, $y = 2$

$\therefore$ A′: $u = -4$, $v = 6$ $\quad\therefore$ A′: $(-4 + 6i)$

B: $z = 4 + 6i$, i.e. $x = 4$, $y = 6$

$\therefore$ B′: $u = 0$, $v = 26$ $\quad\therefore$ B′: $(0 + 26i)$

By a simple application of Pythagoras, we can now calculate the lengths of AB and A′B′, and then determine the magnification factor (A′B′)/(AB).

AB = ...; A′B′ = ...; magnification = ...

24

$$\text{AB} = 4\sqrt{2}; \quad \text{A}'\text{B}' = 4\sqrt{26}; \quad \text{mag} = \sqrt{13}$$

Because

$$\text{AB} = \sqrt{16 + 16} = \sqrt{32} = 4\sqrt{2}$$

$$\text{A}'\text{B}' = \sqrt{16 + 400} = \sqrt{416} = 4\sqrt{26}$$

$$\therefore \quad \text{magnification} = \frac{4\sqrt{26}}{4\sqrt{2}} = \sqrt{13}$$

Also $|a + ib| = |3 + 2i| = \sqrt{9 + 4} = \sqrt{13}$ $\quad\therefore$ mag $= |a + ib|$

Now let us check the rotation.

For AB $\quad \tan\theta_1 = 1 \quad \therefore \ \theta_1 = 45^\circ = 0.7854$ radians

For A′B′ $\quad \tan\theta_2 = 5 \quad \therefore \ \theta_2 = 78^\circ\ 41' = 1.3733$ radians

$$\therefore \quad \text{rotation} = \theta_2 - \theta_1 = 1.3733 - 0.7854 = 0.5879$$

i.e. rotation = 0.5879 radians

Also arg $(a + ib)$ = arg $(3 + 2i)$ = ...

25

0.5879 radians

Because $\arg(3+2i) = \arctan\frac{2}{3} = 33^\circ\ 41' = 0.5879$ radians.

So, in transformation $w = (a+ib)z = (3+2i)z$

(a) AB is magnified by $|a+ib|$, i.e. $\sqrt{13}$

(b) AB is rotated anticlockwise through $\arg(a+ib)$, i.e. $\arg(3+2i)$ i.e. 0.5879 radians.

5 *Combined magnification, rotation and translation*

The work we have just done can be extended to include all three effects of transformation.

In general, a transformation equation $w = az + b$, where a and b are each real or complex, results in

magnification $|a|$; rotation $\arg a$; translation b

Therefore, if $w = (3+i)z + 2 - i$

magnification = ...; rotation = ...;
translation = ...

26

mag $= \sqrt{10} = 3.162$; rotation $= 18^\circ\ 26' = 0.3218$ radians;
translation = 2 units to right, 1 unit downwards

Because

(a) magnification $= |3+i| = \sqrt{9+1} = \sqrt{10} = 3.162$

(b) rotation $= \arg(3+i) = \arctan\dfrac{1}{3} = 18^\circ\ 26' = 0.3218$ radians

(c) translation $= 2 - i$, i.e. 2 to the right, 1 downwards.

Let us work through an example in detail.

Example 1

The straight line joining A $(-2-3i)$ and B $(3+i)$ in the z-plane is subjected to the linear transformation equation

$w = (1+2i)z + 3 - 4i$

Illustrate the mapping onto the w-plane and state the resulting magnification, rotation and translation involved.

The first part is just like examples we have already done. So,

(a) transform the end points A and B onto A′ and B′ in the w-plane

(b) join A′ and B′ with a straight line, since AB is a straight line and the transformation equation is linear.

That can be done without trouble, the final diagram being

....

27

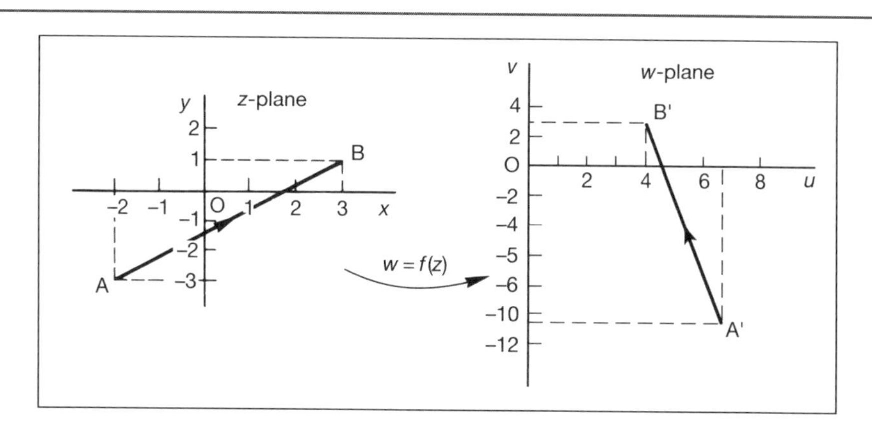

Check the working. $w = (1 + 2i)z + 3 - 4i$

A: $z = x + iy$

$= -2 - 3i$

A′: $w = u + iv = (1 + 2i)(-2 - 3i) + 3 - 4i$

$= -2 - 7i + 6 + 3 - 4i$

$= 7 - 11i$

B: $z = x + iy$

$= 3 + i$

B′: $w = u + iv = (1 + 2i)(3 + i) + 3 - 4i$

$= 3 + 7i - 2 + 3 - 4i$

$= 4 + 3i$

Now for the second part of the problem, we have to state the magnification, rotation and translation when $w = (1 + 2i)z + 3 - 4i$. We remember that the 'tailpiece', i.e. $3 - 4i$, independent of z, represents the

....

28

translation

So, for the moment, we concentrate on $w = (1 + 2i)z$, which determines the magnification and rotation. This tells us that

magnification = ...

rotation = ...

29

$$\text{mag} = |a| = |1+2i| = \sqrt{1+4} = \sqrt{5} = 2.236$$
$$\text{rotation} = \arg a = \arctan \tfrac{2}{1} = 63^\circ\ 26' = 1.107 \text{ radians}$$

The translation is given by $(3 - 4i)$, i.e. 3 units to the right, 4 units downwards.

We can in fact see the intermediate steps if we deal first with the transformation $w = (1 + 2i)z$ and subsequently with the translation $w = 3 - 4i$.

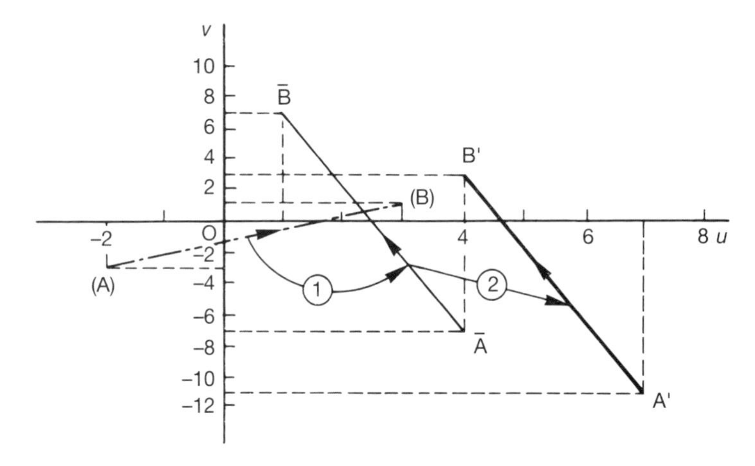

Under $w = (1 + 2i)z$, A and B map onto $\bar{A}$ and $\bar{B}$ where $\bar{A}$ is $w = 4 - 7i$ and $\bar{B}$ is $w = 1 + 7i$.

Then the translation $w = 3 - 4i$ moves all points 3 units to the right and 4 units downwards, so that $\bar{A}$ and $\bar{B}$ now map onto A′ and B′ where A′ is $w = 7 - 11i$ and B′ is $w = 4 + 3i$.

Normally, there is no need to analyze the transformation into intermediate steps.

Now for –

Example 2

Map the straight line joining A $(1 + 2i)$ and B $(4 + i)$ in the z-plane onto the w-plane using the transformation equation

$$w = (2 - 3i)z - 4 + 5i$$

and state the magnification, rotation and translation involved.

There are no snags. Complete the working and check with the next frame.

30

Here is the complete working.

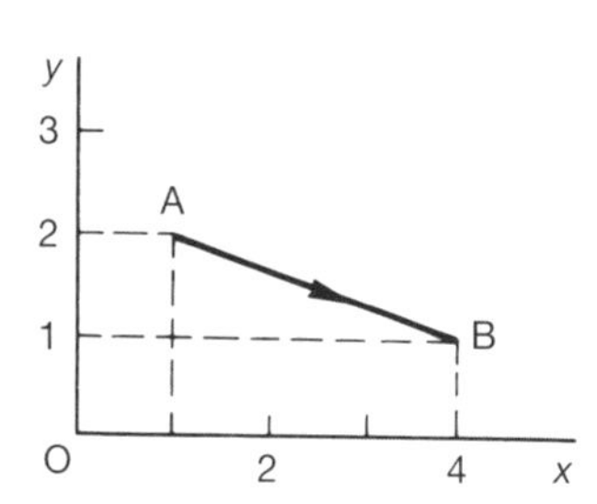

$w = (2 - 3i)z - 4 + 5i$
A: $z = 1 + 2i$
B: $z = 4 + i$

A: $z = 1 + 2i$
A′: $w = (2 - 3i)(1 + 2i) - 4 + 5i = 2 + i + 6 - 4 + 5i = 4 + 6i$
B: $z = 4 + i$
B′: $w = (2 - 3i)(4 + i) - 4 + 5i = 8 - 10i + 3 - 4 + 5i = 7 - 5i$

So we have

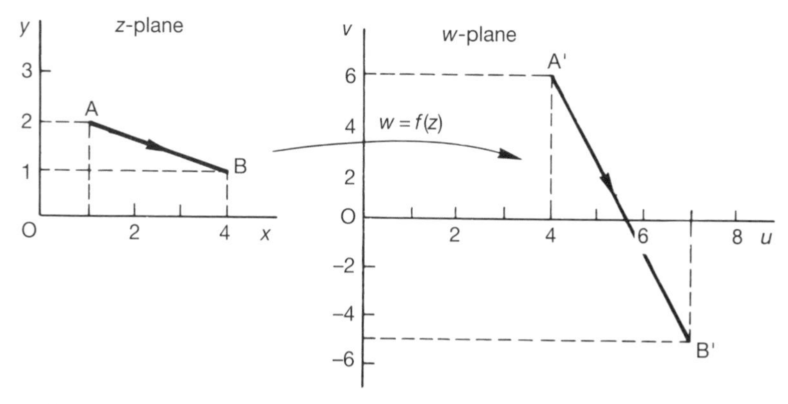

Also we have

(a) magnification $= |2 - 3i| = \sqrt{4 + 9} = \sqrt{13} = 3.606$

(b) rotation $= \arg(2 - 3i) = \arctan(\frac{-3}{2}) = -56^\circ\ 19'$

$= 0.9828$ radians clockwise

(c) translation $= -4 + 5i$ i.e. 4 units to left, 5 units upwards

All very straightforward. Before we move on, here is a short review exercise.

Exercise

Calculate (a) the magnification, (b) the rotation, (c) the translation involved in each of the following transformations.

1 $w = (1 - 2i)z + 2 - 3i$

2 $w = (4 + 3i)z - 2 + 5i$

3 $w = (2 - 3i)z - 1 - i$

4 $w = (i - 4)z + 2i - 3$

5 $w = i2z + 4 - i$

6 $w = (5 + 2i)z + i(3i - 4)$.

Complete all six and then check the results with the next frame.

31

Results:

1 $w = (1 - 2i)z + 2 - 3i$

(a) magnitude $= |1 - 2i| = \sqrt{1+4} = \sqrt{5} = 2.236$

(b) rotation $= \arg(1 - 2i) = \arctan(-2) = -63^\circ\ 26'$

$= 1.107$ radians clockwise

(c) translation $= 2 - 3i$, i.e. 2 units to right, 3 units downwards.

The others are done in the same way and give the following results.

No.	Magnitude	Rotation (rad)	Translation
2	5	0.6435 ac	2L, 5U
3	3.606	0.9828 c	1L, 1D
4	4.123	0.2450 c	3L, 2U
5	2	1.5708 ac	4R, 1D
6	5.385	0.3805 ac	3L, 4D

Now let us start a new section, so on to the next frame

Non-linear transformations

32

So far, we have concentrated on linear transformations of the form $w = az + b$. We can now proceed to something rather more interesting.

1 *Transformation* $w = z^2$ (refer to Frame 12)

The general principles are those we have used before. An example will show the development.

Example 1

The straight line AB in the z-plane as shown is mapped onto the w-plane by $w = z^2$.

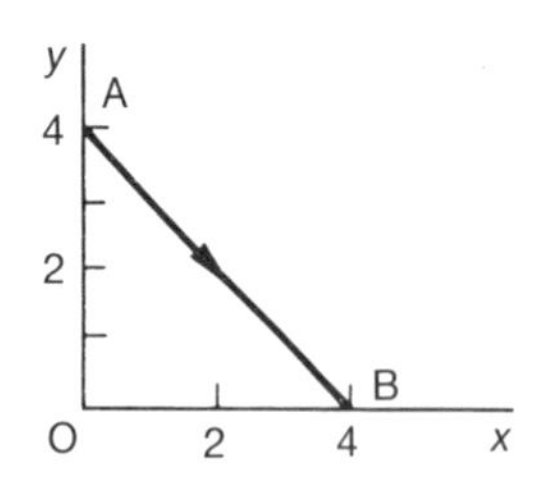

As before, we start by transforming the end points onto A′ and B′ in the w-plane.

A′: $w = \dots\dots\dots\dots$

B′: $w = \dots\dots\dots\dots$

33

$$\boxed{A'\!: \ w = -16; \quad B'\!: \ w = 16}$$

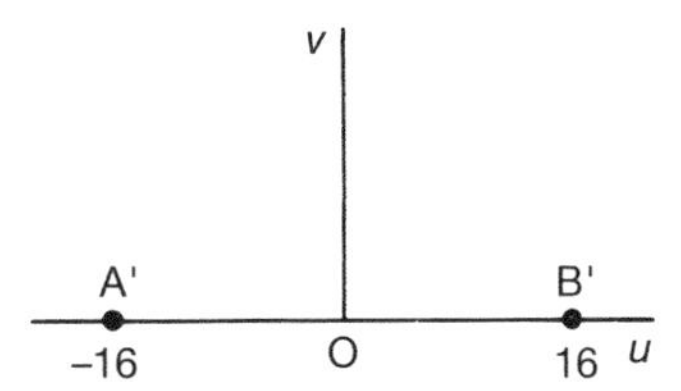

We cannot however assume that AB maps onto the straight line A′B′, since the transformation is not linear. We therefore have to deal with a general point.

$$w = u + iv = z^2 = (x + iy)^2 = x^2 + 2ixy - y^2 = (x^2 - y^2) + 2ixy$$

$$\therefore \ u = x^2 - y^2 \quad \text{and} \quad v = 2xy$$

The Cartesian equation of AB in the z-plane is $y = 4 - x$. So, substituting in the results of the previous line, we can express u and v in terms of x.

$u = \ldots\ldots\ldots\ldots; \quad v = \ldots\ldots\ldots\ldots$

34

$$\boxed{u = 8x - 16; \quad v = 8x - 2x^2}$$

The first gives $x = \dfrac{u+16}{8}$ and substituting this in the expression for v gives …………

35

$$\boxed{v = -\tfrac{1}{32}u^2 + 8}$$

Because

$$v = 8\left(\frac{u+16}{8}\right) - 2\left(\frac{u+16}{8}\right)^2 = u + 16 - \frac{u^2}{32} - u - 8$$

$$\therefore \ v = -\frac{u^2}{32} + 8$$

which is an ‘inverted’ parabola, symmetrical about the v-axis, with $v = 8$ at $u = 0$. The mapping is therefore

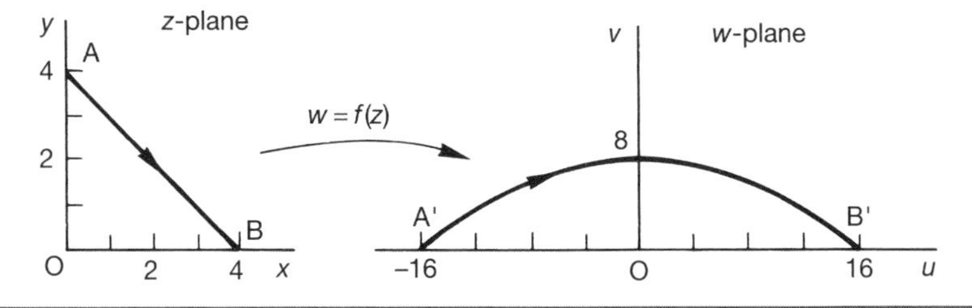

36

Example 2

AB is a straight line in the z-plane joining the origin A to the point B $(a + ib)$. Obtain the mapping of AB onto the w-plane under the transformation $w = z^2$.

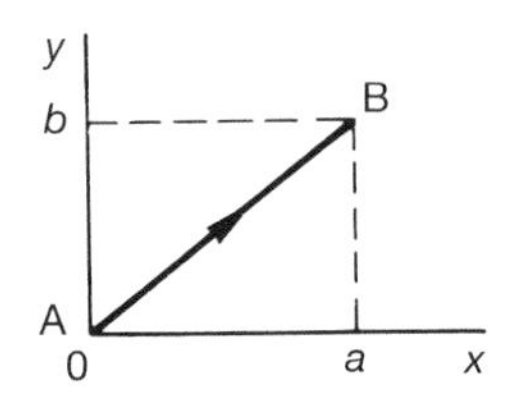

As always, first map the end points.

A′: $w = 0$

B′: $w = (a + ib)^2 = (a^2 - b^2) + 2iab$

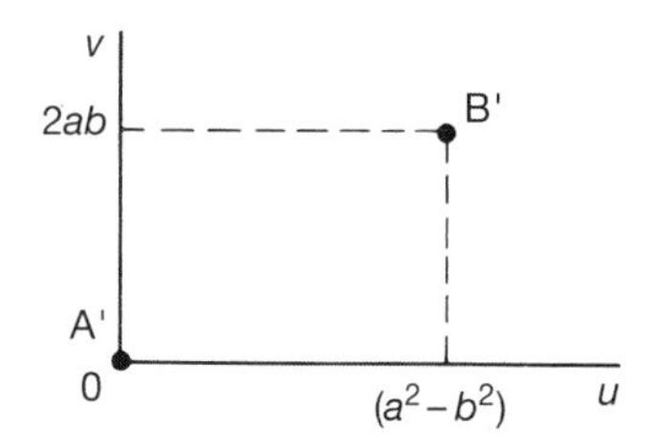

Now to find the path joining A′ and B′, we consider a general point $z = x + iy$.

$$w = u + iv = z^2$$
$$= (x + iy)^2$$
$$= (x^2 - y^2) + 2ixy$$
$$\therefore\ u = x^2 - y^2 \quad \text{and} \quad v = 2xy$$

The equation of AB is $y = \dfrac{b}{a}x$. We can therefore find u and v in terms of x and hence v in terms of u.

$u = \ldots\ldots\ldots\ldots$

$v = \ldots\ldots\ldots\ldots$

$v = f(u) = \ldots\ldots\ldots\ldots$

37

$$u = \left(\frac{a^2 - b^2}{a^2}\right)x^2; \quad v = \left(\frac{2b}{a}\right)x^2; \quad v = \left(\frac{2ab}{a^2 - b^2}\right)u$$

Because

$$u = x^2 - y^2 = x^2 - \left(\frac{b^2}{a^2}\right)x^2 = \left(\frac{a^2 - b^2}{a^2}\right)x^2$$

$$v = 2xy = 2x\left(\frac{b}{a}\right)x = \left(\frac{2b}{a}\right)x^2$$

From the expression for u, $x^2 = \left(\frac{a^2}{a^2 - b^2}\right)u \quad \therefore \; v = \frac{2b}{a}\left(\frac{a^2}{a^2 - b^2}\right)u$

$\therefore \; v = \left(\frac{2ab}{a^2 - b^2}\right)u$ which is of the form $v = ku$.

A′B′ is therefore a straight line through the origin.

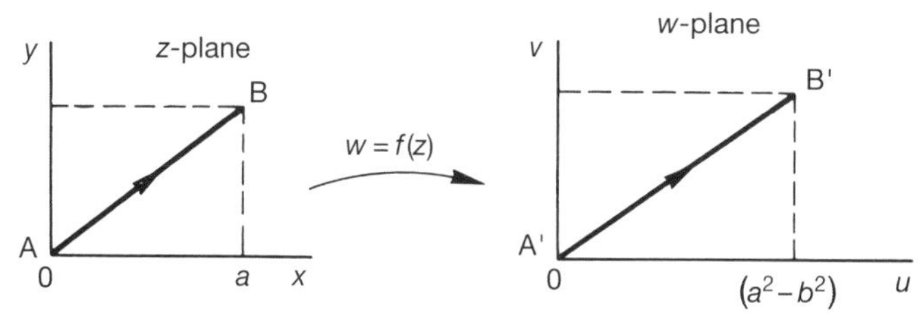

Therefore, under the transformation $w = z^2$, a straight line through the origin in the z-plane maps onto a straight line through the origin in the w-plane, whereas a straight line not passing through the origin maps onto a parabola.

This is worth remembering, so make a note of it

38

Example 3

A triangle consisting of AB, BC, CA in the z-plane is mapped onto the w-plane by the transformation $w = z^2$.

The transformation is $w = z^2$.

$$\therefore \; w = (x + iy)^2 = (x^2 - y^2) + 2ixy$$
$$= u + iv$$
$$\therefore \; u = x^2 - y^2 \quad \text{and} \quad v = 2xy$$

First we can map the end points A, B, C onto A′, B′, C′ in the w-plane.

A′: ....

B′: ....

C′: ....

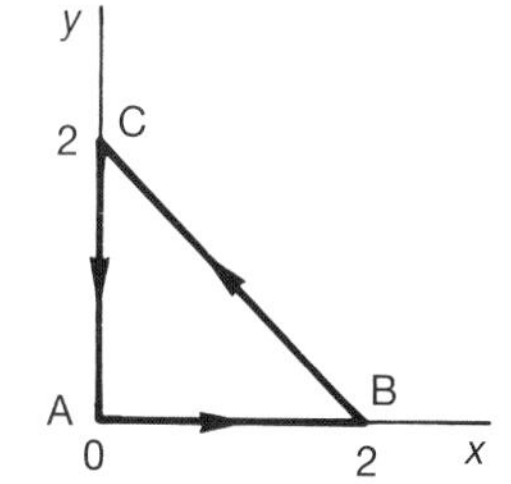

39

$$\text{A}': \ w = 0; \quad \text{B}': \ w = 4; \quad \text{C}': \ w = -4$$

So we establish

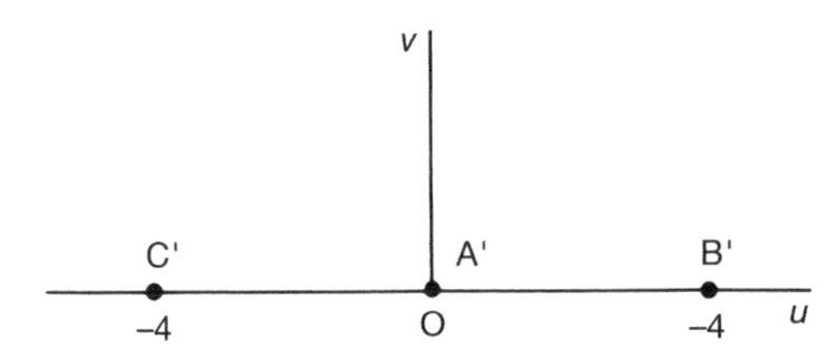

To find the paths joining these three transformed end points, we consider each of the sides of the triangle in turn.

(a) AB: Equation of AB is $y = 0 \quad \therefore \ u = x^2; \quad v = 0$
$\therefore$ Each point in AB maps onto a point between A′ and B′ for which $v = 0$, i.e. part of the u-axis.

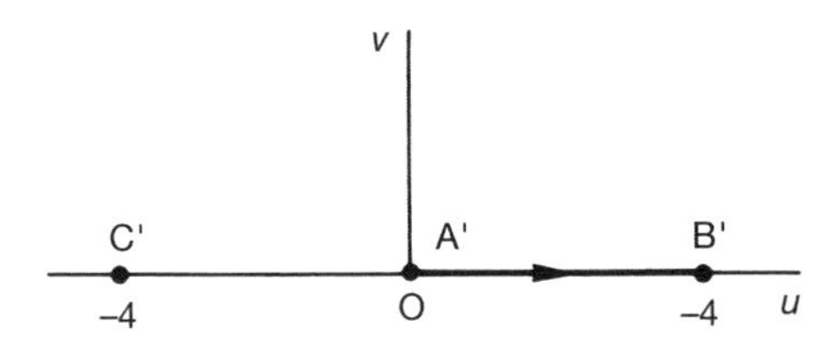

(b) BC: Equation of BC is $y = 2 - x$
Substitute in $u = x^2 - y^2$ and $v = 2xy$ and determine v as a function of u.

$u = \ldots\ldots\ldots\ldots$
$v = \ldots\ldots\ldots\ldots$
$v = f(u) = \ldots\ldots\ldots\ldots$

40

$$u = 4x - 4; \quad v = 4x - 2x^2; \quad v = 2 - \frac{u^2}{8}$$

Because

$$u = x^2 - y^2 = x^2 - (2 - x)^2 = 4x - 4 \quad \therefore \ x = \frac{u + 4}{4}$$

$$v = 2xy = 2x(2 - x) = 4x - 2x^2$$

$$\therefore \ v = 4\left(\frac{u+4}{4}\right) - 2\left(\frac{u+4}{4}\right)^2 = 2 - \frac{u^2}{8}$$

Therefore, the path joining B′ to C′ is an

$\ldots\ldots\ldots\ldots$

41

inverted parabola

$v = 2 - \dfrac{u^2}{8}$ $\therefore$ at $u = 0$, $v = 2$ and the w-plane diagram now becomes

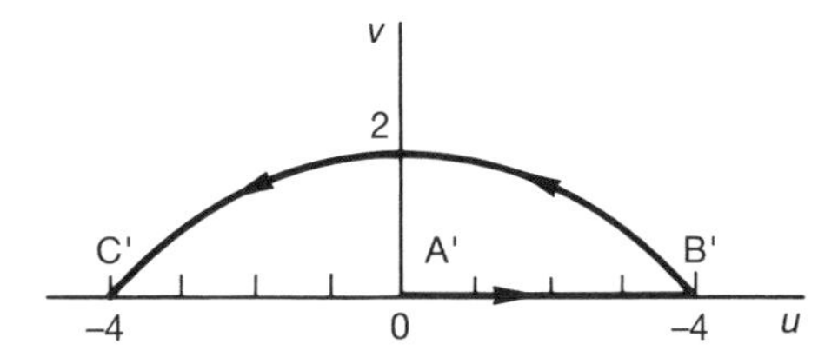

To complete the mapping, we have still to deal with CA. This transforms onto

. . . .

42

the u-axis between C′ and A′

(c) CA: Equation of CA is $x = 0$ $\therefore$ $u = -y^2$, $v = 0$
$\therefore$ Each point between C and A maps onto the negative part of the u-axis between C′ and A′.

So finally we have

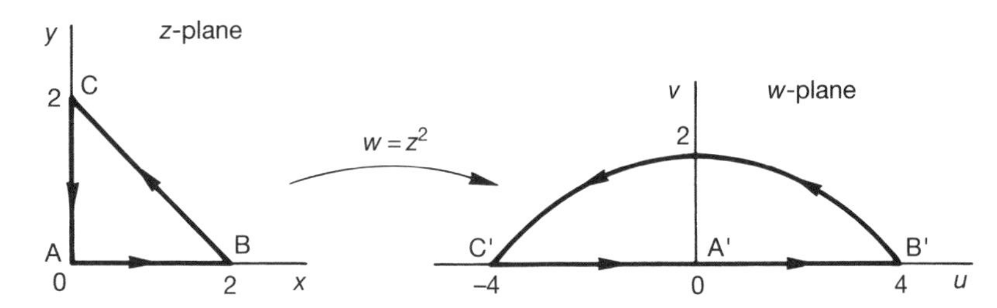

Mapping of regions

In this last example, the three lines AB, BC and CA form the boundary of a triangular region and we have seen how this boundary maps onto the boundary A′B′C′A′ in the w-plane. What we do not know yet is whether the points internal to the triangle map to points internal to the figure in the w-plane or to points external to it.

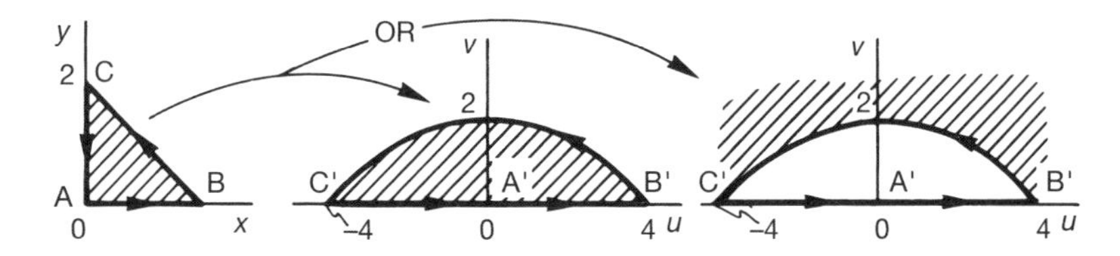

▶

In the z-plane, the region is on the left-hand side as we proceed round the figure in the direction of the arrows ABCA. The region on the left-hand side as we proceed round the figure A′B′C′A′ in the w-plane determines that the transformed region in this case is, in fact, the internal region.

So

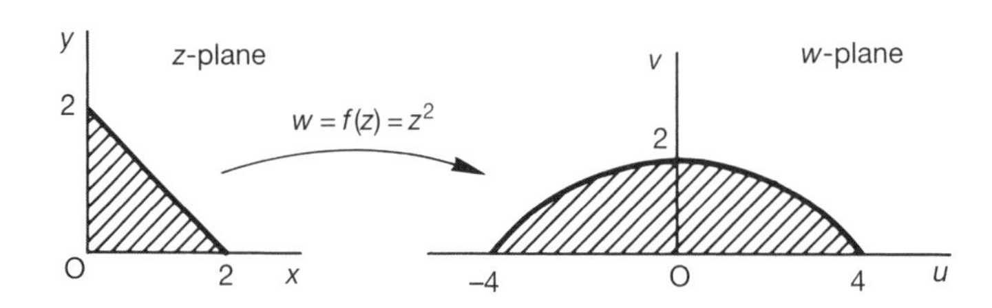

Therefore, every point in the region shaded in the z-plane maps onto a corresponding point in the region shaded in the w-plane.

43

2 *Transformation* $w = \dfrac{1}{z}$ (inversion)

Example 1

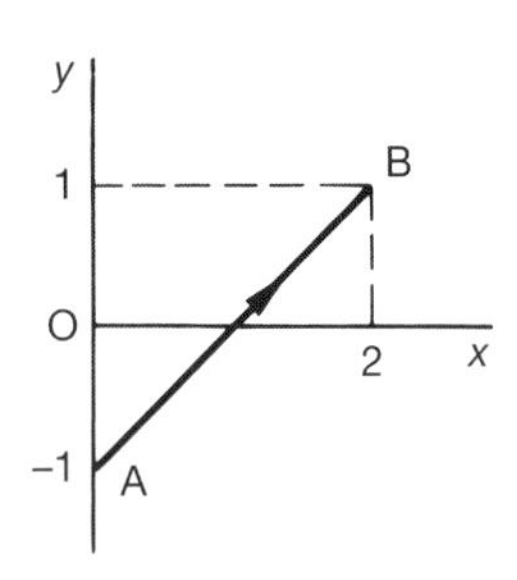

A straight line joining A $(-i)$ and B $(2+i)$ in the z-plane is mapped onto the w-plane by the transformation equation $w = \dfrac{1}{z}$.

Proceeding as before

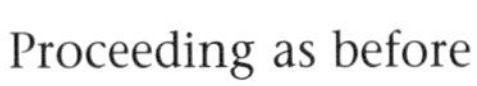

$$w = \frac{1}{z}$$

$$\therefore\ u + iv = \frac{1}{x+iy} = \frac{x-iy}{x^2+y^2}$$

$$\therefore\ u = \frac{x}{x^2+y^2};\quad v = \frac{-y}{x^2+y^2}$$

First we map the end points A and B onto the w-plane.

A′: $w = \ldots\ldots\ldots\ldots$

B′: $w = \ldots\ldots\ldots\ldots$

44

$$\text{A}': \ w = i; \quad \text{B}': w = \frac{2}{5} - \frac{1}{5}i$$

Because

$\text{A}: \ x = 0, \ y = -1 \qquad \therefore \ \text{A}': \ u = 0, \ v = 1 \qquad \therefore \ \text{A}' \text{ is } w = i$

$\text{B}: \ x = 2, \ y = 1 \qquad \therefore \ \text{B}': \ u = \frac{2}{5}, \ v = -\frac{1}{5} \qquad \therefore \ \text{B}' \text{ is } w = \frac{2}{5} - \frac{1}{5}i$

So far then we have

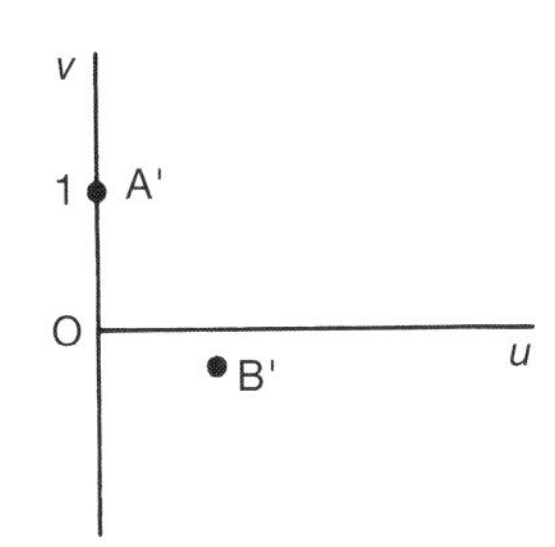

To determine the path A′B′, we can proceed as follows

$$w = \frac{1}{z} \qquad \therefore \ z = \frac{1}{w} \quad \text{i.e.} \quad x + iy = \frac{1}{u + iv} = \frac{u - iv}{u^2 + v^2}$$

$$\therefore \ x = \frac{u}{u^2 + v^2} \quad \text{and} \quad y = \frac{-v}{u^2 + v^2}$$

The equation of AB is $y = x - 1$

$$\therefore \ \frac{-v}{u^2 + v^2} = \frac{u}{u^2 + v^2} - 1$$

which simplifies into ....

45

$$u^2 + v^2 - u - v = 0$$

Because

$$\frac{-v}{u^2 + v^2} = \frac{u}{u^2 + v^2} - 1 \quad \therefore \quad -v = u - u^2 - v^2$$

$$\therefore \ u^2 + v^2 - u - v = 0$$

We can write this as $(u^2 - u) + (v^2 - v) = 0$ and completing the square in each bracket this becomes

$$\left(u - \frac{1}{2}\right)^2 + \left(v - \frac{1}{2}\right)^2 = \frac{1}{2}$$

which we recognize as the equation of a ....

46

$$\text{circle with center } \left(\frac{1}{2}, \frac{1}{2}\right) \text{ and radius } \frac{1}{\sqrt{2}}$$

The path joining A′ and B′ is therefore an arc of this circle.

But we still have problems, for it could be the minor arc or the major arc.

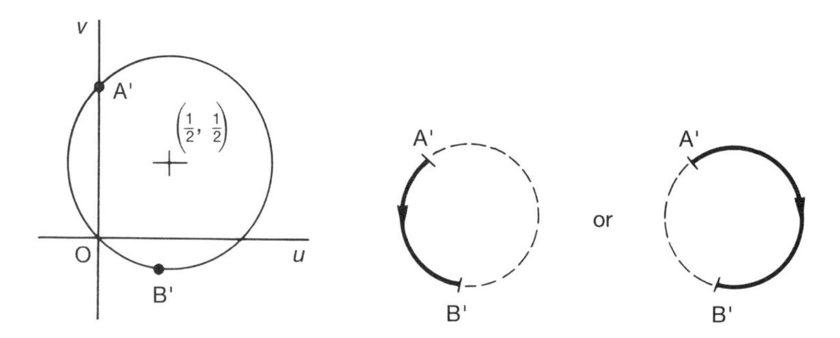

To decide which is correct, we take a further convenient point on the original line AB and determine its image on the w-plane.

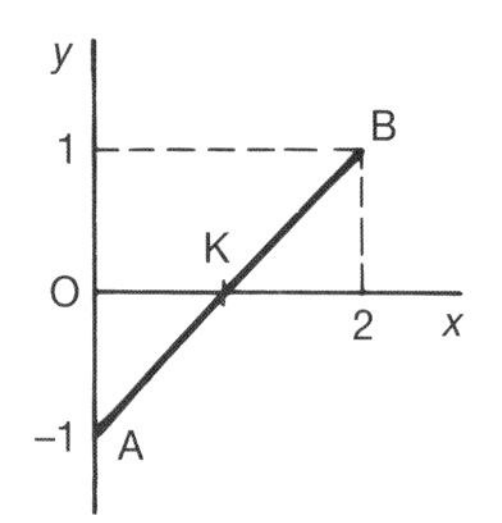

For instance, for K, $x = 1$, $y = 0$

$$\therefore \text{ For K}', \quad u = \frac{x}{x^2 + y^2} = 1$$

$$v = \frac{-y}{x^2 + y^2} = 0$$

$\therefore$ K′ is the point $w = 1$

The path is, therefore, the major arc A′K′B′ developed in the direction indicated.

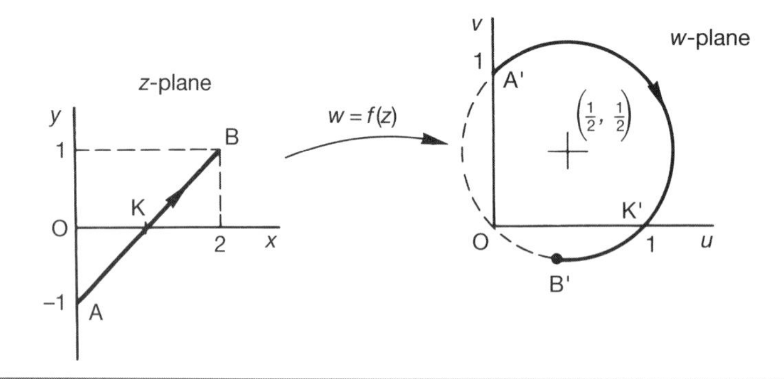

47

If we consider the line AB of the previous example extended to infinity in each direction, its image in the w-plane would then be the complete circle.

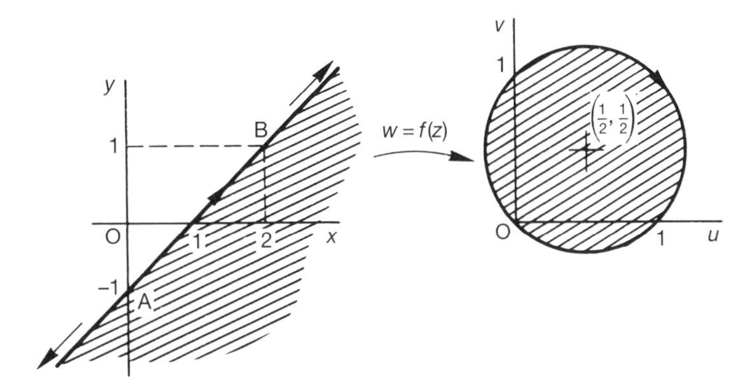

Furthermore, the line AB cuts the entire z-plane into two regions and

(a) the region on the right-hand side of the line relative to the arrowed direction maps onto the region inside the circle in the w-plane

(b) the region on the left-hand side of the line maps onto

....

48

the region outside the circle in the w-plane

Let us now consider a general case.

Example 2

Determine the image in the w-plane of a circle in the z-plane under the inversion transformation $w = \dfrac{1}{z}$.

The general equation of a circle is

$$x^2 + y^2 + 2gx + 2fy + c = 0$$

with center $(-g, -f)$

and radius $\sqrt{g^2 + f^2 - c}$.

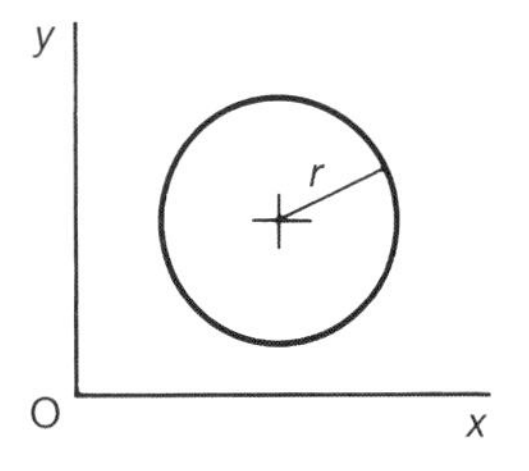

It is convenient at times to write this as

$$A(x^2 + y^2) + Dx + Ey + F = 0$$

in which case

center is .... and radius is ....

49

$$\text{center}\left(-\frac{D}{2A}, -\frac{E}{2A}\right); \quad \text{radius} = \frac{1}{2A}\sqrt{D^2 + E^2 - 4AF}$$

Because

$$g = \frac{D}{2A}, \quad f = \frac{E}{2A}, \quad c = \frac{F}{A}.$$

As before we have $w = \frac{1}{z} \quad \therefore \ z = \frac{1}{w}$

$$\therefore \ x + iy = \frac{1}{u + iv} = \frac{u - iv}{u^2 + v^2} \quad \therefore \ x = \frac{u}{u^2 + v^2}; \quad y = \frac{-v}{u^2 + v^2}$$

Then $A(x^2 + y^2) + Dx + Ey + F = 0$

becomes ....

Simplify it as far as possible

50

$$A + Du - Ev + F(u^2 + v^2) = 0$$

Because we have

$$\frac{A(u^2 + v^2)}{(u^2 + v^2)^2} + \frac{Du}{u^2 + v^2} - \frac{Ev}{u^2 + v^2} + F = 0$$

$$\therefore \ A + Du - Ev + F(u^2 + v^2) = 0$$

Changing the order of terms, this can be written

$$F(u^2 + v^2) + Du - Ev + A = 0$$

which is the equation of a circle with

center ....; radius ....

51

$$\text{center}\left(-\frac{D}{2F}, \frac{E}{2F}\right); \quad \text{radius}\,\frac{1}{2F}\sqrt{D^2 + E^2 - 4FA}$$

Thus any circle in the z-plane transforms, with $w = \frac{1}{z}$, onto another circle in the w-plane.

We have already seen previously that, under inversion, a straight line also maps onto a circle. This may be regarded as a special case of the general result, if we accept a straight line as the circumference of a circle of .... radius.

52

infinite

Because

$$A(x^2+y^2)+Dx+Ey+F=0$$

If $A=0$, this becomes $Dx+Ey+F=0$ i.e. a straight line

and also the center $\left(-\frac{D}{2A}, -\frac{E}{2A}\right)$ becomes infinite

and the radius $\frac{1}{2A}\sqrt{D^2+E^2-4AF}$ becomes infinite.

Therefore, combining the results we have obtained, we have this conclusion:

> Under inversion $w=\frac{1}{z}$, a circle or a straight line in the z-plane transforms onto a circle or a straight line in the w-plane.

Now for one more example.

Example 3

A circle in the z-plane has its center at $z=3$ and a radius of 2 units. Determine its image in the w-plane when transformed by $w=\frac{1}{z}$.

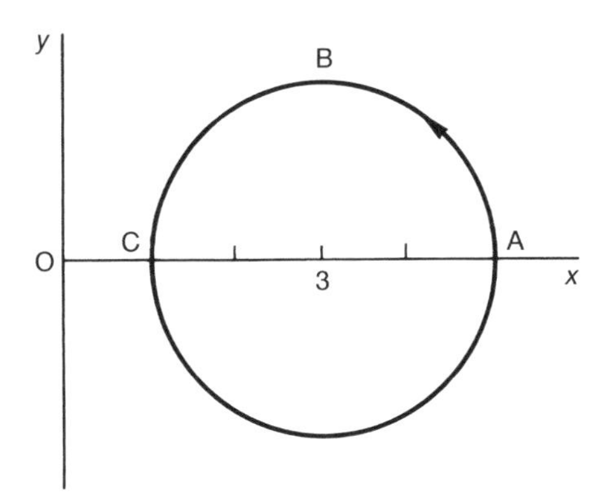

Equation of the circle is

$$(x-3)^2+y^2=4$$
$$x^2-6x+9+y^2=4$$
$$x^2+y^2-6x+5=0.$$

Using $w=\frac{1}{z}$, we can obtain x and y in terms of u and v.

$x=$ . . . . ; $y=$ . . . .

53

$$x = \frac{u}{u^2+v^2}; \quad y = \frac{-v}{u^2+v^2}$$

Because $w = \frac{1}{z}$,

$$\therefore\ z = \frac{1}{w}$$

$$\therefore\ x + iy = \frac{1}{u+iv} = \frac{u-iv}{u^2+v^2}$$

$$\therefore\ x = \frac{u}{u^2+v^2}; \quad y = \frac{-v}{u^2+v^2}$$

Substituting these in the equation of the circle, we get a relationship between u and v, which is

....

54

$$5(u^2+v^2) - 6u + 1 = 0$$

Because the circle is $x^2 + y^2 - 6x + 5 = 0$

$$\therefore\ \frac{u^2}{(u^2+v^2)^2} + \frac{v^2}{(u^2+v^2)^2} - \frac{6u}{u^2+v^2} + 5 = 0$$

$$\frac{1}{u^2+v^2} - \frac{6u}{u^2+v^2} + 5 = 0$$

$$5(u^2+v^2) - 6u + 1 = 0$$

This is of the form $A(u^2+v^2) + Du + Ev + F = 0$
where $A = 5$, $D = -6$, $E = 0$, $F = 1$.

Therefore, the center is ....
and the radius is ....

55

$$\text{center} = \left(\frac{3}{5}, 0\right); \quad \text{radius} = \frac{2}{5}$$

Because the center is $\left(-\frac{D}{2A}, -\frac{E}{2A}\right) = \left(\frac{6}{10}, 0\right)$ i.e. $\left(\frac{3}{5}, 0\right)$

and the radius $= \frac{1}{2A}\sqrt{D^2 + E^2 - 4AF} = \frac{1}{10}\sqrt{36 + 0 - 20} = \frac{2}{5}$.

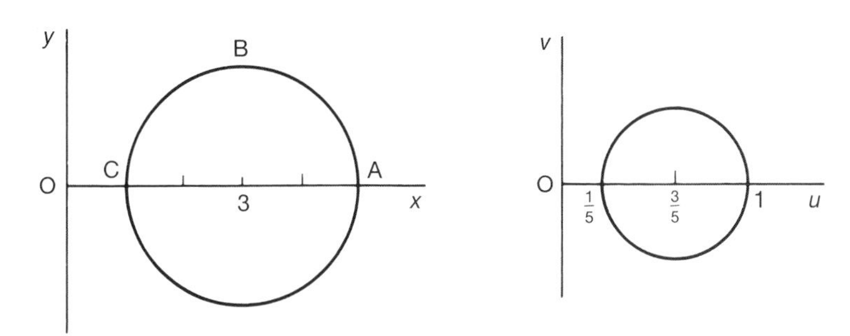

Taking three sample points A, B, C as shown, we can map these onto the w-plane using $u = \frac{x}{x^2 + y^2}$ and $v = \frac{-y}{x^2 + y^2}$.

A′: ....; B′ : ....; C′: ....

56

$$\text{A}': \left(\frac{1}{5}, 0\right); \quad \text{B}': \left(\frac{3}{13}, -\frac{2}{13}\right); \quad \text{C}': (1, 0)$$

So we finally have

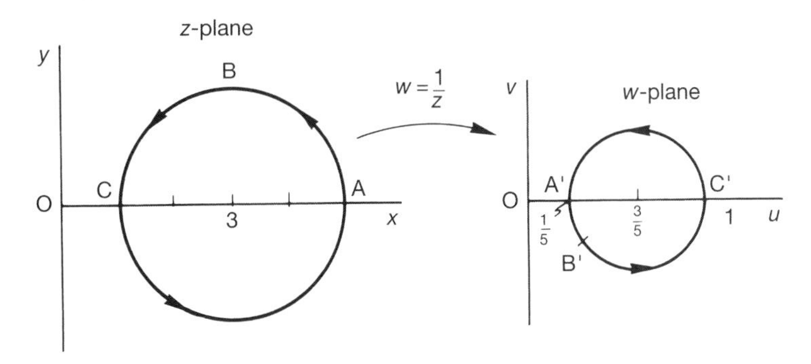

3 *Transformation* $w = \frac{1}{z - a}$

An extension of the method we have just applied occurs with transformations of the form $w = \frac{1}{z - a}$ where a is real or complex.

▶

Example

A circle $|z| = 1$ in the z-plane is mapped onto the w-plane by $w = \dfrac{1}{z-2}$.

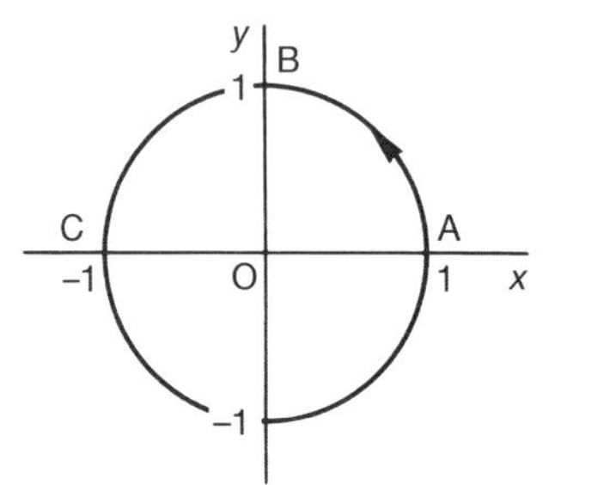

$$w = \frac{1}{z-2} \quad \therefore \; z - 2 = \frac{1}{w}$$

$$x + iy - 2 = \frac{1}{u + iv}$$

$$(x - 2) + iy = \frac{u - iv}{u^2 + v^2}$$

$$\therefore \; x = \frac{u}{u^2 + v^2} + 2; \quad y = \frac{-v}{u^2 + v^2}$$

Cartesian equation of the circle is $x^2 + y^2 = 1$.

We then substitute the expressions for x and y in terms of u and v and obtain the relationship between u and v, which is ....

57

$$\boxed{3(u^2 + v^2) + 4u + 1 = 0}$$

Because we have $\left\{\dfrac{u + 2(u^2 + v^2)}{u^2 + v^2}\right\}^2 + \left\{\dfrac{-v}{u^2 + v^2}\right\}^2 = 1$

$$\{u + 2(u^2 + v^2)\}^2 + v^2 = (u^2 + v^2)^2$$

$$u^2 + 4u(u^2 + v^2) + 4(u^2 + v^2)^2 + v^2 = (u^2 + v^2)^2$$

$$1 + 4u + 4(u^2 + v^2) = u^2 + v^2$$

$$3(u^2 + v^2) + 4u + 1 = 0$$

This can be expressed as

$$u^2 + \frac{4}{3}u + v^2 + \frac{1}{3} = 0$$

$$\left(u + \frac{2}{3}\right)^2 + v^2 = \left(\frac{1}{3}\right)^2$$

which is a circle with center $\left(-\dfrac{2}{3}, 0\right)$ and radius $\dfrac{1}{3}$.

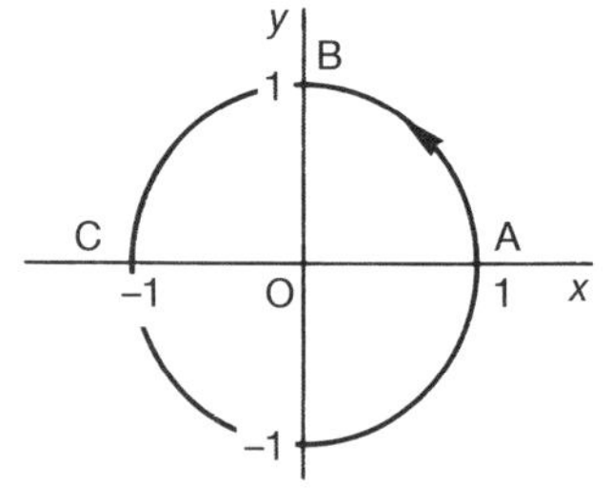

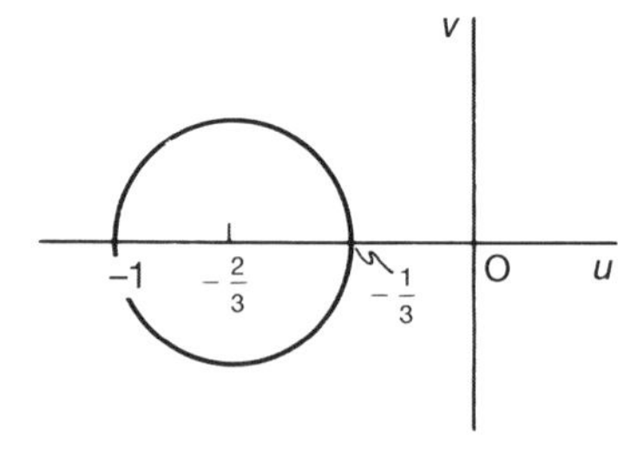

To determine the direction of development relative to the arrowed direction in the z-plane, we consider the mapping of three sample points A, B, C as shown onto the w-plane, giving A′, B′, C′.

A′: ....; B′: ....; C′: ....

58

$$\text{A}': \quad w = (-1, 0); \quad \text{B}': \quad w = \left(-\frac{2}{5}, -\frac{1}{5}\right); \quad \text{C}': \quad w = \left(\frac{1}{3}, 0\right)$$

Because

$$\text{A}: \; z = 1 \qquad \therefore \; w = \frac{1}{z-2} = -1 \qquad \therefore \; \text{A}' = (-1, 0)$$

$$\text{B}: \; z = i \qquad \therefore \; w = \frac{1}{i-2} = \frac{i+2}{-5} \qquad \therefore \; \text{B}' = \left(-\frac{2}{5}, -\frac{1}{5}\right)$$

$$\text{C}: \; z = -1 \qquad \therefore \; w = -\frac{1}{3} \qquad \therefore \; \text{C}' = \left(-\frac{1}{3}, 0\right)$$

Whereupon we have

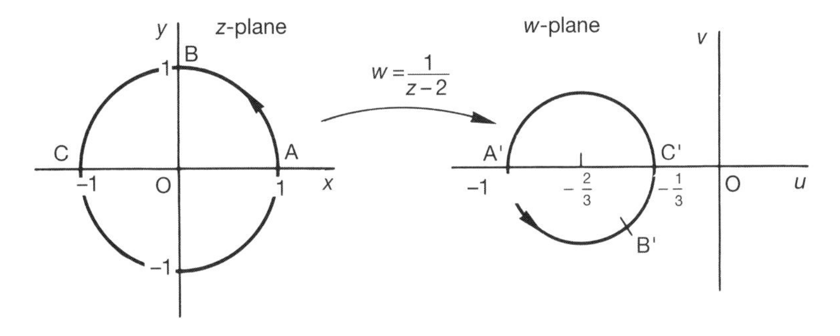

We now have one further transformation which is important, so move on to the next frame for a fresh start

59

4 *Bilinear transformation* $w = \dfrac{az+b}{cz+d}$

Transformation of the form $w = \dfrac{az+b}{cz+d}$ where a, b, c, d are, in general, complex.

Note that

(a) if $cz + d = 1$, $w = az + b$, i.e. the general linear transformation

(b) if $az + b = 1$, $w = \dfrac{1}{cz+d}$, i.e. the form of inversion just considered.

▶

Example

Determine the image in the w-plane of the circle $|z| = 2$ in the z-plane under the transformation $w = \dfrac{z+i}{z-i}$ and show the region in the w-plane onto which the region within the circle is mapped.

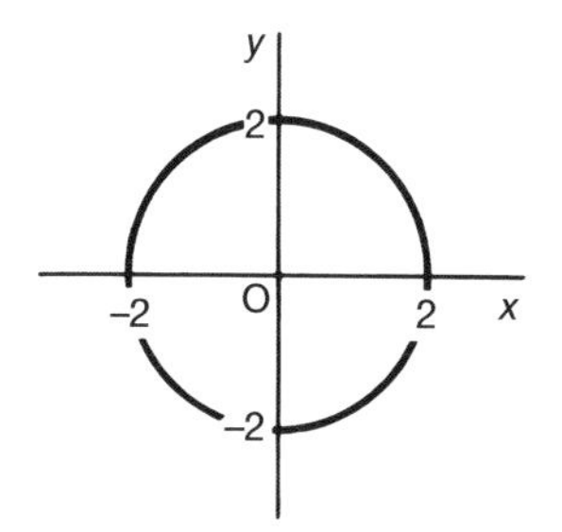

We begin in very much the same way as before by expressing u and v in terms of x and y.

u . . . .; $v =$ . . . .

60

$$u = \frac{x^2+y^2-1}{x^2+y^2-2y+1}; \quad v = \frac{2x}{x^2+y^2-2y+1}$$

Because

$$w = u + iv = \frac{z+i}{z-i} = \frac{x+i(y+1)}{x+i(y-1)}$$

$$= \frac{\{x+i(y+1)\}\{x-i(y-1)\}}{\{x+i(y-1)\}\{x-i(y-1)\}}$$

$$= \frac{x^2+ix(y+1-y+1)+y^2-1}{x^2+(y-1)^2}$$

$$= \frac{x^2+y^2-1+2ix}{x^2+y^2-2y+1}$$

$$\therefore\ u = \frac{x^2+y^2-1}{x^2+y^2-2y+1} \quad \text{and} \quad v = \frac{2x}{x^2+y^2-2y+1}$$

But the equation of the circle is $x^2 + y^2 = 4$, so these expressions simplify to

$u =$ . . . . and $v =$ . . . .

61

$$u = \frac{3}{5-2y}; \quad v = \frac{2x}{5-2y}$$

From these, we can form expressions for x and y in terms of u and v.

$x =$ . . . .; $y =$ . . . .

62

$$x = \frac{3v}{2u}; \quad y = \frac{5u-3}{2u}$$

Because, from the first, $y = \dfrac{5u-3}{2u}$ and substituting in the second gives $x = \dfrac{3v}{2u}$.

But $x^2 + y^2 = 4 \quad \therefore \quad \dfrac{9v^2}{4u^2} + \dfrac{(5u-3)^2}{4u^2} = 4$

which can be simplified to ....

63

$$9(u^2 + v^2) - 30u + 9 = 0$$

Because

$$9v^2 + 25u^2 - 30u + 9 = 16u^2 \quad \therefore \quad 9(u^2 + v^2) - 30u + 9 = 0.$$

Dividing through by 9, we can now rearrange this to

$$\left(u^2 - \frac{30}{9}u\right) + v^2 + 1 = 0$$

i.e. $$\left(u - \frac{5}{3}\right)^2 + v^2 + 1 - \frac{25}{9} = 0$$

$$\left(u - \frac{5}{3}\right)^2 + v^2 = \left(\frac{4}{3}\right)^2$$

which, you will recognize, is a circle in the w-plane with

center .... and radius ....

64

$$\text{center} = \left(\frac{5}{3}, 0\right); \quad \text{radius} = \frac{4}{3}$$

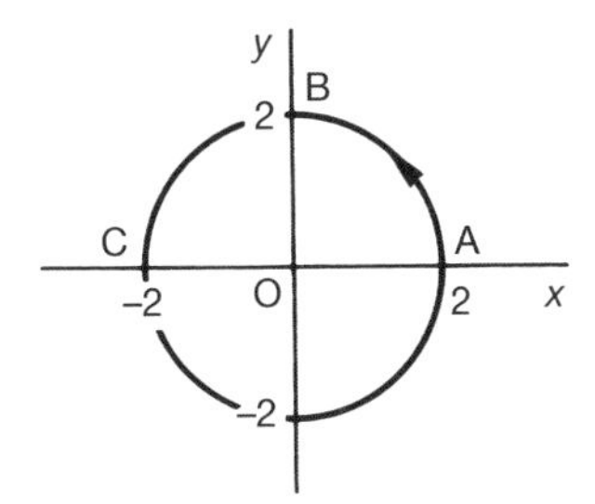

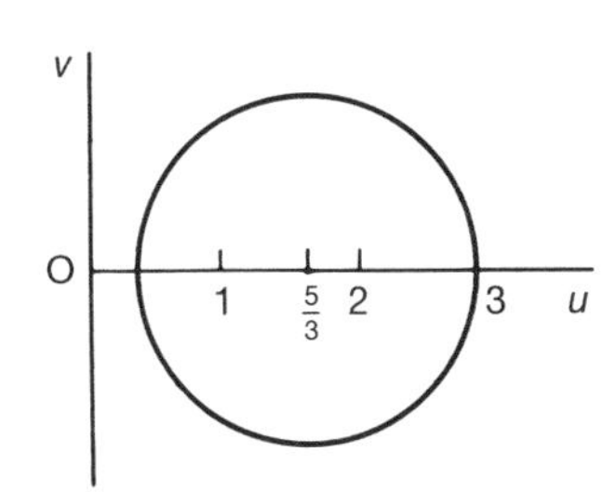

To find the direction of development, we map three sample points A, B, C onto A′, B′, C′ as usual.

A′: ....; B′: ....; C′: ....

65

$$\text{A}': w = \frac{3}{5} + \frac{4}{5}i; \quad \text{B}': w = 3; \quad \text{C}': w = \frac{3}{5} - \frac{4}{5}i$$

Because

A: $z = 2 \quad \therefore \ w = \dfrac{2+i}{2-i} = \dfrac{(2+i)^2}{5} = \dfrac{4+4i-1}{5} = \dfrac{3}{5} + \dfrac{4}{5}i$ i.e. A′

B: $z = 2i \quad \therefore \ w = \dfrac{2i+i}{2i-i} = \dfrac{3i}{i} = 3 \quad \therefore \ w = 3$ i.e. B′

C: $z = -2 \quad \therefore \ w = \dfrac{-2+i}{-2-i} = \dfrac{2-i}{2+i} = \dfrac{(2-i)^2}{5} = \dfrac{3}{5} - \dfrac{4}{5}i$ i.e. C′

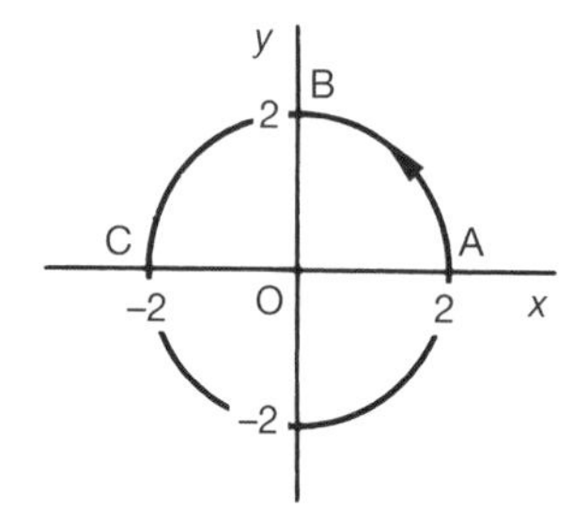

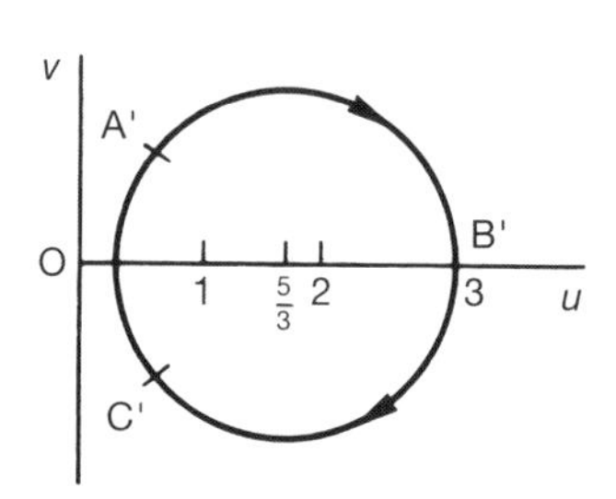

So an anticlockwise progression in the z-plane becomes a clockwise progression in the w-plane with this particular example.

Now we can complete the problem, for the region inside the circle in the z-plane maps onto .... in the w-plane.

66

the region outside the circle

Because the enclosed region in the z-plane is on the left-hand side of the direction of progression. The region on the left-hand side of the direction of progression in the w-plane is thus the region outside the transformed circle.

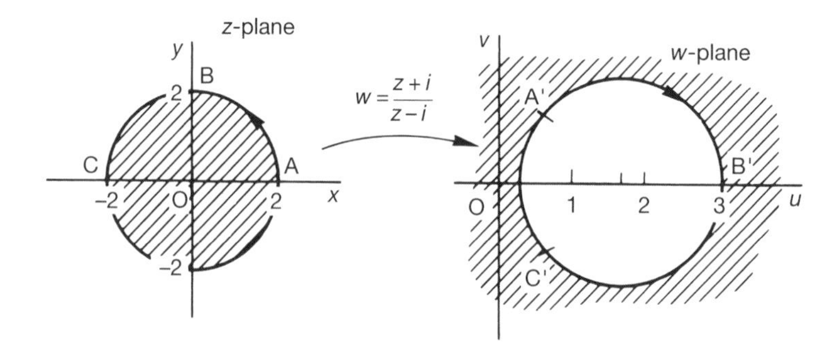

And that brings us successfully to the end of this Program. We shall pursue the topic further in the succeeding Program. Meanwhile, all that remains is to check down the **Summary** and the **Can You?** checklist before working through the **Test exercise**. All very straightforward. The **Further problems** will give you valuable additional practice.

Summary

67

1 *Transformation equation*

$z = x + iy \qquad w = u + iv$

The transformation equation is the relationship between z and w, i.e. $w = f(z)$.

2 *Linear transformation* $w = az + b$ where a and b are real or complex. A straight line in the z-plane maps onto a corresponding straight line in the w-plane.

3 *Types of transformation* $w = az + b$

(a) *magnification* – given by $|a|$

(b) *rotation* – given by arg a

(c) *translation* – given by b.

4 *Non-linear transformation*

(a) $w = z^2$

A straight line through the origin maps onto a corresponding straight line through the origin in the w-plane. A straight line not passing through the origin maps onto a parabola.

(b) $w = \dfrac{1}{z}$ (inversion)

A straight line or a circle maps onto a straight line or a circle in the w-plane.

A straight line may be regarded as a circle of infinite radius.

(c) $w = \dfrac{az + b}{cz + d}$ (bilinear transformation) – with a, b, c, d real or complex.

5 *Mapping of a region* depends on the direction of development. Right-hand regions map onto right-hand regions: left-hand regions onto left-hand regions.

Can You?

68 Checklist 5

Check this list before and after you try the end of Program test.

On a scale of 1 to 5 how confident are you that you can: **Frames**

- Recognize the transformation equation in the form $w = f(z) = u(x, y) + iv(x, y)$?
  Yes ☐ ☐ ☐ ☐ ☐ *No* — 1 and 2
- Illustrate the image of a point in the complex z-plane under a complex mapping onto the w-plane?
  Yes ☐ ☐ ☐ ☐ ☐ *No* — 2 to 7
- Map a straight line in the z-plane onto the w-plane under the transformation $w = f(z)$?
  Yes ☐ ☐ ☐ ☐ ☐ *No* — 7 to 16
- Identify complex mappings that form translations, magnifications, rotations and their combinations?
  Yes ☐ ☐ ☐ ☐ ☐ *No* — 16 to 31
- Deal with the non-linear transformations $w = z^2$, $w = 1/z$, $w = 1/(z - a)$ and $w = (az + b)/(cz + d)$?
  Yes ☐ ☐ ☐ ☐ ☐ *No* — 32 to 66

Test exercise 5

69

1 Map the following points in the z-plane onto the w-plane under the transformation $w = f(z)$.

(a) $z = 3 + 2i$; $w = 2z - 6i$ (c) $z = i(1 - i)$; $w = (2 + i)z - 1$

(b) $z = -2 + i$; $w = 4 + zi$ (d) $z = i - 2$; $w = (1 - i)(z + 3)$.

2 Map the straight line joining A $(2 - i)$ and B $(4 - 3i)$ in the z-plane onto the w-plane using the transformation $w = (1 + 2i)z + 1 - 3i$. State the magnification, rotation and translation involved.

3 A triangle ABC in the z-plane as shown is mapped onto the w-plane under the transformation $w = z^2$.

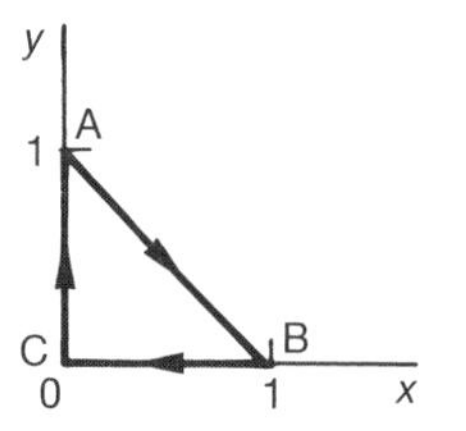

Determine the image in the w-plane and indicate the mapping of the interior triangular region ABC.

4 Map the straight line joining A ($z = i$) and B ($z = 3 + 4i$) in the z-plane onto the w-plane under the inversion transformation $w = \frac{1}{z}$.

Sketch the image of AB in the w-plane.

5 The unit circle $|z| = 1$ in the z-plane is mapped onto the w-plane by $w = \frac{1}{z - 2i}$.

Determine (a) the position of the center and (b) the radius of the circle obtained.

6 The circle $|z| = 2$ is mapped onto the w-plane by the transformation $w = \frac{z + 2i}{z + i}$.

Determine the center and radius of the resulting circle in the w-plane.

Further problems 5

70

1 A triangle ABC in the z-plane with vertices A $(-1 - i)$, B $(2 + 2i)$, C $(-1 + 2i)$ is mapped onto the w-plane under the transformation $w = (1 - i)z + (1 + 2i)$. Determine the image A′B′C′ of ABC in the w-plane.

2 The straight line joining A $(1 + 2i)$ and B $(4 - 3i)$ in the z-plane is mapped onto the w-plane by the transformation equation $w = (2 + 5i)z$. Determine (a) the images of A and B, (b) the magnification, rotation and translation involved.

3 Map the straight line joining A $(-2 + 3i)$ and B $(1 + 2i)$ in the z-plane onto the w-plane using the transformation equation

$w = (-3 + i)z + 2 + 4i.$

State the magnification, rotation and translation occurring in the process.

4

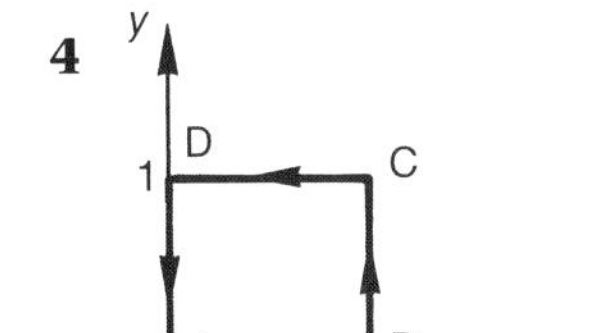

Transform the square ABCD in the z-plane onto the w-plane under the transformation $w = z^2$.

5 Map the square ABCD in the z-plane onto the w-plane using the transformation $w = 2z^2 + 2$.

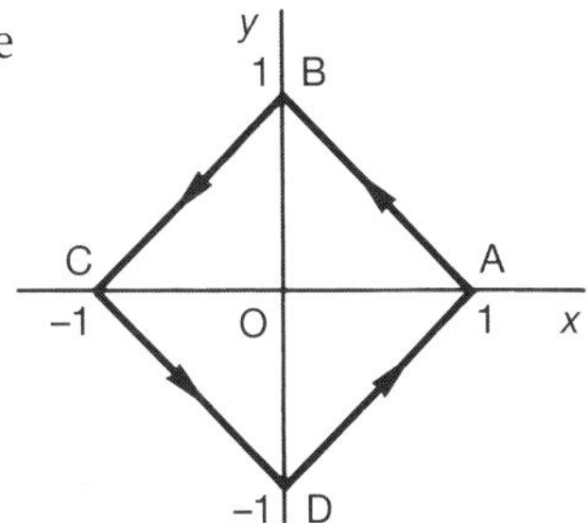

6 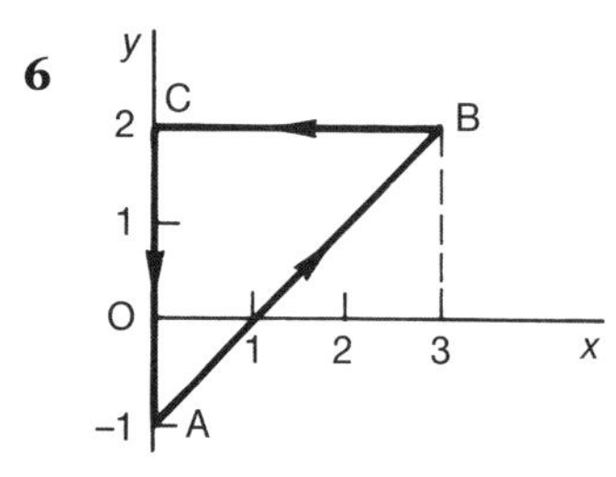

The triangle ABC in the z-plane is mapped onto the w-plane by the transformation $w = 2iz^2 + 1$. Determine the image of ABC in the w-plane.

7 A circle in the z-plane has its center at the point $(-\frac{3}{4} - i)$ and radius $\frac{7}{4}$. Show that its Cartesian equation can be expressed as

$$2(x^2 + y^2) + 3x + 4y - 3 = 0$$

Determine the image of the circle in the w-plane under the inversion transformation $w = \dfrac{1}{z}$.

8 The transformation $w = \dfrac{1}{z-1}$ is applied to the circle $|z| = 2$ in the z-plane. Determine

(a) the image of the circle in the w-plane

(b) the region in the w-plane onto which the region enclosed within the circle in the z-plane is mapped.

9 The circle $|z| = 4$ is described in the z-plane in a counter-clockwise manner. Obtain its image in the w-plane under the transformation $w = \dfrac{z+1}{z-2}$ and state the direction of development.

10 The bilinear transformation $w = \dfrac{z-i}{z+2i}$ is applied to the circle $|z| = 3$ in the z-plane. Determine the equation of the image in the w-plane and state its center and radius.

11 The unit circle $|z| = 1$ in the z-plane is mapped onto the w-plane under the transformation $w = \dfrac{z-1}{z-3}$. Determine the equation of its image and the region onto which the region within the circle is mapped.

12 Obtain the image of the unit circle $|z| = 1$ in the z-plane under the transformation $w = \dfrac{z+3i}{z-2i}$.

13 The circle $|z| = 2$ is mapped onto the w-plane by the transformation $w = \dfrac{z+i}{2z-i}$. Determine

(a) the image of the circle in the w-plane

(b) the mapping of the region enclosed by $|z| = 2$.

14 Show that the transformation equation $w = \dfrac{z-a}{z-b}$ where $z = x + iy$, $a = 1 + 4i$ and $b = 2 + 3i$, transforms the circle $(x-3)^2 + (y-5)^2 = 5$ into a straight line through the origin in the w-plane.

Partial differentiation

Learning outcomes

When you have completed this Program you will be able to:

- Find the first partial derivatives of a function of two real variables
- Find second-order partial derivatives of a function of two real variables
- Calculate errors using partial differentiation

Partial differentiation

1 First partial differentiation

The volume V of a cylinder of radius r and height h is given by

$$V = \pi r^2 h$$

i.e. V depends on two quantities, the values of r and h.

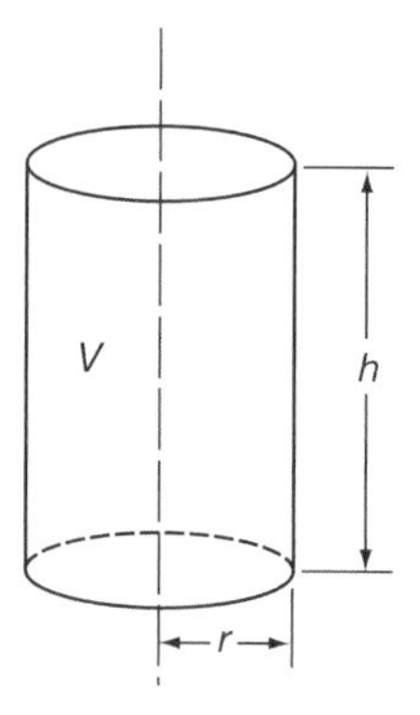

If we keep r constant and increase the height h, the volume V will increase. In these circumstances, we can consider the derivative of V with respect to h – but only if r is kept constant.

$$\text{i.e. } \left[\frac{\mathrm{d}V}{\mathrm{d}h}\right]_{r \text{ constant}} \quad \text{is written } \frac{\partial V}{\partial h}$$

Notice the new type of 'delta'. We already know the meaning of $\frac{\delta y}{\delta x}$ and $\frac{\mathrm{d}y}{\mathrm{d}x}$. Now we have a new one, $\frac{\partial V}{\partial h}$. $\frac{\partial V}{\partial h}$ is called the *partial derivative* of V with respect to h and implies that for our present purpose, the value of r is considered as being kept ....

2

constant

$V = \pi r^2 h$. To find $\frac{\partial V}{\partial h}$, we differentiate the given expression, taking all symbols except V and h as being constant $\therefore \ \frac{\partial V}{\partial h} = \pi r^2 \cdot 1 = \pi r^2$

Of course, we could have considered h as being kept constant, in which case, a change in r would also produce a change in V. We can therefore talk about $\frac{\partial V}{\partial r}$ which simply means that we now differentiate $V = \pi r^2 h$ with respect to r, taking all symbols except V and r as being constant for the time being.

$$\therefore \ \frac{\partial V}{\partial r} = \pi 2rh = 2\pi rh$$

In the statement $V = \pi r^2 h$, V is expressed as a function of two variables, r and h. It therefore has two partial derivatives, one with respect to .... and one with respect to ....

3

One with respect to r; one with respect to h

Another example:

Let us consider the area of the curved surface of the cylinder $A = 2\pi rh$

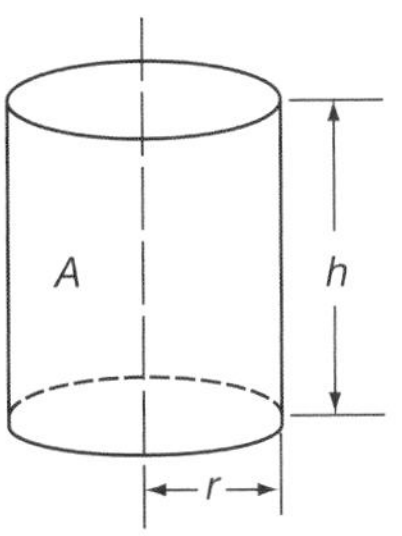

A is a function of r and h, so we can find $\frac{\partial A}{\partial r}$ and $\frac{\partial A}{\partial h}$

To find $\frac{\partial A}{\partial r}$ we differentiate the expression for A with respect to r, keeping all other symbols constant.

To find $\frac{\partial A}{\partial h}$ we differentiate the expression for A with respect to h, keeping all other symbols constant.

So, if $A = 2\pi rh$, then $\frac{\partial A}{\partial r} =$ ....and $\frac{\partial A}{\partial h} =$ ....

4

$\frac{\partial A}{\partial r} = 2\pi h$ and $\frac{\partial A}{\partial h} = 2\pi r$

Of course, we are not restricted to the mensuration of the cylinder. The same will happen with any function which is a function of two independent variables. For example, consider $z = x^2y^3$.

Here z is a function of x and y. We can therefore find $\frac{\partial z}{\partial x}$ and $\frac{\partial z}{\partial y}$.

(a) To find $\frac{\partial z}{\partial x}$, differentiate with respect to x, regarding y as a constant.

$$\therefore \quad \frac{\partial z}{\partial x} = 2xy^3$$

(b) To find $\frac{\partial z}{\partial y}$, differentiate with respect to y, regarding x as a constant.

$$\frac{\partial z}{\partial y} = x^2 3y^2 = 3x^2y^2$$

Partial differentiation is easy! For we regard every independent variable, except the one with respect to which we are differentiating, as being for the time being ....

5

constant

Here are some examples. 'With respect to' is abbreviated to w.r.t.

Example 1

$u = x^2 + xy + y^2$

(a) To find $\frac{\partial u}{\partial x}$, we regard y as being constant.

Partial diff w.r.t. x of $x^2 = 2x$

Partial diff w.r.t. x of $xy = y$ (y is a constant factor)

Partial diff w.r.t. x of $y^2 = 0$ (y^2 is a constant term)

$$\frac{\partial u}{\partial x} = 2x + y$$

(b) To find $\frac{\partial u}{\partial y}$, we regard x as being constant.

Partial diff w.r.t. y of $x^2 = 0$ (x^2 is a constant term)

Partial diff w.r.t. y of $xy = x$ (x is a constant factor)

Partial diff w.r.t. y of $y^2 = 2y$

$$\frac{\partial u}{\partial y} = x + 2y$$

Another example in Frame 6

6

Example 2

$z = x^3 + y^3 - 2x^2y$

$$\frac{\partial z}{\partial x} = 3x^2 + 0 - 4xy = 3x^2 - 4xy$$
$$\frac{\partial z}{\partial y} = 0 + 3y^2 - 2x^2 = 3y^2 - 2x^2$$

And it is all just as easy as that.

Example 3

$z = (2x - y)(x + 3y)$

This is a product, and the usual product rule applies except that we keep y constant when finding $\frac{\partial z}{\partial x}$, and x constant when finding $\frac{\partial z}{\partial y}$.

$$\frac{\partial z}{\partial x} = (2x - y)(1 + 0) + (x + 3y)(2 - 0) = 2x - y + 2x + 6y = 4x + 5y$$
$$\frac{\partial z}{\partial y} = (2x - y)(0 + 3) + (x + 3y)(0 - 1) = 6x - 3y - x - 3y = 5x - 6y$$

▶

Here is one for you to do.

If $z = (4x - 2y)(3x + 5y)$, find $\dfrac{\partial z}{\partial x}$ and $\dfrac{\partial z}{\partial y}$

Find the results and then move on to Frame 7

7

$$\frac{\partial z}{\partial x} = 24x + 14y; \qquad \frac{\partial z}{\partial y} = 14x - 20y$$

Because $z = (4x - 2y)(3x + 5y)$, i.e. product

$$\therefore \frac{\partial z}{\partial x} = (4x - 2y)(3 + 0) + (3x + 5y)(4 - 0)$$
$$= 12x - 6y + 12x + 20y = 24x + 14y$$
$$\frac{\partial z}{\partial y} = (4x - 2y)(0 + 5) + (3x + 5y)(0 - 2)$$
$$= 20x - 10y - 6x - 10y = 14x - 20y$$

There we are. Now what about this one?

Example 4

If $z = \dfrac{2x - y}{x + y}$, find $\dfrac{\partial z}{\partial x}$ and $\dfrac{\partial z}{\partial y}$

Applying the quotient rule, we have:

$$\frac{\partial z}{\partial x} = \frac{(x + y)(2 - 0) - (2x - y)(1 + 0)}{(x + y)^2} = \frac{3y}{(x + y)^2}$$

and $$\frac{\partial z}{\partial y} = \frac{(x + y)(0 - 1) - (2x - y)(0 + 1)}{(x + y)^2} = \frac{-3x}{(x + y)^2}$$

That was not difficult. Now you do this one:

If $z = \dfrac{5x + y}{x - 2y}$, find $\dfrac{\partial z}{\partial x}$ and $\dfrac{\partial z}{\partial y}$

When you have finished, on to the next frame

8

$$\frac{\partial z}{\partial x} = \frac{-11y}{(x-2y)^2}; \quad \frac{\partial z}{\partial y} = \frac{11x}{(x-2y)^2}$$

Here is the working:

(a) To find $\frac{\partial z}{\partial x}$, we regard y as being constant.

$$\therefore \frac{\partial z}{\partial x} = \frac{(x-2y)(5+0)-(5x+y)(1-0)}{(x-2y)^2}$$
$$= \frac{5x-10y-5x-y}{(x-2y)^2} = \frac{-11y}{(x-2y)^2}$$

(b) To find $\frac{\partial z}{\partial y}$, we regard x as being constant.

$$\therefore \frac{\partial z}{\partial y} = \frac{(x-2y)(0+1)-(5x+y)(0-2)}{(x-2y)^2}$$
$$= \frac{x-2y+10x+2y}{(x-2y)^2} = \frac{11y}{(x-2y)^2}$$

In practice, we do not write down the zeros that occur in the working, but this is how we think.

Let us do one more example, so move on to the next frame

9

Example 5

If $z = \sin(3x+2y)$ find $\frac{\partial z}{\partial x}$ and $\frac{\partial z}{\partial y}$

Here we have what is clearly a 'function of a function'. So we apply the usual procedure, except to remember that when we are finding:

(a) $\frac{\partial z}{\partial x}$, we treat y as constant, and

(b) $\frac{\partial z}{\partial y}$, we treat x as constant.

Here goes then.

$$\frac{\partial z}{\partial x} = \cos(3x+2y) \times \frac{\partial}{\partial x}(3x+2y) = \cos(3x+2y) \times 3 = 3\cos(3x+2y)$$
$$\frac{\partial z}{\partial y} = \cos(3x+2y) \times \frac{\partial}{\partial y}(3x+2y) = \cos(3x+2y) \times 2 = 2\cos(3x+2y)$$

There it is. So in partial differentiation, we can apply all the ordinary rules of normal differentiation, except that we regard the independent variables other than the one we are using, as being for the time being . . . .

10

constant

Fine. Now here is a short exercise for you to do by way of review.

In each of the following cases, find $\frac{\partial z}{\partial x}$ and $\frac{\partial z}{\partial y}$:

1 $z = 4x^2 + 3xy + 5y^2$

2 $z = (3x + 2y)(4x - 5y)$

3 $z = \tan(3x + 4y)$

4 $z = \frac{\sin(3x + 2y)}{xy}$

Finish them all, then move on to Frame 11 for the results

11

Here are the answers:

1 $z = 4x^2 + 3xy + 5y^2$ $\quad \frac{\partial z}{\partial x} = 8x + 3y \quad \frac{\partial z}{\partial y} = 3x + 10y$

2 $z = (3x + 2y)(4x - 5y)$ $\quad \frac{\partial z}{\partial x} = 24x - 7y \quad \frac{\partial z}{\partial y} = -7x - 20y$

3 $z = \tan(3x + 4y)$ $\quad \frac{\partial z}{\partial x} = 3\sec^2(3x + 4y) \quad \frac{\partial z}{\partial y} = 4\sec^2(3x + 4y)$

4 $z = \frac{\sin(3x + 2y)}{xy}$

$$\frac{\partial z}{\partial x} = \frac{3x\cos(3x + 2y) - \sin(3x + 2y)}{x^2 y} \qquad \frac{\partial z}{\partial y} = \frac{2y\cos(3x + 2y) - \sin(3x + 2y)}{xy^2}$$

If you have got *all* the answers correct, turn straight on to Frame 15. If you have not got all these answers, or are at all uncertain, move to Frame 12.

12

Let us work through these examples in detail.

1 $z = 4x^2 + 3xy + 5y^2$

To find $\frac{\partial z}{\partial x}$, regard y as a constant:

$\therefore \frac{\partial z}{\partial x} = 8x + 3y + 0$, i.e. $8x + 3y$ $\qquad \therefore \frac{\partial z}{\partial x} = 8x + 3y$

Similarly, regarding x as constant:

$\frac{\partial z}{\partial y} = 0 + 3x + 10y$, i.e. $3x + 10y$ $\qquad \therefore \frac{\partial z}{\partial y} = 3x + 10y$

2 $z = (3x + 2y)(4x - 5y)$ $\qquad$ Product rule

$$\frac{\partial z}{\partial x} = (3x + 2y)(4) + (4x - 5y)(3)$$

$$= 12x + 8y + 12x - 15y = 24x - 7y$$

$$\frac{\partial z}{\partial y} = (3x + 2y)(-5) + (4x - 5y)(2)$$

$$= -15x - 10y + 8x - 10y = -7x - 20y$$

*Move on for the solutions to **3** and **4***

13

3 $z = \tan(3x + 4y)$

$$\frac{\partial z}{\partial x} = \sec^2(3x + 4y)(3) = 3\sec^2(3x + 4y)$$

$$\frac{\partial z}{\partial y} = \sec^2(3x + 4y)(4) = 4\sec^2(3x + 4y)$$

4 $z = \dfrac{\sin(3x + 2y)}{xy}$

$$\frac{\partial z}{\partial x} = \frac{xy\cos(3x + 2y)(3) - \sin(3x + 2y)(y)}{x^2y^2}$$

$$= \frac{3x\cos(3x + 2y) - \sin(3x + 2y)}{x^2y}$$

Now have another go at finding $\dfrac{\partial z}{\partial y}$ in the same way.

Then check it with Frame 14

14

Here it is:

$$z = \frac{\sin(3x + 2y)}{xy}$$

$$\therefore \frac{\partial z}{\partial y} = \frac{xy\cos(3x + 2y)\cdot(2) - \sin(3x + 2y)\cdot(x)}{x^2y^2}$$

$$= \frac{2y\cos(3x + 2y) - \sin(3x + 2y)}{xy^2}$$

That should have cleared up any troubles. This business of partial differentiation is perfectly straightforward. All you have to remember is that for the time being, all the independent variables except the one you are using are kept constant – and behave like constant factors or constant terms according to their positions.

On you go now to Frame 15 and continue the Program

15 Second partial differentiation

Right. Now let us move on a step.

Consider $z = 3x^2 + 4xy - 5y^2$

Then $\dfrac{\partial z}{\partial x} = 6x + 4y$ and $\dfrac{\partial z}{\partial y} = 4x - 10y$

The expression $\dfrac{\partial z}{\partial x} = 6x + 4y$ is itself a function of x and y. We could therefore find its partial derivatives with respect to x or to y.

(a) If we differentiate it partially w.r.t. x, we get:

$\frac{\partial}{\partial x}\left\{\frac{\partial z}{\partial x}\right\}$ and this is written $\frac{\partial^2 z}{\partial x^2}$ (much like an ordinary second derivative, but with the partial ∂)

$$\therefore\ \frac{\partial^2 z}{\partial x^2} = \frac{\partial}{\partial x}(6x + 4y) = 6$$

This is called the second partial derivative of z with respect to x.

(b) If we differentiate partially w.r.t. y, we get:

$\frac{\partial}{\partial y}\left\{\frac{\partial z}{\partial x}\right\}$ and this is written $\frac{\partial^2 z}{\partial y \cdot \partial x}$

Note that the operation now being performed is given by the left-hand of the two symbols in the denominator.

$$\frac{\partial^2 z}{\partial y \cdot \partial x} = \frac{\partial}{\partial y}\left\{\frac{\partial z}{\partial x}\right\} = \frac{\partial}{\partial y}\{6x + 4y\} = 4$$

16

So we have this:

$$z = 3x^2 + 4xy - 5y^2$$

$$\frac{\partial z}{\partial x} = 6x + 4y \qquad \frac{\partial z}{\partial y} = 4x - 10y$$

$$\frac{\partial^2 z}{\partial x^2} = 6$$

$$\frac{\partial^2 z}{\partial y \cdot \partial x} = 4$$

Of course, we could carry out similar steps with the expression for $\frac{\partial z}{\partial y}$ on the right. This would give us:

$$\frac{\partial^2 z}{\partial y^2} = -10$$

$$\frac{\partial^2 z}{\partial x \cdot \partial y} = 4$$

Note that $\frac{\partial^2 z}{\partial y \cdot \partial x}$ means $\frac{\partial}{\partial y}\left\{\frac{\partial z}{\partial x}\right\}$ so $\frac{\partial^2 z}{\partial x \cdot \partial y}$ means ....

17

$$\boxed{\frac{\partial^2 z}{\partial x \cdot \partial y} \text{ means } \frac{\partial}{\partial x}\left\{\frac{\partial z}{\partial y}\right\}}$$

Collecting our previous results together then, we have:

$$z = 3x^2 + 4xy - 5y^2$$

$$\frac{\partial z}{\partial x} = 6x + 4y \qquad \frac{\partial z}{\partial y} = 4x - 10y$$

$$\frac{\partial^2 z}{\partial x^2} = 6 \qquad \frac{\partial^2 z}{\partial y^2} = -10$$

$$\frac{\partial^2 z}{\partial y \cdot \partial x} = 4 \qquad \frac{\partial^2 z}{\partial x \cdot \partial y} = 4$$

We see in this case, that $\frac{\partial^2 z}{\partial y \cdot \partial x} = \frac{\partial^2 z}{\partial x \cdot \partial y}$. There are then, *two* first derivatives and *four* second derivatives, though the last two seem to have the same value.

Here is one for you to do.

If $z = 5x^3 + 3x^2y + 4y^3$, find $\frac{\partial z}{\partial x}, \frac{\partial z}{\partial y}, \frac{\partial^2 z}{\partial x^2}, \frac{\partial^2 z}{\partial y^2}, \frac{\partial^2 z}{\partial x \cdot \partial y}$ and $\frac{\partial^2 z}{\partial y \cdot \partial x}$

When you have completed all that, move to Frame 18

18

Here are the results:

$$z = 5x^3 + 3x^2y + 4y^3$$

$$\frac{\partial z}{\partial x} = 15x^2 + 6xy \qquad \frac{\partial z}{\partial y} = 3x^2 + 12y^2$$

$$\frac{\partial^2 z}{\partial x^2} = 30x + 6y \qquad \frac{\partial^2 z}{\partial y^2} = 24y$$

$$\frac{\partial^2 z}{\partial y \cdot \partial x} = 6x \qquad \frac{\partial^2 z}{\partial x \cdot \partial y} = 6x$$

Again in this example also, we see that $\frac{\partial^2 z}{\partial y \cdot \partial x} = \frac{\partial^2 z}{\partial x \cdot \partial y}$. Now do this one.

It looks more complicated, but it is done in just the same way. Do not rush at it; take your time and all will be well. Here it is. Find all the first and second partial derivatives of $z = x\cos y - y\cos x$.

Then to Frame 19

19

Check your results with these.

$z = x\cos y - y\cos x$

When differentiating w.r.t. x, y is constant (and therefore $\cos y$ also).
When differentiating w.r.t. y, x is constant (and therefore $\cos x$ also).

So we get:

$$\frac{\partial z}{\partial x} = \cos y + y \cdot \sin x \qquad \frac{\partial z}{\partial y} = -x \cdot \sin y - \cos x$$

$$\frac{\partial^2 z}{\partial x^2} = y \cdot \cos x \qquad \frac{\partial^2 z}{\partial y^2} = -x \cdot \cos y$$

$$\frac{\partial^2 z}{\partial y \cdot \partial x} = -\sin y + \sin x \qquad \frac{\partial^2 z}{\partial x \cdot \partial y} = -\sin y + \sin x$$

And again, $\dfrac{\partial^2 z}{\partial y \cdot \partial x} = \dfrac{\partial^2 z}{\partial x \cdot \partial y}$

In fact this will always be so for the functions you are likely to meet, so that there are really *three* different second partial derivatives (and not four). In practice, if you have found $\dfrac{\partial^2 z}{\partial y \cdot \partial x}$ it is a useful check to find $\dfrac{\partial^2 z}{\partial x \cdot \partial y}$ separately. They should give the same result, of course.

20

What about this one?

If $V = \ln(x^2 + y^2)$, prove that $\dfrac{\partial^2 V}{\partial x^2} + \dfrac{\partial^2 V}{\partial y^2} = 0$

This merely entails finding the two second partial derivatives and substituting them in the left-hand side of the statement. So here goes:

$$V = \ln(x^2 + y^2)$$

$$\frac{\partial V}{\partial x} = \frac{1}{(x^2 + y^2)} 2x$$

$$= \frac{2x}{x^2 + y^2}$$

$$\frac{\partial^2 V}{\partial x^2} = \frac{(x^2 + y^2)2 - 2x \cdot 2x}{(x^2 + y^2)^2}$$

$$= \frac{2x^2 + 2y^2 - 4x^2}{(x^2 + y^2)^2} = \frac{2y^2 - 2x^2}{(x^2 + y^2)^2} \qquad \text{(a)}$$

Now you find $\dfrac{\partial^2 V}{\partial y^2}$ in the same way and hence prove the given identity.

When you are ready, move on to Frame 21

21

We had found that $\dfrac{\partial^2 V}{\partial x^2} = \dfrac{2y^2 - 2x^2}{(x^2+y^2)^2}$

So making a fresh start from $V = \ln(x^2 + y^2)$, we get:

$$\frac{\partial V}{\partial y} = \frac{1}{(x^2+y^2)} \cdot 2y = \frac{2y}{x^2+y^2}$$

$$\frac{\partial^2 V}{\partial y^2} = \frac{(x^2+y^2)2 - 2y \cdot 2y}{(x^2+y^2)^2}$$

$$= \frac{2x^2 + 2y^2 - 4y^2}{(x^2+y^2)^2} = \frac{2x^2 - 2y^2}{(x^2+y^2)^2} \qquad \text{(b)}$$

Substituting now the two results in the identity, gives:

$$\frac{\partial^2 V}{\partial x^2} + \frac{\partial^2 V}{\partial y^2} = \frac{2y^2 - 2x^2}{(x^2+y^2)^2} + \frac{2x^2 - 2y^2}{(x^2+y^2)^2}$$

$$= \frac{2y^2 - 2x^2 + 2x^2 - 2y^2}{(x^2+y^2)^2} = 0$$

Now on to Frame 22

22

Here is another kind of example that you should see.

Example 1

If $V = f(x^2 + y^2)$, show that $x\dfrac{\partial V}{\partial y} - y\dfrac{\partial V}{\partial x} = 0$

Here we are told that V is a function of $(x^2 + y^2)$ but the precise nature of the function is not given. However, we can treat this as a 'function of a function' and write $f'(x^2 + y^2)$ to represent the derivative of the function w.r.t. its own combined variable $(x^2 + y^2)$.

$$\therefore \quad \frac{\partial V}{\partial x} = f'(x^2+y^2) \times \frac{\partial}{\partial x}(x^2+y^2) = f'(x^2+y^2) \cdot 2x$$

$$\frac{\partial V}{\partial y} = f'(x^2+y^2) \cdot \frac{\partial}{\partial y}(x^2+y^2) = f'(x^2+y^2) \cdot 2y$$

$$\therefore \quad x\frac{\partial V}{\partial y} - y\frac{\partial V}{\partial x} = x \cdot f'(x^2+y^2) \cdot 2y - y \cdot f'(x^2+y^2) \cdot 2x$$

$$= 2xy \cdot f'(x^2+y^2) - 2xy \cdot f'(x^2+y^2)$$

$$= 0$$

Let us have another one of that kind in the next frame

23

Example 2

If $z = f\left\{\frac{y}{x}\right\}$, show that $x\frac{\partial z}{\partial x} + y\frac{\partial z}{\partial y} = 0$

Much the same as before:

$$\frac{\partial z}{\partial x} = f'\left\{\frac{y}{x}\right\} \cdot \frac{\partial}{\partial x}\left\{\frac{y}{x}\right\} = f'\left\{\frac{y}{x}\right\}\left(-\frac{y}{x^2}\right) = -\frac{y}{x^2}f'\left\{\frac{y}{x}\right\}$$

$$\frac{\partial z}{\partial y} = f'\left\{\frac{y}{x}\right\} \cdot \frac{\partial}{\partial y}\left\{\frac{y}{x}\right\} = f'\left\{\frac{y}{x}\right\} \cdot \frac{1}{x} = \frac{1}{x}f'\left\{\frac{y}{x}\right\}$$

$$\therefore\ x\frac{\partial z}{\partial x} + y\frac{\partial z}{\partial y} = x\left(-\frac{y}{x^2}\right)f'\left\{\frac{y}{x}\right\} + y\frac{1}{x}f'\left\{\frac{y}{x}\right\}$$

$$= -\frac{y}{x}f'\left\{\frac{y}{x}\right\} + \frac{y}{x}f'\left\{\frac{y}{x}\right\}$$

$$= 0$$

And one for you, just to get your hand in:

If $V = f(ax + by)$, show that $b\frac{\partial V}{\partial x} - a\frac{\partial V}{\partial y} = 0$

When you have done it, check your working against that in Frame 24

24

Here is the working; this is how it goes.

$$V = f(ax + by)$$

$$\therefore\ \frac{\partial V}{\partial x} = f'(ax + by) \cdot \frac{\partial}{\partial x}(ax + by)$$

$$= f'(ax + by) \cdot a = a \cdot f'(ax + by) \qquad \text{(a)}$$

$$\frac{\partial z}{\partial y} = f'(ax + by) \cdot \frac{\partial}{\partial y}(ax + by)$$

$$= f'(ax + by) \cdot b = b \cdot f'(ax + by) \qquad \text{(b)}$$

$$\therefore\ b\frac{\partial V}{\partial x} - a\frac{\partial V}{\partial y} = ab \cdot f'(ax + by) - ab \cdot f'(ax + by)$$

$$= 0$$

Move on to Frame 25

25

So to sum up so far.

Partial differentiation is easy, no matter how complicated the expression to be differentiated may seem.

To differentiate partially w.r.t. x, all independent variables other than x are constant for the time being.

To differentiate partially w.r.t. y, all independent variables other than y are constant for the time being.

So that, if z is a function of x and y, i.e. if $z = f(x, y)$, we can find:

$$\frac{\partial z}{\partial x} \qquad \frac{\partial z}{\partial y}$$

$$\frac{\partial^2 z}{\partial x^2} \qquad \frac{\partial^2 z}{\partial y^2}$$

$$\frac{\partial^2 z}{\partial y \cdot \partial x} \qquad \frac{\partial^2 z}{\partial x \cdot \partial y}$$

And also: $$\frac{\partial^2 z}{\partial y \cdot \partial x} = \frac{\partial^2 z}{\partial x \cdot \partial y}$$

Now for a review exercise

Review exercise

26

1 Find all first and second partial derivatives for each of the following functions:

(a) $z = 3x^2 + 2xy + 4y^2$

(b) $z = \sin xy$

(c) $z = \dfrac{x+y}{x-y}$

2 If $z = \ln(e^x + e^y)$, show that $\dfrac{\partial z}{\partial x} + \dfrac{\partial z}{\partial y} = 1$.

3 If $z = x \cdot f(xy)$, express $x\dfrac{\partial z}{\partial x} - y\dfrac{\partial z}{\partial y}$ in its simplest form.

When you have finished, check with the solutions in Frame 27

27

1 (a) $z = 3x^2 + 2xy + 4y^2$

$$\frac{\partial z}{\partial x} = 6x + 2y \qquad \frac{\partial z}{\partial y} = 2x + 8y$$

$$\frac{\partial^2 z}{\partial x^2} = 6 \qquad \frac{\partial^2 z}{\partial y^2} = 8$$

$$\frac{\partial^2 z}{\partial y \cdot \partial x} = 2 \qquad \frac{\partial^2 z}{\partial x \cdot \partial y} = 2$$

(b) $z = \sin xy$

$$\frac{\partial z}{\partial x} = y\cos xy \qquad \frac{\partial z}{\partial y} = x\cos xy$$

$$\frac{\partial^2 z}{\partial x^2} = -y^2 \sin xy \qquad \frac{\partial^2 z}{\partial y^2} = -x^2 \sin xy$$

$$\frac{\partial^2 z}{\partial y \cdot \partial x} = y(-x\sin xy) + \cos xy = \cos xy - xy\sin xy$$

$$\frac{\partial^2 z}{\partial x \cdot \partial y} = x(-y\sin xy) + \cos xy = \cos xy - xy\sin xy$$

(c) $z = \dfrac{x+y}{x-y}$

$$\frac{\partial z}{\partial x} = \frac{(x-y)1 - (x+y)1}{(x-y)^2} = \frac{-2y}{(x-y)^2}$$

$$\frac{\partial z}{\partial y} = \frac{(x-y)1 - (x+y)(-1)}{(x-y)^2} = \frac{2x}{(x-y)^2}$$

$$\frac{\partial^2 z}{\partial x^2} = (-2y)\frac{(-2)}{(x-y)^3} = \frac{4y}{(x-y)^3}$$

$$\frac{\partial^2 z}{\partial y^2} = 2x\frac{(-2)}{(x-y)^3}(-1) = \frac{4x}{(x-y)^3}$$

$$\frac{\partial^2 z}{\partial y \cdot \partial x} = \frac{(x-y)^2(-2) - (-2y)2(x-y)(-1)}{(x-y)^4}$$

$$= \frac{-2(x-y)^2 - 4y(x-y)}{(x-y)^4}$$

$$= \frac{-2}{(x-y)^2} - \frac{4y}{(x-y)^3}$$

$$= \frac{-2x+2y-4y}{(x-y)^3} = \frac{-2x-2y}{(x-y)^3}$$

$$\frac{\partial^2 z}{\partial x \cdot \partial y} = \frac{(x-y)^2(2) - 2x \cdot 2(x-y)1}{(x-y)^4}$$

$$= \frac{2(x-y)^2 - 4x(x-y)}{(x-y)^4}$$

$$= \frac{2}{(x-y)^2} - \frac{4x}{(x-y)^3}$$

$$= \frac{2x-2y-4x}{(x-y)^3} = \frac{-2x-2y}{(x-y)^3}$$

2 $z = \ln(e^x + e^y)$

$$\frac{\partial z}{\partial x} = \frac{1}{e^x + e^y} \cdot e^x \qquad \frac{\partial z}{\partial y} = \frac{1}{e^x + e^y} \cdot e^y$$

$$\frac{\partial z}{\partial x} + \frac{\partial z}{\partial y} = \frac{e^x}{e^x + e^y} + \frac{e^y}{e^x + e^y}$$

$$= \frac{e^x + e^y}{e^x + e^y} = 1$$

$$\frac{\partial z}{\partial x} + \frac{\partial z}{\partial y} = 1$$

3 $z = x \cdot f(xy)$

$$\frac{\partial z}{\partial x} = x \cdot f'(xy) \cdot y + f(xy)$$

$$\frac{\partial z}{\partial y} = x \cdot f'(xy) \cdot x$$

$$x\frac{\partial z}{\partial x} - y\frac{\partial z}{\partial y} = x^2yf'(xy) + xf(xy) - x^2yf'(xy)$$

$$x\frac{\partial z}{\partial x} - y\frac{\partial z}{\partial y} = xf(xy) = z$$

That was a pretty good review test. Do not be unduly worried if you made a slip or two in your working. Try to avoid doing so, of course, but you are doing fine. Now on to the next part of the Program.

So far we have been concerned with the technique of partial differentiation. Now let us look at one of its applications.

So move on to Frame 28

Small increments

28

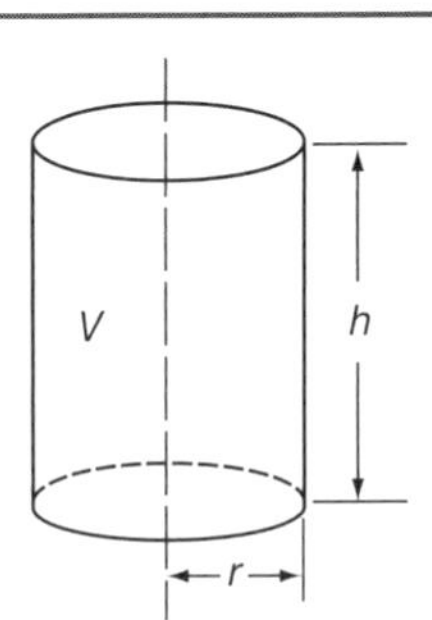

If we return to the volume of the cylinder with which we started this Program, we have once again that $V = \pi r^2 h$. We have seen that we can find $\frac{\partial V}{\partial r}$ with h constant, and $\frac{\partial V}{\partial h}$ with r constant.

$$\frac{\partial V}{\partial r} = 2\pi rh; \quad \frac{\partial V}{\partial h} = \pi r^2$$

Now let us see what we get if r and h both change simultaneously.

If r becomes $r+\delta r$, and h becomes $h+\delta h$, let V become $V+\delta V$. Then the new volume is given by:

$$\begin{aligned} V+\delta V &= \pi(r+\delta r)^2(h+\delta h) \\ &= \pi(r^2+2r\delta r+[\delta r]^2)(h+\delta h) \\ &= \pi(r^2h+2rh\delta r+h[\delta r]^2+r^2\delta h+2r\delta r\delta h+[\delta r]^2\delta h) \end{aligned}$$

Subtract $V=\pi r^2 h$ from each side, giving:

$$\delta V = \pi(2rh\delta r+h[\delta r]^2+r^2\delta h+2r\delta r\delta h+[\delta r]^2\delta h)$$

$$\approx \pi(2rh\delta r+r^2\delta h)$$ since δr and δh are small and all the remaining terms are of a higher degree of smallness.

Therefore

$$\delta V \approx 2\pi rh\delta r+\pi r^2\delta h,$$ that is:

$$\delta V \approx \frac{\partial V}{\partial r}\delta r+\frac{\partial V}{\partial h}\delta h$$

Let us now do a numerical example to see how it all works out.

On to Frame 29

29

A cylinder has dimensions $r=5$ cm, $h=10$ cm. Find the approximate increase in volume when r increases by 0.2 cm and h decreases by 0.1 cm. Well now

$$V=\pi r^2 h \text{ so } \frac{\partial V}{\partial r}=2\pi rh \qquad \frac{\partial V}{\partial h}=\pi r^2$$

In this case, when $r=5$ cm, $h=10$ cm so

$$\frac{\partial V}{\partial r}=2\pi 5\cdot 10=100\pi \qquad \frac{\partial V}{\partial h}=\pi r^2=\pi 5^2=25\pi$$

$\delta r=0.2$ and $\delta h=-0.1$ (minus because h is decreasing)

$$\therefore\ \delta V \approx \frac{\partial V}{\partial r}.\delta r+\frac{\partial V}{\partial h}.\delta h$$

$$\begin{aligned} \delta V &= 100\pi(0.2)+25\pi(-0.1) \\ &= 20\pi-2.5\pi=17.5\pi \end{aligned}$$

$$\therefore\ \delta V \approx 54.98\ \text{cm}^3$$

i.e. the volume increases by $54.98\ \text{cm}^3$

Just like that!

30

This kind of result applies not only to the volume of the cylinder, but to any function of two independent variables. Here is an example:

If z is a function of x and y, i.e. $z = f(x, y)$ and if x and y increase by small amounts δx and δy, the increase δz will also be relatively small. If we expand δz in powers of δx and δy, we get:

$$\delta z = A\delta x + B\delta y + \text{higher powers of } \delta x \text{ and } \delta y,$$

where A and B are functions of x and y.

If y remains constant, so that $\delta y = 0$, then:

$$\delta z = A\delta x + \text{higher powers of } \delta x$$

$$\therefore \; \frac{\delta z}{\delta x} = A. \text{ So that if } \delta x \to 0, \text{ this becomes } A = \frac{\partial z}{\partial x}$$

Similarly, if x remains constant, making $\delta y \to 0$ gives $B = \dfrac{\partial z}{\partial y}$

$$\therefore \; \delta z = \frac{\partial z}{\partial x}\delta x + \frac{\partial z}{\partial y}\delta y + \text{ higher powers of very small quantities which can be ignored}$$

$$\delta z = \frac{\partial z}{\partial x}\delta x + \frac{\partial z}{\partial y}\delta y$$

31

So, if $z = f(x, y)$

$$\delta z = \frac{\partial z}{\partial x}\delta x + \frac{\partial z}{\partial y}\delta y$$

This is the key to all the forthcoming applications and will be quoted over and over again.

The result is quite general and a similar result applies for a function of three independent variables. For example:

If $z = f(x, y, w)$

$$\text{then } \delta z = \frac{\partial z}{\partial x}\delta x + \frac{\partial z}{\partial y}\delta y + \frac{\partial z}{\partial w}\delta w$$

If we remember the rule for a function of two independent variables, we can easily extend it when necessary.

Here it is once again:

$$\text{If } z = f(x, y) \text{ then } \delta z = \frac{\partial z}{\partial x}\delta x + \frac{\partial z}{\partial y}\delta y$$

Copy this result into your record book in a prominent position, such as it deserves!

32

Now for a couple of examples

Example 1

If $I = \frac{V}{R}$, and $V = 250$ volts and $R = 50$ ohms, find the change in I resulting from an increase of 1 volt in V and an increase of 0.5 ohm in R.

$$I = f(V, R) \qquad \therefore \ \delta I = \frac{\partial I}{\partial V}\delta V + \frac{\partial I}{\partial R}\delta R$$

$$\frac{\partial I}{\partial V} = \frac{1}{R} \text{ and } \frac{\partial I}{\partial R} = -\frac{V}{R^2}$$

$$\therefore \ \delta I = \frac{1}{R}\delta V - \frac{V}{R^2}\delta R$$

So when $R = 50$, $V = 250$, $\delta V = 1$ and $\delta R = 0.5$:

$$\delta I = \frac{1}{50}(1) - \frac{250}{2500}(0.5)$$
$$= \frac{1}{50} - \frac{1}{20}$$
$$= 0.02 - 0.05 = -0.03$$

i.e. I decreases by 0.03 amperes

33

Here is another example.

Example 2

If $y = \frac{ws^3}{d^4}$, find the percentage increase in y when w increases by 2 per cent, s decreases by 3 per cent and d increases by 1 per cent.

Notice that, in this case, y is a function of three variables, w, s and d. The formula therefore becomes:

$$\delta y = \frac{\partial y}{\partial w}\delta w + \frac{\partial y}{\partial s}\delta s + \frac{\partial y}{\partial d}\delta d$$

We have

$$\frac{\partial y}{\partial w} = \frac{s^3}{d^4}; \quad \frac{\partial y}{\partial s} = \frac{3ws^2}{d^4}; \quad \frac{\partial y}{\partial d} = -\frac{4ws^3}{d^5}$$

$$\therefore \ \delta y = \frac{s^3}{d^4}\delta w + \frac{3ws^2}{d^4}\delta s + \frac{-4ws^3}{d^5}\delta d$$

Now then, what are the values of $\delta w, \delta s$ and δd?

Is it true to say that $\delta w = \frac{2}{100}; \quad \delta s = \frac{-3}{100}; \quad \delta d = \frac{1}{100}$?

If not, why not?

Next frame

34

No. It is not correct

Because δw is not $\dfrac{2}{100}$ of a unit, but 2 per cent of w, i.e. $\delta w = \dfrac{2}{100}$ of $w = \dfrac{2w}{100}$

Similarly, $\delta s = \dfrac{-3}{100}$ of $s = \dfrac{-3s}{100}$ and $\delta d = \dfrac{d}{100}$. Now that we have cleared that point up, we can continue with the problem.

$$\delta y = \frac{s^3}{d^4}\left(\frac{2w}{100}\right) + \frac{3ws^2}{d^4}\left(\frac{-3s}{100}\right) - \frac{4ws^3}{d^5}\left(\frac{d}{100}\right)$$

$$= \frac{ws^3}{d^4}\left(\frac{2}{100}\right) - \frac{ws^3}{d^4}\left(\frac{9}{100}\right) - \frac{ws^3}{d^4}\left(\frac{4}{100}\right)$$

$$= \frac{ws^3}{d^4}\left\{\frac{2}{100} - \frac{9}{100} - \frac{4}{100}\right\}$$

$$= y\left\{-\frac{11}{100}\right\} = -11 \text{ per cent of } y$$

i.e. y decreases by 11 per cent

Remember that where the increment of w is given as 2 per cent, it is *not* $\dfrac{2}{100}$ of a unit, but $\dfrac{2}{100}$ of w, and the symbol w must be included.

Move on to Frame 35

35

Now here is an exercise for you to do.

$P = w^2hd$. If errors of up to 1 per cent (plus or minus) are possible in the measured values of w, h and d, find the maximum possible percentage error in the calculated values of P.

This is very much like the previous example, so you will be able to deal with it without any trouble. Work it right through and then go on to Frame 36 and check your result.

36

$$P = w^2hd. \quad \therefore \ \delta P = \frac{\partial P}{\partial w}\cdot\delta w + \frac{\partial P}{\partial h}\cdot\delta h + \frac{\partial P}{\partial d}\cdot\delta d$$

$$\frac{\partial P}{\partial w} = 2whd; \quad \frac{\partial P}{\partial h} = w^2d; \quad \frac{\partial P}{\partial d} = w^2h$$

$$\delta P = 2whd\cdot\delta w + w^2d\cdot\delta h + w^2h\cdot\delta d$$

Now $\quad \delta w = \pm\dfrac{w}{100}; \quad \delta h = \pm\dfrac{h}{100}, \quad \delta d = \pm\dfrac{d}{100}$

$$\delta P = 2whd\left(\pm\frac{w}{100}\right) + w^2d\left(\pm\frac{h}{100}\right) + w^2h\left(\pm\frac{d}{100}\right)$$

$$= \pm\frac{2w^2hd}{100} \pm \frac{w^2dh}{100} \pm \frac{w^2hd}{100}$$

The greatest possible error in P will occur when the signs are chosen so that they are all of the same kind, i.e. all plus or minus. If they were mixed, they would tend to cancel each other out.

▶

$$\therefore \quad \delta P = \pm w^2 hd\left\{\frac{2}{100}+\frac{1}{100}+\frac{1}{100}\right\} = \pm P\left(\frac{4}{100}\right)$$

$\therefore$ Maximum possible error in P is 4 per cent of P

Finally, here is one last example for you to do. Work right through it and then check your results with those in Frame 37.

The two sides forming the right-angle of a right-angled triangle are denoted by a and b. The hypotenuse is h. If there are possible errors of ± 0.5 per cent in measuring a and b, find the maximum possible error in calculating (a) the area of the triangle and (b) the length of h.

37

(a) $\delta A = 1$ per cent of A
(b) $\delta h = 0.5$ per cent of h

Here is the working in detail:

(a) $A = \dfrac{a \cdot b}{2} \qquad \delta A = \dfrac{\partial A}{\partial a}\cdot \delta a + \dfrac{\partial A}{\partial b}\cdot \delta b$

$\dfrac{\partial A}{\partial a} = \dfrac{b}{2}; \quad \dfrac{\partial A}{\partial b} = \dfrac{a}{2}; \quad \delta a = \pm\dfrac{a}{200}; \quad \delta b = \pm\dfrac{b}{200}$

$$\delta A = \frac{b}{2}\left(\pm\frac{a}{200}\right) + \frac{a}{2}\left(\pm\frac{b}{200}\right)$$

$$= \pm\frac{a\cdot b}{2}\left[\frac{1}{200}+\frac{1}{200}\right] = \pm A\cdot\frac{1}{100}$$

$\therefore \quad \delta A = 1$ per cent of A

(b) $h = \sqrt{a^2+b^2} = (a^2+b^2)^{\frac{1}{2}}$

$$\delta h = \frac{\partial h}{\partial a}\delta a + \frac{\partial h}{\partial b}\delta b$$

$$\frac{\partial h}{\partial a} = \frac{1}{2}(a^2+b^2)^{-\frac{1}{2}}(2a) = \frac{a}{\sqrt{a^2+b^2}}$$

$$\frac{\partial h}{\partial b} = \frac{1}{2}(a^2+b^2)^{-\frac{1}{2}}(2b) = \frac{b}{\sqrt{a^2+b^2}}$$

Also $\quad \delta a = \pm\dfrac{a}{200}; \quad \delta b = \pm\dfrac{b}{200}$

$$\therefore \quad \delta h = \frac{a}{\sqrt{a^2+b^2}}\left(\pm\frac{a}{200}\right) + \frac{b}{\sqrt{a^2+b^2}}\left(\pm\frac{b}{200}\right)$$

$$= \pm\frac{1}{200}\frac{a^2+b^2}{\sqrt{a^2+b^2}}$$

$$= \pm\frac{1}{200}\sqrt{a^2+b^2} = \pm\frac{1}{200}(h)$$

$\therefore \quad \delta h = 0.5$ per cent of h

▶

That brings us to the end of this particular Program. We shall meet partial differentiation again in the next Program when we shall consider some more of its applications. But for the time being, there remain only the **Can You?** checklist and the **Test exercise**.

So on now to Frames 38 and 39

Can You?

38 Checklist 6

Check this list before and after you try the end of Program test.

On a scale of 1 to 5 how confident are you that you can: **Frames**

- Find the first partial derivatives of a function of two real variables? 1 to 14
  Yes ☐ ☐ ☐ ☐ ☐ *No*
- Find second-order partial derivatives of a function of two real variables? 15 to 25
  Yes ☐ ☐ ☐ ☐ ☐ *No*
- Calculate errors using partial differentiation? 28 to 37
  Yes ☐ ☐ ☐ ☐ ☐ *No*

Test exercise 6

39

Take your time over the questions; do them carefully.

1 Find all first and second partial derivatives of the following:

(a) $z = 4x^3 - 5xy^2 + 3y^3$

(b) $z = \cos(2x + 3y)$

(c) $z = e^{x^2 - y^2}$

(d) $z = x^2 \sin(2x + 3y)$

2 (a) If $V = x^2 + y^2 + z^2$, express in its simplest form

$$x\frac{\partial V}{\partial x} + y\frac{\partial V}{\partial y} + z\frac{\partial V}{\partial z}.$$

(b) If $z = f(x + ay) + F(x - ay)$, find $\dfrac{\partial^2 z}{\partial x^2}$ and $\dfrac{\partial^2 z}{\partial y^2}$ and hence prove that

$$\frac{\partial^2 z}{\partial y^2} = a^2 \cdot \frac{\partial^2 z}{\partial x^2}.$$

▶

3 The power P dissipated in a resistor is given by $P = \dfrac{E^2}{R}$.

If $E = 200$ volts and $R = 8$ ohms, find the change in P resulting from a drop of 5 volts in E and an increase of 0.2 ohm in R.

4 If $\theta = kHLV^{-\frac{1}{2}}$, where k is a constant, and there are possible errors of ± 1 per cent in measuring H, L and V, find the maximum possible error in the calculated value of θ.

That's it

Further problems 6

40

1 If $z = \dfrac{1}{x^2 + y^2 - 1}$, show that $x\dfrac{\partial z}{\partial x} + y\dfrac{\partial z}{\partial y} = -2z(1+z)$.

2 Prove that, if $V = \ln(x^2 + y^2)$, then $\dfrac{\partial^2 V}{\partial x^2} + \dfrac{\partial^2 V}{\partial y^2} = 0$.

3 If $z = \sin(3x + 2y)$, verify that $3\dfrac{\partial^2 z}{\partial y^2} - 2\dfrac{\partial^2 z}{\partial x^2} = 6z$.

4 If $u = \dfrac{x + y + z}{(x^2 + y^2 + z^2)^{\frac{1}{2}}}$, show that $x\dfrac{\partial u}{\partial x} + y\dfrac{\partial u}{\partial y} + z\dfrac{\partial u}{\partial z} = 0$.

5 Show that the equation $\dfrac{\partial^2 z}{\partial x^2} + \dfrac{\partial^2 z}{\partial y^2} = 0$, is satisfied by

$$z = \ln\sqrt{x^2 + y^2} + \frac{1}{2}\tan^{-1}\left(\frac{y}{x}\right)$$

6 If $z = e^x(x\cos y - y\sin y)$, show that $\dfrac{\partial^2 z}{\partial x^2} + \dfrac{\partial^2 z}{\partial y^2} = 0$.

7 If $u = (1 + x)\sinh(5x - 2y)$, verify that $4\dfrac{\partial^2 u}{\partial x^2} + 20\dfrac{\partial^2 u}{\partial x \cdot \partial y} + 25\dfrac{\partial^2 u}{\partial y^2} = 0$.

8 If $z = f\left(\dfrac{y}{x}\right)$, show that $x^2\dfrac{\partial^2 z}{\partial x^2} + 2xy\dfrac{\partial^2 z}{\partial x \cdot \partial y} + y^2\dfrac{\partial^2 z}{\partial y^2} = 0$.

9 If $z = (x + y) \cdot f\left(\dfrac{y}{x}\right)$, where f is an arbitrary function, show that $x\dfrac{\partial z}{\partial x} + y\dfrac{\partial z}{\partial y} = z$.

10 In the formula $D = \dfrac{Eh^3}{12(1 - v^2)}$, h is given as 0.1 ± 0.002 and v as 0.3 ± 0.02. Express the approximate maximum error in D in terms of E.

11 The formula $z = \dfrac{a^2}{x^2 + y^2 - a^2}$ is used to calculate z from observed values of x and y. If x and y have the same percentage error p, show that the percentage error in z is approximately $-2p(1 + z)$.

▶

12 In a balanced bridge circuit, $R_1 = R_2R_3/R_4$. If R_2, R_3, R_4 have known tolerances of $\pm x$ per cent, $\pm y$ per cent, $\pm z$ per cent respectively, determine the maximum percentage error in R_1, expressed in terms of x, y and z.

13 The deflection y at the center of a circular plate suspended at the edge and uniformly loaded is given by $y = \dfrac{kwd^4}{t^3}$, where w = total load, d = diameter of plate, t = thickness and k is a constant.

Calculate the approximate percentage change in y if w is increased by 3 per cent, d is decreased by $2\frac{1}{2}$ per cent and t is increased by 4 per cent.

14 The coefficient of rigidity (n) of a wire of length (L) and uniform diameter (d) is given by $n = \dfrac{AL}{d^4}$, where A is a constant. If errors of ± 0.25 per cent and ± 1 per cent are possible in measuring L and d respectively, determine the maximum percentage error in the calculated value of n.

15 If $k/k_0 = (T/T_0)^n \cdot p/760$, show that the change in k due to small changes of a per cent in T and b per cent in p is approximately $(na + b)$ per cent.

16 The deflection y at the center of a rod is known to be given by $y = \dfrac{kwl^3}{d^4}$, where k is a constant. If w increases by 2 per cent, l by 3 per cent, and d decreases by 2 per cent, find the percentage increase in y.

17 The displacement y of a point on a vibrating stretched string, at a distance x from one end, at time t, is given by

$$\frac{\partial^2 y}{\partial t^2} = c^2 \cdot \frac{\partial^2 y}{\partial x^2}$$

Show that one solution of this equation is $y = A \sin\dfrac{px}{c} \cdot \sin(pt + a)$, where A, p, c and a are constants.

18 If $y = A\sin(px + a)\cos(qt + b)$, find the error in y due to small errors δx and δt in x and t respectively.

19 Show that $\phi = Ae^{-kt/2}\sin pt \cos qx$, satisfies the equation

$$\frac{\partial^2 \phi}{\partial x^2} = \frac{1}{c^2}\left\{\frac{\partial^2 \phi}{\partial t^2} + k\frac{\partial \phi}{\partial t}\right\}, \text{ provided that } p^2 = c^2q^2 - \frac{k^2}{4}.$$

20 Show that (a) the equation $\dfrac{\partial^2 V}{\partial x^2} + \dfrac{\partial^2 V}{\partial y^2} + \dfrac{\partial^2 V}{\partial z^2} = 0$ is satisfied by

$V = \dfrac{1}{\sqrt{x^2 + y^2 + z^2}}$, and that (b) the equation $\dfrac{\partial^2 V}{\partial x^2} + + \dfrac{\partial^2 V}{\partial y^2} = 0$ is satisfied by $V = \tan^{-1}\left(\dfrac{y}{x}\right)$.

Integration

Learning outcomes

When you have completed this Program you will be able to:

- Evaluate double and triple integrals and apply them to the determination of the areas of plane figures and the volumes of solids
- Understand the role of the differential of a function of two or more real variables
- Determine exact differentials in two real variables and their integrals
- Evaluate the area enclosed by a closed curve by contour integration
- Evaluate line integrals and appreciate their properties
- Evaluate line integrals around closed curves within a simply connected region
- Link line integrals to integrals along the x-axis
- Link line integrals to integrals along a contour given in parametric form
- Discuss the dependence of a line integral between two points on the path of integration
- Determine exact differentials in three real variables and their integrals
- Demonstrate the validity and use of Green's theorem

Introduction

1

You will no doubt recognize the following.

1 *Double integrals*

$$\int_{y_1}^{y_2}\int_{x_1}^{x_2} f(x, y)\,\mathrm{d}x\,\mathrm{d}y$$

is a double integral and is evaluated from the inside outwards, i.e.

$$\left[\int_{y_1}^{y_2}\left[\int_{x_1}^{x_2} f(x, y)\,\mathrm{d}x\right]_{①}\mathrm{d}y\right]_{②}$$

A double integral is sometimes expressed in the form

$$\int_{y_1}^{y_2}\mathrm{d}y\int_{x_1}^{x_2} f(x, y)\,\mathrm{d}x$$

in which case, we evaluate from the right-hand end, i.e.

$$\int_{y_1}^{y_2}\mathrm{d}y\left[\int_{x_1}^{x_2} f(x, y)\,\mathrm{d}x\right]_{①}$$

then $$\left[\int_{y_1}^{y_2}\left[\int_{x_1}^{x_2} f(x, y)\,\mathrm{d}x\right]\mathrm{d}y\right]_{②}$$

2 *Triple integrals*

Triple integrals follow the same procedure.

$\int_{z_1}^{z_2}\int_{y_1}^{y_2}\int_{x_1}^{x_2} f(x, y, z)\,\mathrm{d}x\,\mathrm{d}y\,\mathrm{d}z$ is evaluated in the order

$$\left[\int_{z_1}^{z_2}\left[\int_{y_1}^{y_2}\left[\int_{x_1}^{x_2} f(x,y,z)\,\mathrm{d}x\right]_{①}\mathrm{d}y\right]_{②}\mathrm{d}z\right]_{③}$$

▶

3 *Applications*

(a) *Areas of plane figures*

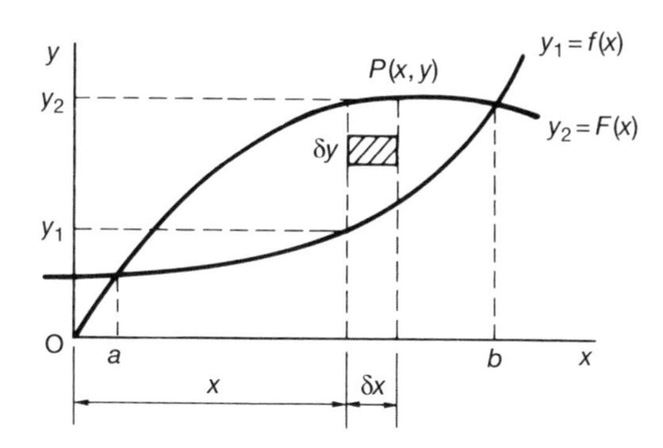

Area of element $\delta A = \delta x \delta y$

Area of strip $\approx \sum_{y=y_1}^{y=y_2} \delta x \delta y$

Area of all such strips $\approx \sum_{x=a}^{x=b} \left\{ \sum_{y=y_1}^{y=y_2} \delta x \delta y \right\}$

If $\delta x \to 0$ and $\delta y \to 0$, $A = \int_a^b \int_{y_1}^{y_2} \mathrm{d}y \, \mathrm{d}x$

(b) *Areas of plane figures bounded by a polar curve* $r = f(\theta)$ and radius vectors at $\theta = \theta_1$ and $\theta = \theta_2$

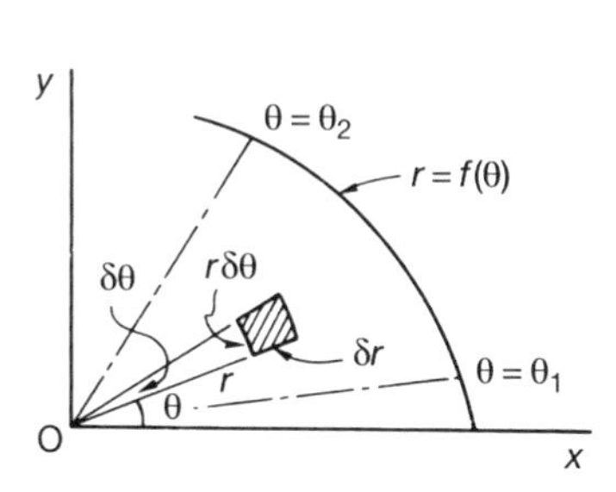

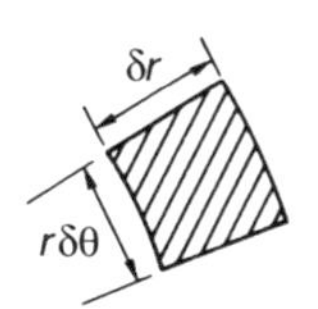

Small arc of circle of radius r, subtending angle $\delta\theta$ at centre.

$\therefore$ Arc $= r\delta\theta$

Area of element $\delta A \approx r\delta\theta \, \delta r$

Area of thin sector $\approx \sum_{r=0}^{r=f(\theta)} r \, \delta\theta \, \delta r$

$\therefore$ Total area of all such sectors $\approx \sum_{\theta=\theta_1}^{\theta=\theta_2} \left\{ \sum_{r=0}^{r=f(\theta)} r \, \delta r \, \delta\theta \right\}$

$\therefore$ If $\delta r \to 0$ and $\delta\theta \to 0$, $A = \int_{\theta_1}^{\theta_2} \int_0^{r=f(\theta)} r \, \mathrm{d}r \, \mathrm{d}\theta$

▶

(c) *Volume of solids*

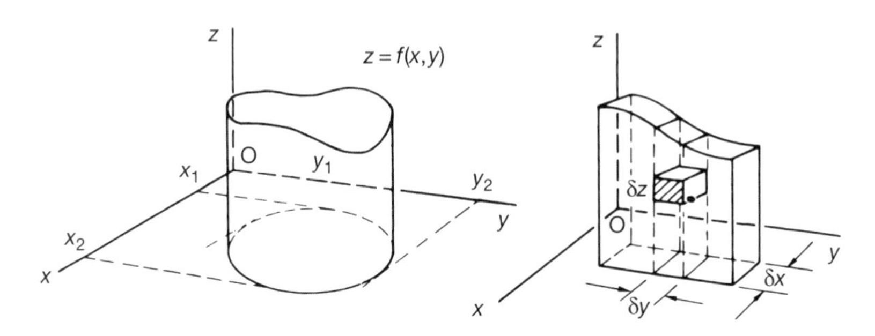

Volume of element $\delta V = \delta x \, \delta y \, \delta z$

$$\text{Volume of column} \approx \sum_{z=0}^{z=f(x,\,y)} \delta x \, \delta y \, \delta z$$

$$\text{Volume of slice} \approx \sum_{y=y_1}^{y=y_2} \left\{ \sum_{z=0}^{z=f(x,\,y)} \delta x \, \delta y \, \delta z \right\}$$

$\therefore$ Total volume $V \approx$ sum of all such slices

$$\text{i.e.} \quad V \approx \sum_{x=x_1}^{x=x_2} \sum_{y=y_1}^{y=y_2} \sum_{z=0}^{z=f(x,\,y)} \delta x \, \delta y \, \delta z$$

Then, if $\delta x \to 0$, $\delta y \to 0$, $\delta z \to 0$,

$$V = \int_{x_1}^{x_2} \int_{y_1}^{y_2} \int_{0}^{z=f(x,\,y)} \mathrm{d}z \, \mathrm{d}y \, \mathrm{d}x$$

If $z = f(x, y)$, this becomes

$$V = \int_{x_1}^{x_2} \int_{y_1}^{y_2} f(x, y) \, \mathrm{d}y \, \mathrm{d}x$$

2

4 *Review examples* As a means of 'warming up', let us work through one or two straightforward examples on the previous work.

Example 1

Find the area of the plane figure bounded by the curves $y_1 = (x-1)^2$ and $y_2 = 4-(x-3)^2$.

The first thing, as always, is to sketch the curves – each of which is a parabola – and to determine their points of intersection.

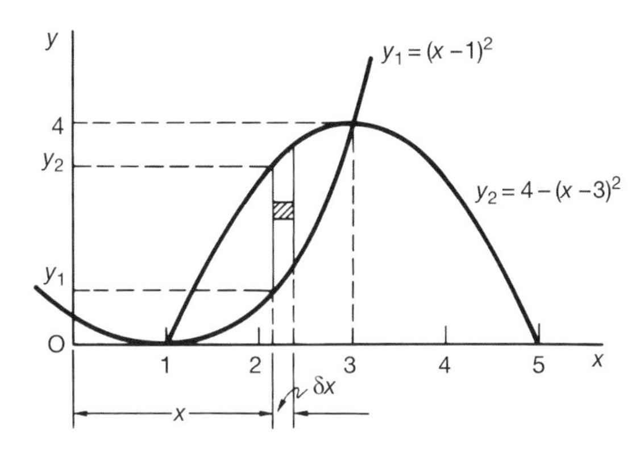

Points of intersection: $(x-1)^2 = 4-(x-3)^2$

$$x^2 - 2x + 1 = 4 - x^2 + 6x - 9 \quad \text{i.e. } x^2 - 4x + 3 = 0$$

$$\therefore \ (x-1)(x-3) = 0 \quad \therefore \ x = 1 \ \text{ or } \ x = 3.$$

Now we have all the information to determine the required area, which is

....

3

$$\boxed{A = 2\tfrac{2}{3} \text{ square units}}$$

Because

$$A = \int_{x=1}^{x=3}\int_{y_1}^{y_2} \mathrm{d}y\,\mathrm{d}x = \int_{x=1}^{x=3}\int_{y=(x-1)^2}^{y=4-(x-3)^2} \mathrm{d}y\,\mathrm{d}x$$

$$= \int_1^3 \{4-(x-3)^2-(x-1)^2\}\,\mathrm{d}x = -2\int_1^3 (x^2-4x+3)\,\mathrm{d}x$$

$$= -2\left[\frac{x^3}{3} - 2x^2 + 3x\right]_1^3 = 2\tfrac{2}{3} \text{ square units}$$

Now for another.

Example 2

A rectangular plate is bounded by the x and y axes and the lines $x = 6$ and $y = 4$. The thickness t of the plate at any point is proportional to the square of the distance of the point from the origin. Determine the total volume of the plate.

First of all draw the figure and build up the appropriate double integral. Do not evaluate it yet. The expression is therefore

$$V = \dots\dots\dots\dots$$

4

$$V = \int_{x=0}^{x=6}\int_{y=0}^{y=4} k(x^2 + y^2)\,\mathrm{d}y\,\mathrm{d}x$$

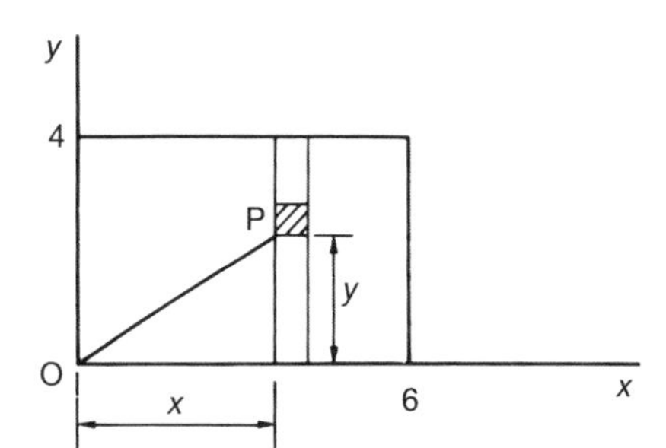

Thickness t of plate at P is

$$t = k\,\mathrm{OP}^2 = k(x^2 + y^2)$$

Element of area $= \delta y\,\delta x$

$\therefore$ Element of volume at P $\approx k(x^2 + y^2)\,\delta y\,\delta x$

$\therefore$ Total volume $V = \displaystyle\int_{x=0}^{x=6}\int_{y=0}^{y=4} k(x^2 + y^2)\,\mathrm{d}y\,\mathrm{d}x$

Now we can evaluate the integral. We start from the inside with

$$\int_{y=2}^{y=4} k(x^2 + y^2)\,\mathrm{d}y,$$

remembering that for this integral (volume of the strip) x is constant. This gives $\dots\dots\dots\dots$

5

$$k\left(4x^2 + \frac{64}{3}\right)$$

Because

$$k\int_0^4 (x^2 + y^2)\,\mathrm{d}y = k\left[x^2 y + \frac{y^3}{3}\right]_{y=0}^{y=4} = k\left(4x^2 + \frac{64}{3}\right)$$

Then $\quad V = k\displaystyle\int_0^6 \left(4x^2 + \frac{64}{3}\right)\mathrm{d}x = \dots\dots\dots\dots$

6

$$V = 416k \text{ cubic units}$$

That was easy enough. Notice that an alternative interpretation of this problem could be that of a uniform lamina with a variable density $\rho = k(x^2 + y^2)$ at any point (x, y). Now for one in polar coordinates.

Example 3

Express as a double integral the area enclosed by one loop of the curve $r = 3\cos 2\theta$ and evaluate the integral.

Consider the half loop shown.

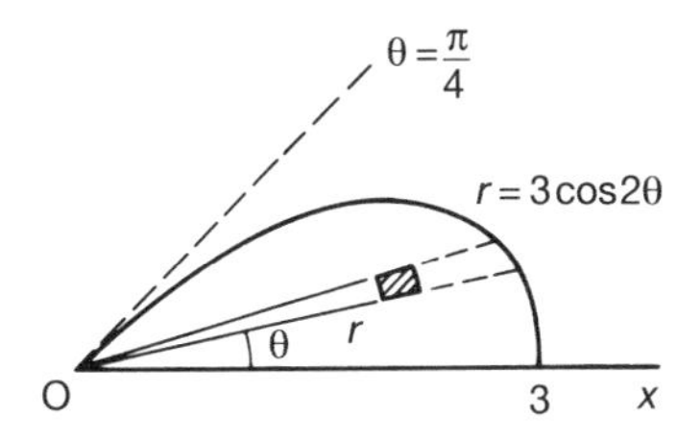

First set up the double integral which is ....

$$A = \int_{\theta=0}^{\theta=\pi/4} \int_{r=0}^{r=3\cos 2\theta} r\,\mathrm{d}r\,\mathrm{d}\theta$$

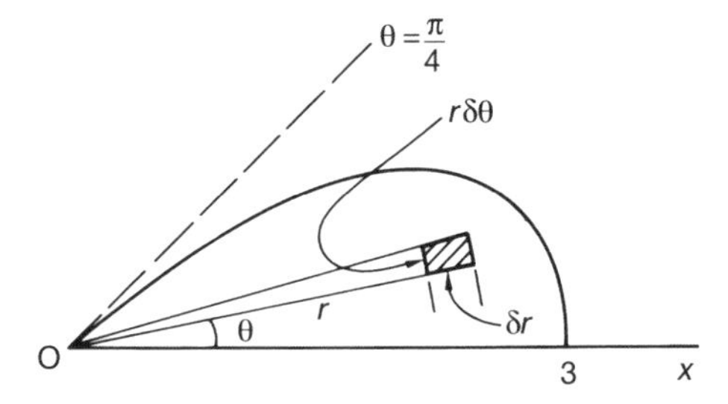

Area of element $= r\,\delta r\,\delta\theta$

$$\therefore \text{ Area of sector} \approx \sum_{r=0}^{r=3\cos 2\theta} r\,\delta r\,\delta\theta$$

$$\therefore \text{ Area of half loop} \approx \sum_{\theta=0}^{\theta=\pi/4} \sum_{r=0}^{r=3\cos 2\theta} r\,\delta r\,\delta\theta$$

If $\delta r \to 0$ and $\delta\theta \to 0$,

$$A = \int_{\theta=0}^{\theta=\pi/4} \int_{r=0}^{r=3\cos 2\theta} r\,\mathrm{d}r\,\mathrm{d}\theta$$

Now finish it off to find the area of the whole loop, which is

....

8

$$\frac{9\pi}{8} \text{ square units}$$

Because

$$\begin{aligned}
A &= \int_{\theta=0}^{\theta=\pi/4} \int_{r=0}^{r=3\cos 2\theta} r \, dr \, d\theta \\
&= \int_0^{\pi/4} \left[\frac{r^2}{2}\right]_0^{3\cos 2\theta} d\theta \\
&= \frac{9}{2}\int_0^{\pi/4} \cos^2 2\theta \, d\theta \\
&= \frac{9}{4}\int_0^{\pi/4} (1 + \cos 4\theta) \, d\theta \\
&= \frac{9}{4}\left[\theta + \frac{\sin 4\theta}{4}\right]_0^{\pi/4} \\
&= \frac{9\pi}{16}
\end{aligned}$$

This is the area of a half loop.

$$\text{Required area} = \frac{9\pi}{8} \text{ square units}$$

Now here is another.

Example 4

Find the volume of the solid bounded by the planes $z = 0$, $x = 1$, $x = 3$, $y = 1$, $y = 2$ and the surface $z = x^2 y^2$.

As always, we start off by sketching the figure. When you have done that, check the result with the next frame.

9

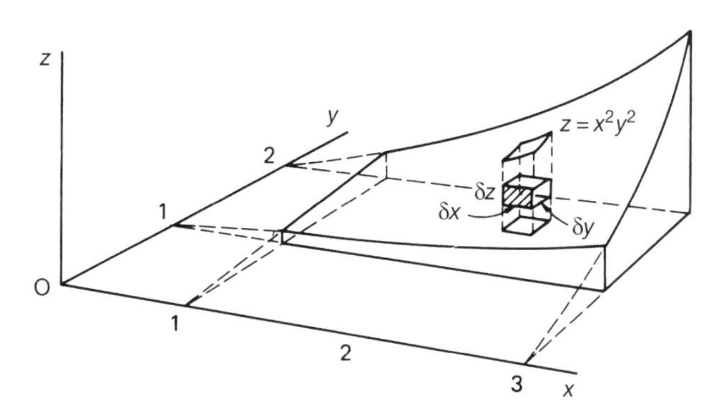

We now build up the integral which will give us the volume of the solid.

Element of volume $\delta V = \delta x\, \delta y\, \delta z$

$$\text{Volume of column} \approx \sum_{z=0}^{z=x^2y^2} \delta x\, \delta y\, \delta z$$

$$\text{Volume of slice} \approx \sum_{y=1}^{y=2} \left\{ \sum_{z=0}^{z=x^2y^2} \delta x\, \delta y\, \delta z \right\}$$

$$\text{Volume of solid} \approx \sum_{x=1}^{x=3} \left\{ \sum_{y=1}^{y=2} \sum_{z=0}^{z=x^2y^2} \delta x\, \delta y\, \delta z \right\}$$

When $\delta x \to 0$, $\delta y \to 0$, $\delta z \to 0$,

$$V = \int_{x=1}^{x=3} \int_{y=1}^{y=2} \int_{z=0}^{z=x^2y^2} dz\, dy\, dx$$

Evaluating this, $V = \ldots\ldots\ldots\ldots$

10

$$\boxed{V = 20\tfrac{2}{9} \text{ cubic units}}$$

Because, starting with the innermost integral

$$V = \int_{x=1}^{x=3} \int_{y=1}^{y=2} \Big[z \Big]_0^{x^2y^2} dy\, dx = \int_1^3 \int_1^2 x^2y^2\, dy\, dx$$

$$= \int_1^3 \left[\frac{x^2y^3}{3} \right]_{y=1}^{y=2} dx \qquad = \int_1^3 \frac{7x^2}{3}\, dx = 20\tfrac{2}{9}$$

Now that we have reviewed the basics, let us move on to something rather different

Differentials

11

It is convenient in various branches of the calculus to denote small increases in value of a variable by the use of *differentials*. The method is particularly useful in dealing with the effects of small finite changes and shortens the writing of calculus expressions.

We are already familiar with the diagram from which finite changes δy and δx in a function $y = f(x)$ are depicted.

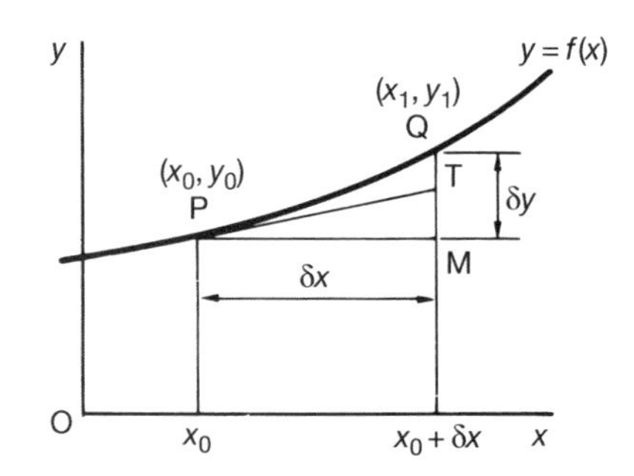

The increase in y from P to Q $=$ MQ $= \delta y = f(x_0 + \delta x) - f(x_0)$

If PT is the tangent at P, then MQ $=$ MT $+$ TQ. Also $\dfrac{\text{MT}}{\delta x} = f'(x_0)$

$\therefore\ \text{MT} = f'(x_0)\delta x$

$\therefore\ \text{MQ} = \delta y = f'(x_0) \cdot \delta x + \text{TQ}$

and, if Q is close to P, then $\delta y \approx f'(x_0)\delta x$

We define the differentials $\mathrm{d}y$ and $\mathrm{d}x$ as finite quantities such that

$$\mathrm{d}y = f'(x_0)\,\mathrm{d}x$$

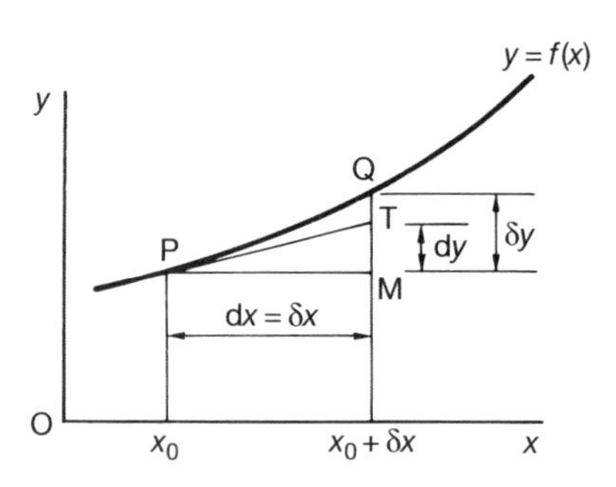

Note that the differentials $\mathrm{d}y$ and $\mathrm{d}x$ are finite quantities – not necessarily zero – and can therefore exist alone.

Note too that $\mathrm{d}x = \delta x$.

From the diagram, we can see that

δy is the increase in y as we move from P to Q along the curve.
$\mathrm{d}y$ is the increase in y as we move from P to T along the tangent.

As Q approaches P, the difference between δy and $\mathrm{d}y$ decreases to zero. The use of differentials simplifies the writing of many relationships and is based on the general statement $\mathrm{d}y = f'(x)\,\mathrm{d}x$.

For example

(a) $y = x^5$ then $\mathrm{d}y = 5x^4\,\mathrm{d}x$
(b) $y = \sin 3x$ then $\mathrm{d}y = 3\cos 3x\,\mathrm{d}x$
(c) $y = e^{4x}$ then $\mathrm{d}y = 4\,e^{4x}\,\mathrm{d}x$
(d) $y = \cosh 2x$ then $\mathrm{d}y = 2\,\sinh 2x\,\mathrm{d}x$

Note that when the left-hand side is a differential $\mathrm{d}y$ the right-hand side must also contain a differential. Remember therefore to include the '$\mathrm{d}x$' on the right-hand side.

The product and quotient rules can also be expressed in differentials.

$$\frac{\mathrm{d}}{\mathrm{d}x}(uv) = u\frac{\mathrm{d}v}{\mathrm{d}x} + v\frac{\mathrm{d}u}{\mathrm{d}x} \quad \text{becomes} \quad \mathrm{d}(uv) = u\,\mathrm{d}v + v\,\mathrm{d}u$$

$$\frac{\mathrm{d}}{\mathrm{d}x}\left(\frac{u}{v}\right) = \frac{v\dfrac{\mathrm{d}u}{\mathrm{d}x} - u\dfrac{\mathrm{d}v}{\mathrm{d}x}}{v^2} \quad \text{becomes} \quad \mathrm{d}\left(\frac{u}{v}\right) = \frac{v\,\mathrm{d}u - u\,\mathrm{d}v}{v^2}$$

So, if $y = e^{2x}\sin 4x$, $\mathrm{d}y = \dots\dots\dots\dots$

and if $y = \dfrac{\cos 2t}{t^2}$ $\mathrm{d}y = \dots\dots\dots\dots$

12

$$y = e^{2x}\sin 4x, \quad \mathrm{d}y = 2e^{2x}(2\cos 4x + \sin 4x)\,\mathrm{d}x$$

$$y = \frac{\cos 2t}{t^2}, \quad \mathrm{d}y = -\frac{2}{t^3}\{t\sin 2t + \cos 2t\}\,\mathrm{d}t$$

That was easy enough. Let us now consider a function of two independent variables, $z = f(x, y)$.

If $z = f(x, y)$ then $z + \delta z = f(x + \delta x, y + \delta y)$

$$\therefore\ \delta z = f(x + \delta x,\ y + \delta y) - f(x, y)$$

Expanding δz in terms of δx and δy, gives

$\delta z = A\delta x + B\delta y +$ higher powers of δx and δy,
where A and B are functions of x and y.

If y remains constant, i.e. $\delta y = 0$, then

$$\delta z = A\,\delta x + \text{higher powers of } \delta x \quad \therefore\ \frac{\delta z}{\delta x} \approx A$$

$$\therefore\ \text{If } \delta x \to 0, \text{ then } A = \frac{\partial z}{\partial x}$$

▶

Similarly, if x remains constant, i.e. $\delta x = 0$, then

$$\delta z = B\,\delta y + \text{higher powers of } \delta y \quad \therefore \quad \frac{\delta z}{\delta y} \approx B$$

$$\therefore \text{ If } \delta y \to 0, \text{ then } B = \frac{\partial z}{\partial y}$$

$$\therefore \quad \delta z = \frac{\partial z}{\partial x}\delta x + \frac{\partial z}{\partial y}\delta y + \text{higher powers of small quantities}$$

$$\therefore \quad \delta z = \frac{\partial z}{\partial x}\delta x + \frac{\partial z}{\partial y}\delta y$$

In terms of differentials, this result can be written

$$\text{If } z = f(x, y), \text{ then } \mathrm{d}z = \frac{\partial z}{\partial x}\mathrm{d}x + \frac{\partial z}{\partial y}\mathrm{d}y$$

The result can be extended to functions of more than two independent variables.

$$\text{If } z = f(x, y, w), \quad \mathrm{d}z = \frac{\partial z}{\partial x}\mathrm{d}x + \frac{\partial z}{\partial y}\mathrm{d}y + \frac{\partial z}{\partial w}\mathrm{d}w$$

Make a note of these results in differential form as shown.

Exercise

Determine the differential dz for each of the following functions.

1 $z = x^2 + y^2$

2 $z = x^3 \sin 2y$

3 $z = (2x - 1)\,e^{3y}$

4 $z = x^2 + 2y^2 + 3w^2$

5 $z = x^3y^2w$.

Finish all five and then check the results.

13

1 $\mathrm{d}z = 2(x\,\mathrm{d}x + y\,\mathrm{d}y)$

2 $\mathrm{d}z = x^2(3\sin 2y\,\mathrm{d}x + 2x\cos 2y\,\mathrm{d}y)$

3 $\mathrm{d}z = e^{3y}\{2\,\mathrm{d}x + (6x - 3)\mathrm{d}y\}$

4 $\mathrm{d}z = 2(x\,\mathrm{d}x + 2y\,\mathrm{d}y + 3w\,\mathrm{d}w)$

5 $\mathrm{d}z = x^2y(3yw\,\mathrm{d}x + 2xw\,\mathrm{d}y + xy\,\mathrm{d}w)$

Now move on

Exact differential

14

We have just established that if $z = f(x, y)$

$$dz = \frac{\partial z}{\partial x}dx + \frac{\partial z}{\partial y}dy$$

We now work in reverse.

Any expression $dz = P\,dx + Q\,dy$, where P and Q are functions of x and y, is an *exact differential* if it can be integrated to determine z.

$$\therefore\ P = \frac{\partial z}{\partial x} \quad \text{and} \quad Q = \frac{\partial z}{\partial y}$$

Now $\dfrac{\partial P}{\partial y} = \dfrac{\partial^2 z}{\partial y\,\partial x}$ and $\dfrac{\partial Q}{\partial x} = \dfrac{\partial^2 z}{\partial x\,\partial y}$ and we know that $\dfrac{\partial^2 z}{\partial y\,\partial x} = \dfrac{\partial^2 z}{\partial x\,\partial y}$.

Therefore, for dz to be an exact differential $\dfrac{\partial P}{\partial y} = \dfrac{\partial Q}{\partial x}$ and this is the test we apply.

Example 1

$dz = (3x^2 + 4y^2)\,dx + 8xy\,dy$.

If we compare the right-hand side with $P\,dx + Q\,dy$, then

$$P = 3x^2 + 4y^2 \quad \therefore\ \frac{\partial P}{\partial y} = 8y$$

$$Q = 8xy \quad \therefore\ \frac{\partial Q}{\partial x} = 8y$$

$$\frac{\partial P}{\partial y} = \frac{\partial Q}{\partial x} \qquad \therefore\ \text{dz is an exact differential}$$

Similarly, we can test this one.

Example 2

$dz = (1 + 8xy)\,dx + 5x^2\,dy$.

From this we find …………

15

dz is *not* an exact differential

Because $dz = (1 + 8xy)\,dx + 5x^2\,dy$

$$\therefore\ P = 1 + 8xy \quad \therefore\ \frac{\partial P}{\partial y} = 8x$$

$$Q = 5x^2 \quad \therefore\ \frac{\partial Q}{\partial x} = 10x$$

$$\frac{\partial P}{\partial y} \neq \frac{\partial Q}{\partial x} \qquad \therefore\ \text{dz is not an exact differential.}$$

Exercise

Determine whether each of the following is an exact differential.

1. $dz = 4x^3y^3\,dx + 3x^4y^2\,dy$
2. $dz = (4x^3y + 2xy^3)\,dx + (x^4 + 3x^2y^2)\,dy$
3. $dz = (15y^2e^{3x} + 2xy^2)\,dx + (10ye^{3x} + x^2y)\,dy$
4. $dz = (3x^2e^{2y} - 2y^2e^{3x})\,dx + (2x^3e^{2y} - 2ye^{3x})\,dy$
5. $dz = (4y^3\cos 4x + 3x^2\cos 2y)\,dx + (3y^2\sin 4x - 2x^3\sin 2y)\,dy.$

16

1 Yes	**2** Yes	**3** No	**4** No	**5** Yes

We have just tested whether certain expressions are, in fact, exact differentials – and we said previously that, by definition, an exact differential can be integrated. But how exactly do we go about it? The following examples will show.

Integration of exact differentials

$$dz = P\,dx + Q\,dy \quad \text{where } P = \frac{\partial z}{\partial x} \text{ and } Q = \frac{\partial z}{\partial y}$$

$$\therefore\ z = \int P\,dx \quad \text{and also} \quad z = \int Q\,dy$$

Example 1

$dz = (2xy + 6x)\,dx + (x^2 + 2y^3)\,dy.$

$$P = \frac{\partial z}{\partial x} = 2xy + 6x \qquad \therefore\ z = \int (2xy + 6x)\,dx$$

$\therefore\ z = x^2y + 3x^2 + f(y)$ where $f(y)$ is an arbitrary function of y only, and is akin to the constant of integration in a normal integral.

$$\text{Also } Q = \frac{\partial z}{\partial y} = x^2 + 2y^3 \qquad \therefore\ z = \int (x^2 + 2y^3)\,dy$$

$$\therefore\ z = \ldots\ldots\ldots\ldots$$

17

$$z = x^2y + \frac{y^4}{2} + F(x) \text{ where } F(x) \text{ is an arbitrary function of } x \text{ only}$$

So the two results tell us

$$z = x^2y + 3x^2 + f(y) \qquad (1)$$

and $$z = x^2y + \frac{y^4}{2} + F(x) \qquad (2)$$

For these two expressions to represent the same function, then

$f(y)$ in (1) must be $\frac{y^4}{2}$ already in (2)

and $F(x)$ in (2) must be $3x^2$ already in (1)

$$\therefore\ z = x^2y + 3x^2 + \frac{y^4}{2}$$

Example 2

Integrate $\mathrm{d}z = (8e^{4x} + 2xy^2)\,\mathrm{d}x + (4\cos 4y + 2x^2y)\,\mathrm{d}y$.

Argue through the working in just the same way, from which we obtain

$z = \ldots\ldots\ldots\ldots$

18

$$z = 2e^{4x} + x^2y^2 + \sin 4y$$

Here it is. $\mathrm{d}z = (8e^{4x} + 2xy^2)\,\mathrm{d}x + (4\cos 4y + 2x^2y)\,\mathrm{d}y$

$$P = \frac{\partial z}{\partial x} = 8e^{4x} + 2xy^2 \qquad \therefore\ z = \int (8e^{4x} + 2xy^2)\mathrm{d}x$$

$$\therefore\ z = 2e^{4x} + x^2y^2 + f(y) \qquad (1)$$

$$Q = \frac{\partial z}{\partial y} = 4\cos 4y + 2x^2y \qquad \therefore\ z = \int (4\cos 4y + 2x^2y)\,\mathrm{d}y$$

$$\therefore\ z = \sin 4y + x^2y^2 + F(x) \qquad (2)$$

For (1) and (2) to agree, $f(y) = \sin 4y$ and $F(x) = 2e^{4x}$

$$\therefore\ z = 2\,e^{4x} + x^2y^2 + \sin 4y$$

They are all done in the same way, so you will have no difficulty with the short exercise that follows. *On you go.*

Exercise

Integrate the following exact differentials to obtain the function z.

1 $\mathrm{d}z = (6x^2 + 8xy^3)\,\mathrm{d}x + (12x^2y^2 + 12y^3)\,\mathrm{d}y$

2 $\mathrm{d}z = (3x^2 + 2xy + y^2)\,\mathrm{d}x + (x^2 + 2xy + 3y^2)\,\mathrm{d}y$

3 $\mathrm{d}z = 2(y+1)e^{2x}\,\mathrm{d}x + (e^{2x} - 2y)\,\mathrm{d}y$

4 $\mathrm{d}z = (3y^2\cos 3x - 3\sin 3x)\,\mathrm{d}x + (2y\sin 3x + 4)\,\mathrm{d}y$

5 $\mathrm{d}z = (\sinh y + y\sinh x)\mathrm{d}x + (x\cosh y + \cosh x)\,\mathrm{d}y$.

Finish all five before checking with the next frame.

19

1	$z = 2x^3 + 4x^2y^3 + 3y^4$
2	$z = x^3 + x^2y + xy^2 + y^3$
3	$z = e^{2x}(1 + y) - y^2$
4	$z = y^2 \sin 3x + \cos 3x + 4y$
5	$z = x \sinh y + y \cosh x.$

In the last one, of course, we find that the two expressions for z agree without any further addition of $f(y)$ or $F(x)$.

We shall be meeting exact differentials again later on, but for the moment let us deal with something different. On then to the next frame

Area enclosed by a closed curve

20

One of the earliest applications of integration is finding the area of a plane figure bounded by the x-axis, the curve $y = f(x)$ and ordinates at $x = x_1$ and $x = x_2$.

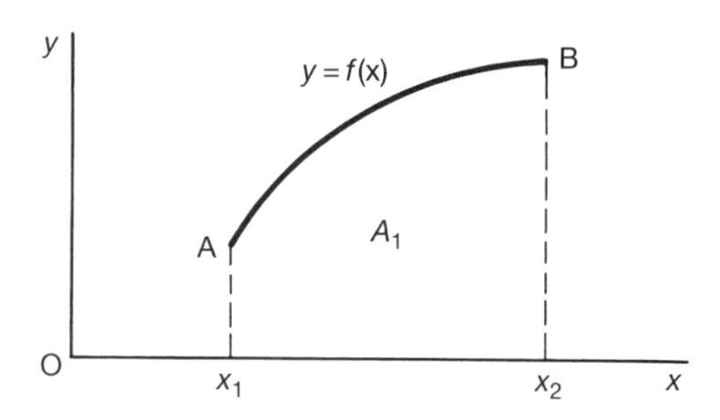

$$A_1 = \int_{x_1}^{x_2} y \, dx = \int_{x_1}^{x_2} f(x) \, dx$$

If points A and B are joined by another curve, $y = F(x)$

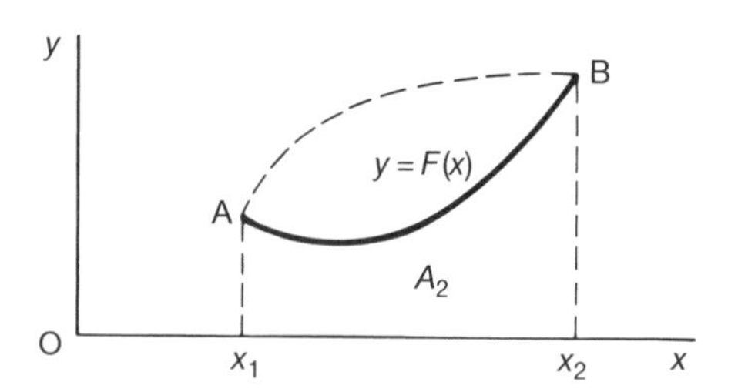

$$A_2 = \int_{x_1}^{x_2} F(x) \, dx$$

▶

Combining the two figures, we have

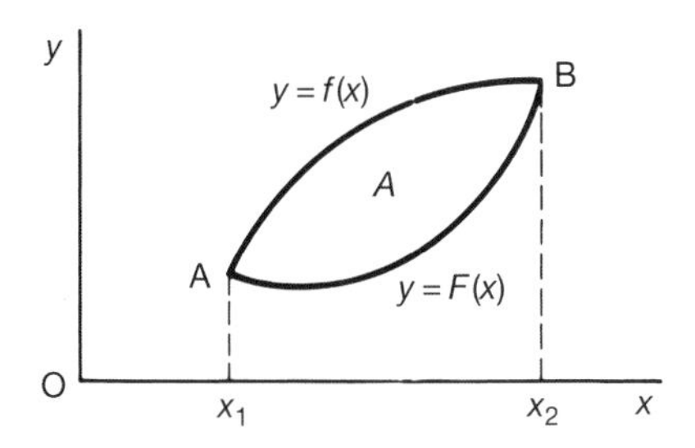

$$A = A_1 - A_2$$

$$\therefore\ A = \int_{x_1}^{x_2} f(x)\,dx - \int_{x_1}^{x_2} F(x)\,dx$$

It is convenient on occasions to arrange the limits so that the integration follows the path round the enclosed area in a regular order.

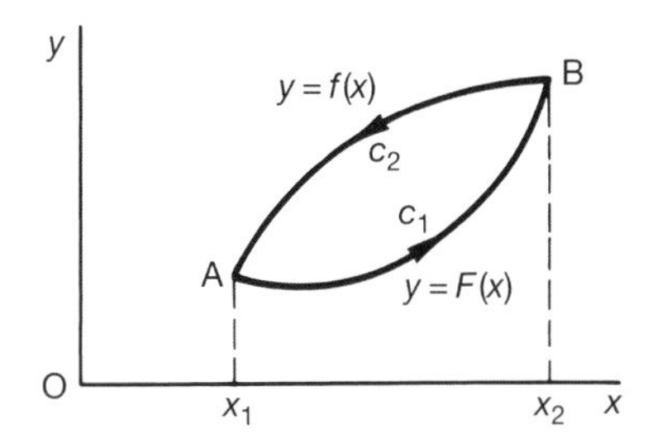

For example

$\int_{x_1}^{x_2} F(x)\,dx$ gives A_2 as before, but integrating from B to A along c_2 with $y = f(x)$, i.e. $\int_{x_2}^{x_1} f(x)\,dx$, is the integral for A_1 with the sign changed, i.e. $\int_{x_2}^{x_1} f(x)\,dx = -\int_{x_1}^{x_2} f(x)\,dx$

$\therefore$ The result $A = A_1 - A_2 = \int_{x_1}^{x_2} f(x)\,dx - \int_{x_1}^{x_2} F(x)\,dx$ becomes

$A = \ldots\ldots\ldots\ldots$

21

$$A = -\int_{x_1}^{x_2} F(x)\,dx - \int_{x_2}^{x_1} f(x)\,dx$$

i.e.
$$A = -\left\{\int_{x_1}^{x_2} F(x)\,dx + \int_{x_2}^{x_1} f(x)\,dx\right\}$$

If we proceed round the boundary in a *counter-clockwise manner*, the enclosed area is kept on the *left-hand side* and the resulting area is considered *positive*. If we proceed round the boundary in a *clockwise manner*, the enclosed area remains on the *right-hand side* and the resulting area is *negative*.

▶

The final result above can be written in the form

$$A = -\oint y \, dx$$

where the symbol $\oint$ indicates that the integral is to be evaluated round the closed boundary in the positive (i.e. counter-clockwise) direction

$$\therefore\ A = -\oint y \, dx = -\left\{ \int_{x_1}^{x_2} F(x)\,dx + \int_{x_2}^{x_1} f(x)\,dx \right\}$$

(along c_1) (along c_2)

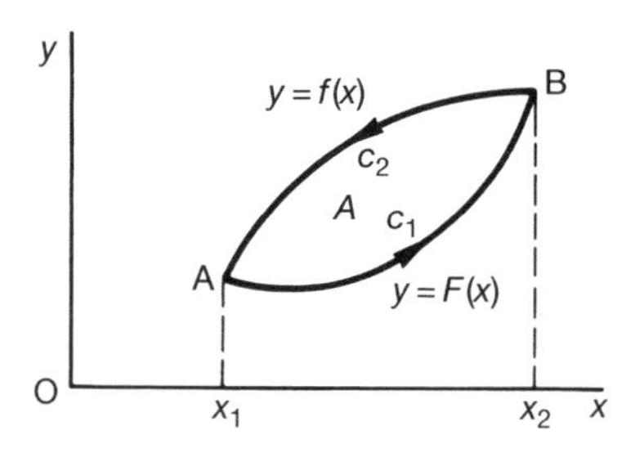

Let us apply this result to a very simple case.

Example 1

Determine the area enclosed by the graphs of $y = x^3$ and $y = 4x$ for $x \geq 0$. First we need to know the points of intersection. These are

. . . .

22

$x = 0$ and $x = 2$

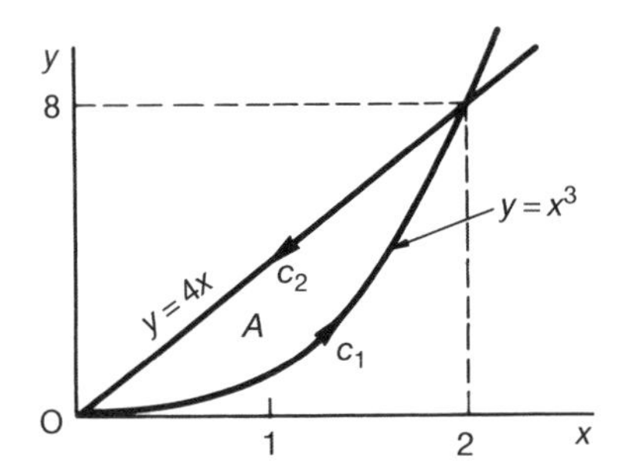

We integrate in a counter-clockwise manner

c_1: $y = x^3$, limits $x = 0$ to $x = 2$

c_2: $y = 4x$, limits $x = 2$ to $x = 0$.

$$A = -\oint y \, dx = \dots\dots\dots\dots$$

23

$$A = 4 \text{ square units}$$

Because

$$A = -\oint y\,\mathrm{d}x = -\left\{\int_0^2 x^3\,\mathrm{d}x + \int_2^0 4x\,\mathrm{d}x\right\}$$

$$= -\left\{\left[\frac{x^4}{4}\right]_0^2 + \left[2x^2\right]_2^0\right\} = 4$$

Another example.

Example 2

Find the area of the triangle with vertices (0, 0), (5, 3) and (2, 6).

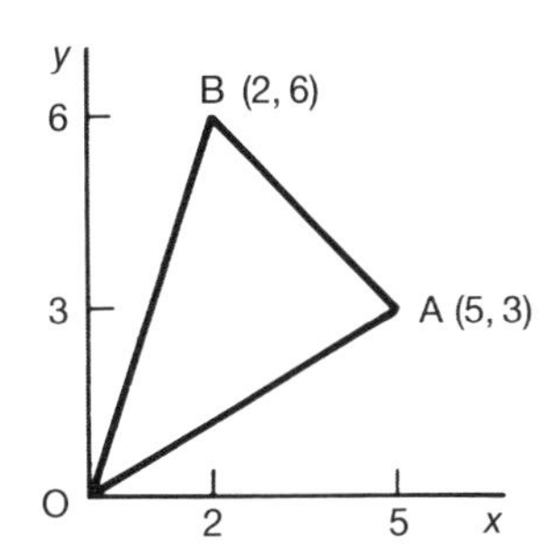

The equation of

OA is ....

BA is ....

OB is ....

24

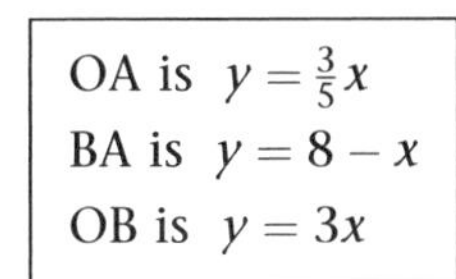

OA is $y = \frac{3}{5}x$
BA is $y = 8 - x$
OB is $y = 3x$

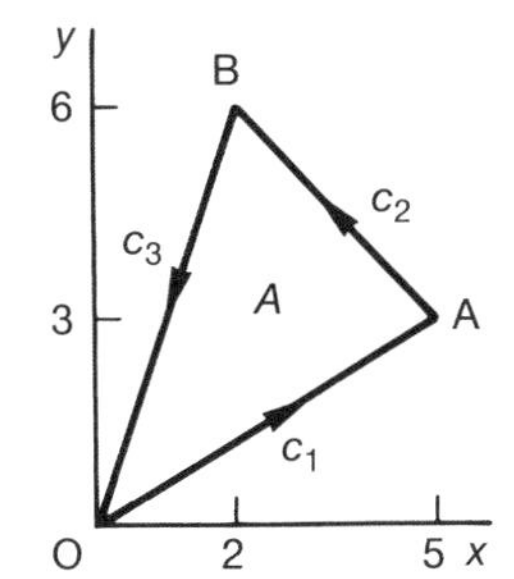

Then $A = -\oint y\,\mathrm{d}x$

$= \ldots\ldots\ldots\ldots$

Write down the component integrals with appropriate limits.

25

$$A = -\oint y\,\mathrm{d}x = -\left\{\int_0^5 \frac{3}{5}x\,\mathrm{d}x + \int_5^2 (8 - x)\,\mathrm{d}x + \int_2^0 3x\,\mathrm{d}x\right\}$$

The limits chosen must progress the integration round the boundary of the figure in a *counter-clockwise manner*. Finishing off the integration, we have

$A = \ldots\ldots\ldots\ldots$

26

$A = 12$ square units

The actual integration is easy enough.

The work we have just done leads us on to consider ***line integrals****, so let us make a fresh start in the next frame*

Line integrals

27

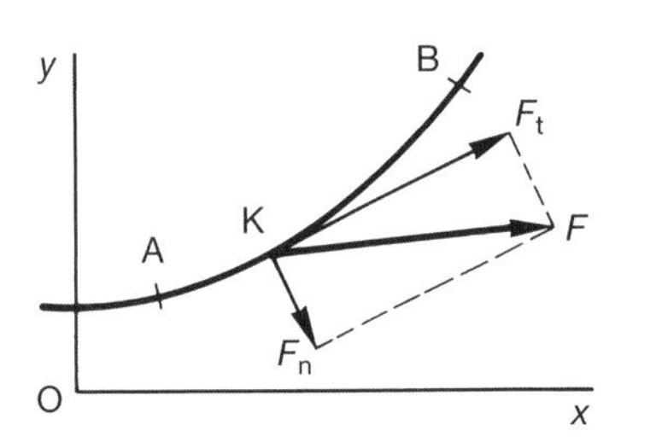

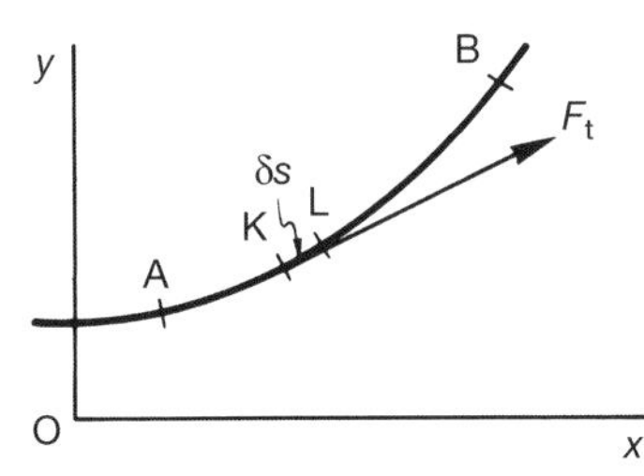

If a field exists in the x–y plane, producing a force F on a particle at K, then F can be resolved into two components

F_t along the tangent to the curve AB at K
F_n along the normal to the curve AB at K.

The work done in moving the particle through a small distance δs from K to L along the curve is then approximately $F_t\,\delta s$. So the total work done in moving a particle along the curve from A to B is given by

....

28

$$\lim_{\delta s \to 0} \sum F_t\,\delta s = \int F_t\,ds \text{ from A to B}$$

This is normally written $\int_{AB} F_t\,ds$ where A and B are the end points of the curve, or as $\int_c F_t\,ds$ where the curve c connecting A and B is defined.

Such an integral thus formed is called a *line integral* since integration is carried out along the path of the particular curve c joining A and B.

$$\therefore\ I = \int_{AB} F_t\,ds = \int_c F_t\,ds$$

where c is the curve $y = f(x)$ between A (x_1, y_1) and B (x_2, y_2).

There is in fact an alternative form of the integral which is often useful, so let us also consider that

Alternative form of a line integral

29

It is often more convenient to integrate with respect to x or y than to take arc length as the variable.

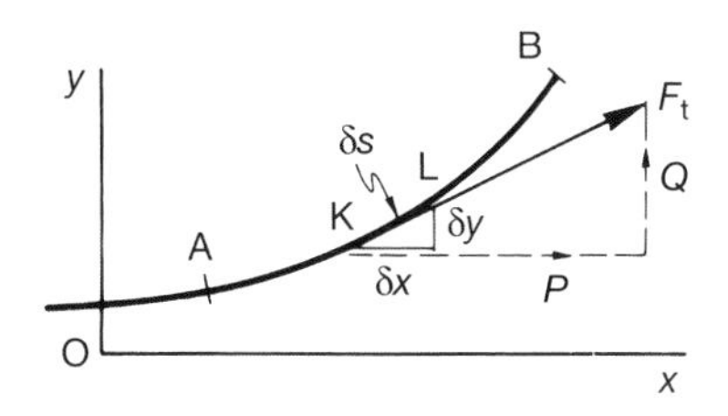

If F_t has a component

P in the x-direction

Q in the y-direction

then the work done from K to L can be stated as $P\,\delta x + Q\,\delta y$.

$$\therefore \quad \int_{AB} F_t \, ds = \int_{AB} (P\,dx + Q\,dy)$$

where P and Q are functions of x and y.

In general then, the line integral can be expressed as

$$I = \int_c F_t \, ds = \int_c (P\,dx + Q\,dy)$$

where c is the prescribed curve and F, or P and Q, are functions of x and y.

Make a note of these results – then we will apply them to one or two examples

30

Example 1

Evaluate $\int_c (x + 3y)dx$ from A (0, 1) to B (2, 5) along the curve $y = 1 + x^2$.

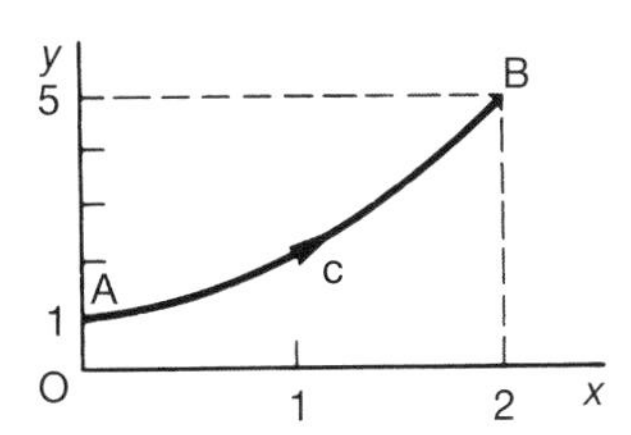

The line integral is of the form

$$\int_c (P\,dx + Q\,dy)$$

where, in this case, $Q = 0$ and c is the curve $y = 1 + x^2$.

It can be converted at once into an ordinary integral by substituting for y and applying the appropriate limits of x.

$$I = \int_c (P\,dx + Q\,dy) = \int_c (x + 3y)\,dx = \int_0^2 (x + 3 + 3x^2)\,dx$$

$$= \left[\frac{x^2}{2} + 3x + x^3\right]_0^2 = 16$$

Now for another, so move on

31

Example 2

Evaluate $I = \int_c (x^2 + y)\,dx + (x - y^2)\,dy$ from A (0, 2) to B (3, 5) along the curve $y = 2 + x$.

$$I = \int_c (P\,dx + Q\,dy)$$
$$P = x^2 + y = x^2 + 2 + x = x^2 + x + 2$$
$$Q = x - y^2 = x - (4 + 4x + x^2)$$
$$= -(x^2 + 3x + 4)$$

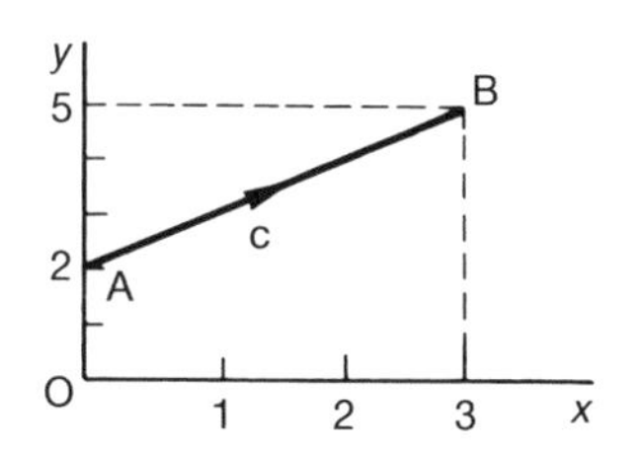

Also $y = 2 + x \quad \therefore\ dy = dx$ and the limits are $x = 0$ to $x = 3$.

$\therefore\ I = \ldots\ldots\ldots\ldots$

32

$$\boxed{I = -15}$$

Because

$$I = \int_0^3 \{(x^2 + x + 2)\,dx - (x^2 + 3x + 4)\,dx\}$$
$$\int_0^3 -(2x + 2)dx = \Big[x^2 - 2x\Big]_0^3 = -15$$

Here is another.

Example 3

Evaluate $I = \int_c \{(x^2 + 2y)\,dx + xy\,dy\}$ from O (0, 0) to B (1, 4) along the curve $y = 4x^2$.

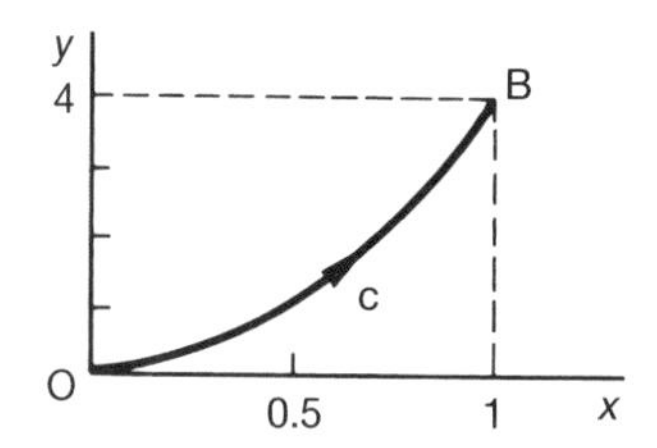

In this case, c is the curve $y = 4x^2$.

$\therefore\ dy = 8x\,dx$

Substitute for y in the integral and apply the limits.

Then $I = \ldots\ldots\ldots\ldots$

Finish it off: it is quite straightforward.

33

$$I = 9.4$$

Because

$$I = \int_c \{(x^2 + 2y)\,dx + xy\,dy\} \qquad y = 4x^2 \qquad \therefore\ dy = 8x\,dx$$

Also $x^2 + 2y = x^2 + 8x^2 = 9x^2$; $\qquad xy = 4x^3$

$$\therefore\ I = \int_0^1 \{9x^2\,dx + 32x^4\,dx\} = \int_0^1 (9x^2 + 32x^4)\,dx = 9.4$$

They are all done in very much the same way.

Move on for Example 4

34

Example 4

Evaluate $I = \int_c \{(x^2 + 2y)\,dx + xy\,dy\}$ from O (0, 0) to A (1, 0) along line $y = 0$ and then from A (1, 0) to B (1, 4) along the line $x = 1$.

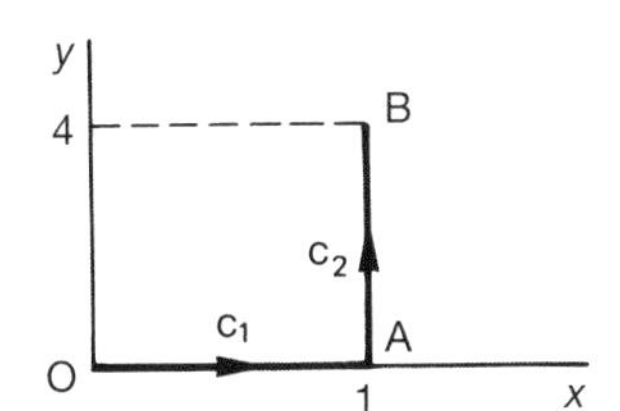

(1) OA: c_1 is the line $y = 0$ $\therefore$ $dy = 0$. Substituting $y = 0$ and $dy = 0$ in the given integral gives

$$I_{OA} = \int_0^1 x^2\,dx = \left[\frac{x^3}{3}\right]_0^1 = \frac{1}{3}$$

(2) AB: Here c_2 is the line $x = 1$ $\qquad \therefore\ dx = 0$

$$\therefore\ I_{AB} = \ldots\ldots\ldots\ldots$$

35

$$I_{AB} = 8$$

Because

$$I_{AB} = \int_0^4 \{(1 + 2y)(0) + y\,dy\}$$

$$= \int_0^4 y\,dy$$

$$= \left[\frac{y^2}{2}\right]_0^4 = 8$$

Then $I = I_{OA} + I_{AB} = \frac{1}{3} + 8 = 8\frac{1}{3}$ $\quad \therefore\ I = 8\frac{1}{3}$

If we now look back to Examples 3 and 4 just completed, we find that we have evaluated the same integral between the same two end points, but $\ldots\ldots\ldots\ldots$

36

along different paths of integration

If we combine the two diagrams, we have

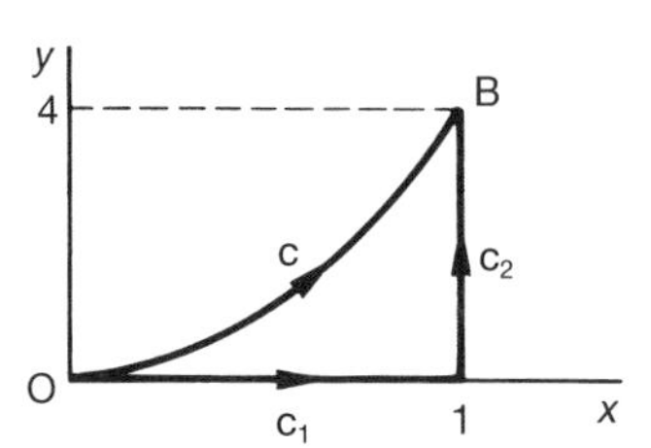

where c is the curve $y = 4x^2$ and $c_1 + c_2$ are the lines $y = 0$ and $x = 1$.

The results obtained were

$I_c = 9\frac{2}{5}$ and $I_{c_1+c_2} = 8\frac{1}{3}$

Notice therefore that integration along two distinct paths joining the same two end points does not necessarily give the same results.

37

Let us pause here a moment and list the main properties of line integrals.

Properties of line integrals

1 $\int_c F \mathrm{d}s = \int_c \{P\,\mathrm{d}x + Q\,\mathrm{d}y\}$

2 $\int_{AB} F \mathrm{d}s = -\int_{BA} F\,\mathrm{d}s$ and $\int_{AB} \{P\,\mathrm{d}x + Q\,\mathrm{d}y\} = -\int_{BA} \{P\,\mathrm{d}x + Q\,\mathrm{d}y\}$
i.e. the sign of a line integral is reversed when the direction of the integration along the path is reversed.

3 (a) For a path of integration parallel to the y-axis, i.e. $x = k$,

$$\mathrm{d}x = 0. \quad \therefore \int_c P\,\mathrm{d}x = 0 \quad \therefore I_c = \int_c Q\,\mathrm{d}y.$$

(b) For a path of integration parallel to the x-axis, i.e. $y = k$,

$$\mathrm{d}y = 0. \quad \therefore \int_c Q\,\mathrm{d}y = 0 \quad \therefore I_c = \int_c P\,\mathrm{d}x.$$

4 If the path of integration c joining A to B is divided into two parts AK and KB, then $I_c = I_{AB} = I_{AK} + I_{KB}$.

5 In all cases, the function $y = f(x)$ that describes the path of integration involved must be continuous and single-valued – or dealt with as in item **6** below.

6 If the function $y = f(x)$ that describes the path of integration c is not single-valued for part of its extent, the path is divided into two sections.

$y = f_1(x)$ from A to K
$y = f_2(x)$ from K to B.

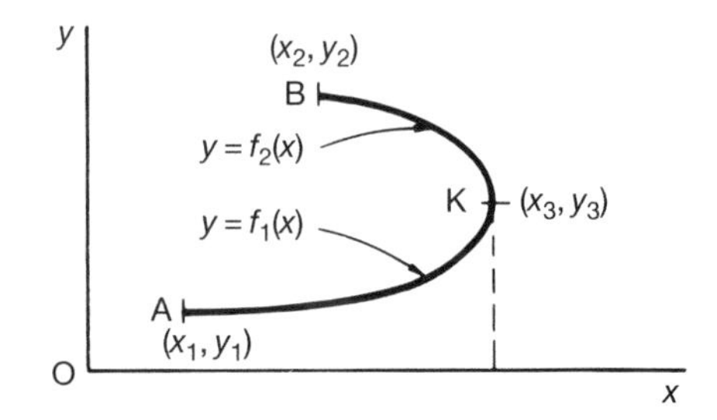

Make a note of this list for future reference and review

38

Example

Evaluate $I = \int_c (x + y)\,dx$ from A (0, 1) to B (0, − 1) along the semi-circle $x^2 + y^2 = 1$ for $x \geq 0$.

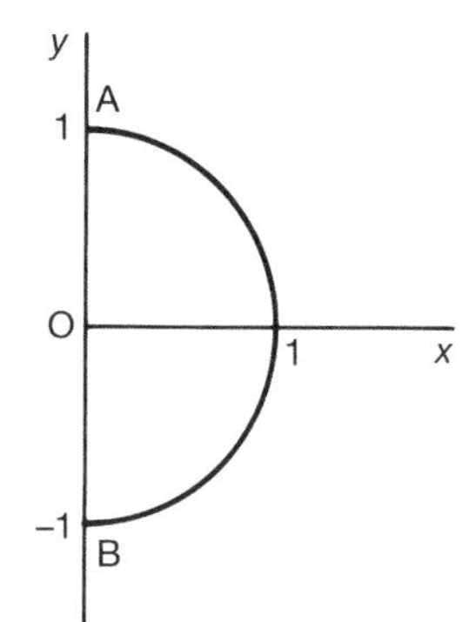

The first thing we notice is that

…………

39

> the function $y = f(x)$ that describes the path of integration c is *not* single-valued

For any value of x, $y = \pm\sqrt{1 - x^2}$. Therefore, we divide c into two parts

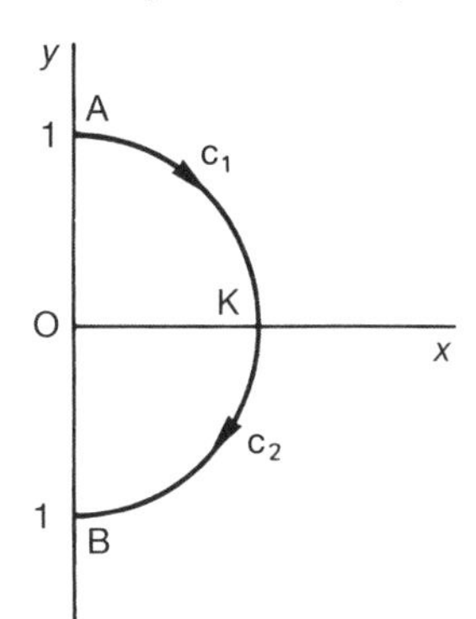

(1) $y = \sqrt{1 - x^2}$ from A to K

(2) $y = -\sqrt{1 - x^2}$ from K to B.

As usual, $I = \int_c (P\,dx + Q\,dy)$ and in this particular case, $Q =$ …………

40

> $Q = 0$

$$\therefore\ I = \int_c P\,dx = \int_0^1 \left(x + \sqrt{1 - x^2}\right) dx + \int_1^0 (x - \sqrt{1 - x^2})\,dx$$

$$= \int_0^1 (x + \sqrt{1 - x^2} - x + \sqrt{1 - x^2})\,dx = 2\int_0^1 \sqrt{1 - x^2}\,dx$$

Now substitute $x = \sin\theta$ and finish it off.

$I =$ …………

41

$$\boxed{I = \frac{\pi}{2}}$$

Because

$$I = 2\int_0^1 \sqrt{1-x^2}\,dx \quad x = \sin\theta \quad \therefore\ dx = \cos\theta\,d\theta$$

$$\sqrt{1-x^2} = \cos\theta$$

Limits: $x = 0,\ \theta = 0;\ \ x = 1,\ \theta = \frac{\pi}{2}$

$$\therefore\ I = 2\int_0^{\pi/2} \cos^2\theta\,d\theta = \int_0^{\pi/2} (1+\cos 2\theta)\,d\theta$$

$$= \left[\theta + \frac{\sin 2\theta}{2}\right]_0^{\pi/2}$$

$$= \frac{\pi}{2}$$

Now let us extend this line of development a stage further.

42 Regions enclosed by closed curves

A region is said to be *simply connected* if a path joining A and B can be deformed to coincide with any other line joining A and B without going outside the region.

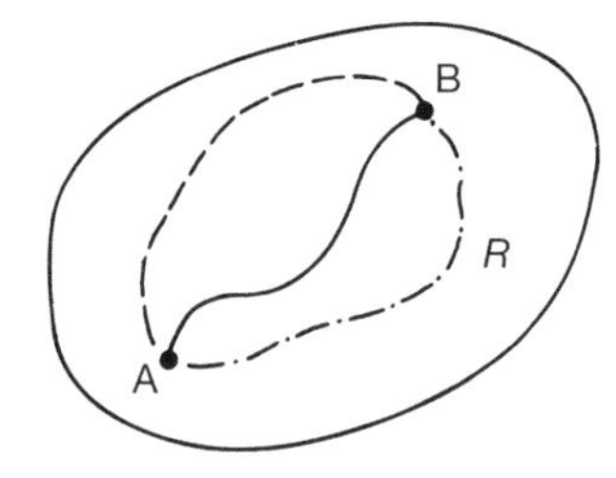

Another definition is that a region is simply connected if any closed path in the region can be contracted to a single point without leaving the region.

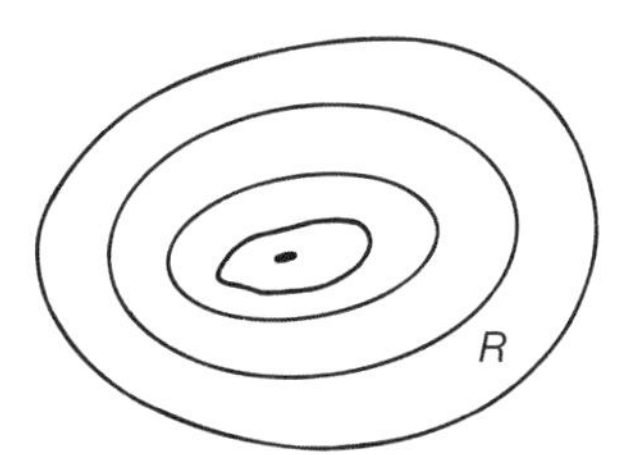

Clearly, this would not be satisfied in the case where the region R contains one or more 'holes'.

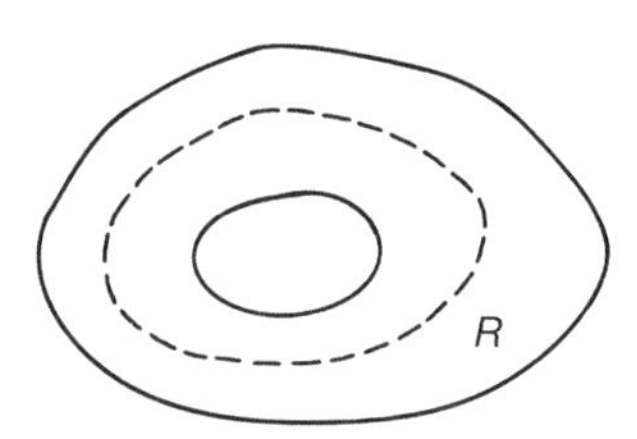

The closed curves involved in problems in this Program all relate to simply connected regions, so no difficulties will arise.

Line integrals round a closed curve

43

We have already introduced the symbol $\oint$ to indicate that an integral is to be evaluated round a closed curve in the positive (counter-clockwise) direction.

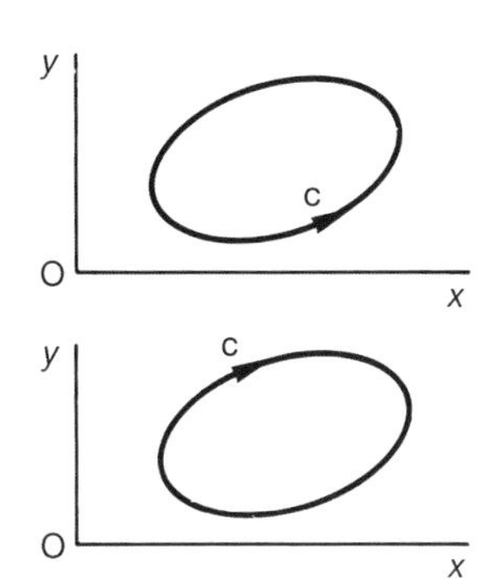

Positive direction (counter-clockwise) line integral denoted by $\oint$.

Negative direction (clockwise) line integral denoted by $-\oint$.

With a closed curve, the y-values on the path c cannot be single-valued. Therefore, we divide the path into two or more parts and treat each separately.

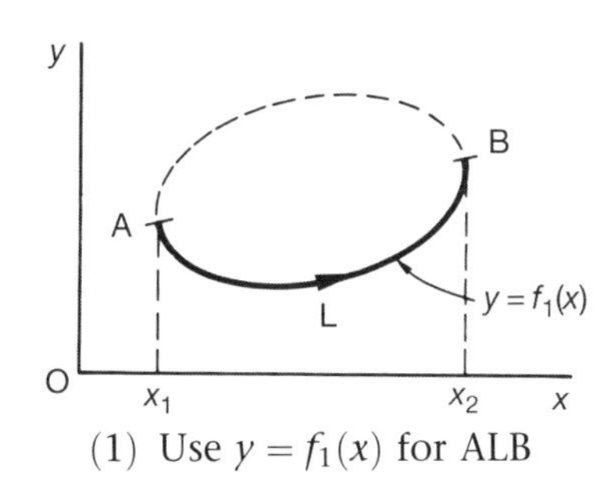

(1) Use $y = f_1(x)$ for ALB

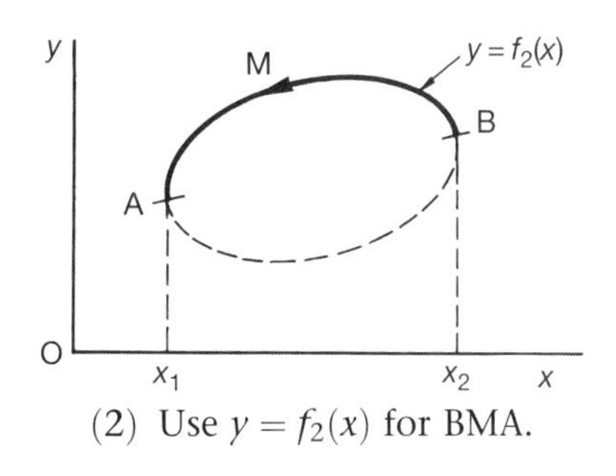

(2) Use $y = f_2(x)$ for BMA.

Unless specially required otherwise, we always proceed round the closed curve in a ....

44

counter-clockwise direction

Example 1

Evaluate the line integral $I = \oint_c (x^2\,dx - 2xy\,dy)$ where c comprises the three sides of the triangle joining O (0, 0), A (1, 0) and B (0, 1).

First draw the diagram and mark in c_1, c_2 and c_3, the proposed directions of integration. Do just that.

45

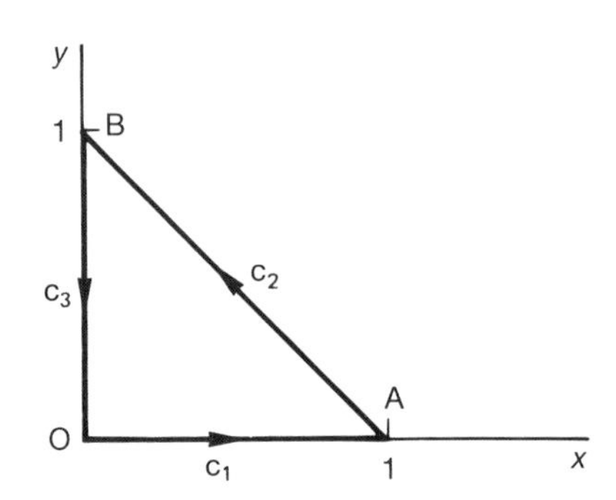

The three sections of the path of integration must be arranged in a counter-clockwise manner round the figure. Now we deal with each part separately.

(a) OA: c_1 is the line $y = 0 \quad \therefore \; dy = 0$.

Then $I = \oint (x^2 \, dx - 2xy \, dy)$ for this part becomes

$$I_1 = \int_0^1 x^2 \, dx = \left[\frac{x^3}{3}\right]_0^1 = \frac{1}{3} \qquad \therefore \; I_1 = \frac{1}{3}$$

(b) AB: c_2 is the line $y = 1 - x \quad \therefore \; dy = -dx$

$I_2 = \dots\dots\dots\dots$ (evaluate it)

46

$$\boxed{I_2 = -\tfrac{2}{3}}$$

Because c_2 is the line $y = 1 - x \quad \therefore \; dy = -dx$.

$$I_2 = \int_1^0 \{x^2 \, dx + 2x(1 - x) \, dx\} = \int_1^0 (x^2 + 2x - 2x^2) \, dx$$

$$= \int_1^0 (2x - x^2) dx = \left[x^2 - \frac{x^3}{3}\right]_1^0 = -\frac{2}{3} \qquad \therefore \; I_2 = -\frac{2}{3}$$

Note that counter-clockwise progression is obtained by arranging the limits in the appropriate order.

Now we have to determine I_3 for BO.

(c) BO: c_3 is the line $x = 0$

$I_3 = \dots\dots\dots\dots$

47

$$\boxed{I_3 = 0}$$

Because for c_3, $\quad x = 0 \quad \therefore \; dx = 0 \quad \therefore \; I_3 = \int 0 \, dy = 0 \quad \therefore \; I_3 = 0$

Finally, $I = I_1 + I_2 + I_3 = \frac{1}{3} - \frac{2}{3} + 0 = -\frac{1}{3} \qquad \therefore \; I = -\frac{1}{3}$

Let us work through another example.

Example 2

Evaluate $\oint_c y\,dx$ when c is the circle $x^2 + y^2 = 4$.

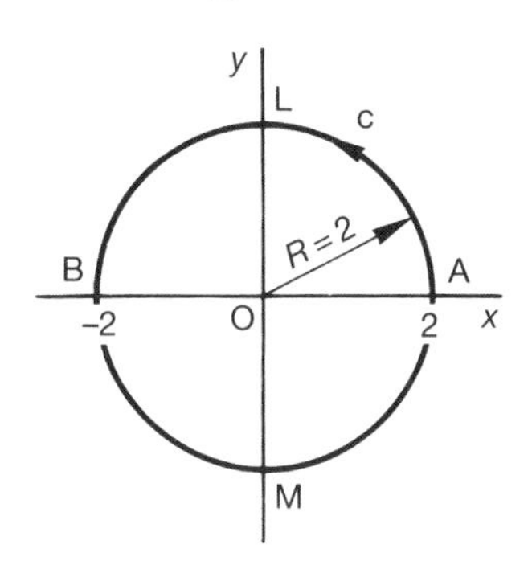

$$x^2 + y^2 = 4 \quad \therefore\; y = \pm\sqrt{4 - x^2}$$

y is thus not single-valued. Therefore use $y = \sqrt{4 - x^2}$ for ALB between $x = 2$ and $x = -2$ and $y = -\sqrt{4 - x^2}$ for BMA between $x = -2$ and $x = 2$.

$$\begin{aligned}\therefore\; I &= \int_2^{-2} \sqrt{4 - x^2}\,dx + \int_{-2}^{2} \{-\sqrt{4 - x^2}\}\,dx \\ &= 2\int_2^{-2} \sqrt{4 - x^2}\,dx = -2\int_{-2}^{2} \sqrt{4 - x^2}\,dx \\ &= -4\int_0^2 \sqrt{4 - x^2}\,dx.\end{aligned}$$

To evaluate this integral, substitute $x = 2\sin\theta$ and finish it off.

$I = \ldots\ldots\ldots\ldots$

48

$$\boxed{I = -4\pi}$$

Because

$$x = 2\sin\theta \quad \therefore\; dx = 2\cos\theta\,d\theta \quad \therefore\; \sqrt{4 - x^2} = 2\cos\theta$$

limits: $x = 0,\ \theta = 0;\ x = 2,\ \theta = \dfrac{\pi}{2}$

$$\begin{aligned}\therefore\; I &= -4\int_0^{\pi/2} 2\cos\theta\, 2\cos\theta\,d\theta = -16\int_0^{\pi/2} \cos^2\theta\,d\theta \\ &= -8\int_0^{\pi/2} (1 + \cos 2\theta)\,d\theta = -8\left[\theta + \frac{\sin 2\theta}{2}\right]_0^{\pi/2} = -4\pi\end{aligned}$$

Now for one more

Example 3

Evaluate $I = \oint_c \{xy\,dx + (1 + y^2)\,dy\}$ where c is the boundary of the rectangle joining A (1, 0), B (3, 0), C (3, 2) and D (1, 2).

First draw the diagram and insert c_1, c_2, c_3, c_4.

That gives $\ldots\ldots\ldots\ldots$

49

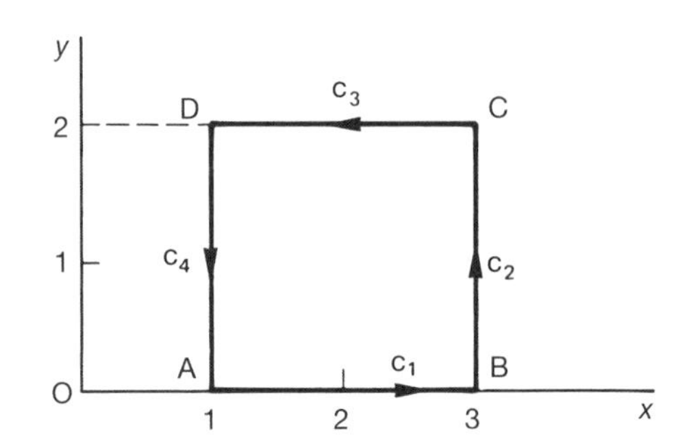

Now evaluate I_1 for AB; I_2 for BC; I_3 for CD; I_4 for DA; and finally I.

Complete the working and then check with the next frame

50

$$I_1 = 0; \quad I_2 = 4\tfrac{2}{3}; \quad I_3 = -8; \quad I_4 = -4\tfrac{2}{3}; \quad I = -8$$

Here is the complete working.

$$I = \oint_c \{xy\,dx + (1+y^2)\,dy\}$$

(a) AB: c_1 is $y = 0$ $\therefore$ $dy = 0$ $\therefore$ $I_1 = 0$

(b) BC: c_2 is $x = 3$ $\therefore$ $dx = 0$

$$\therefore\ I_2 = \int_0^2 (1+y^2)dy = \left[y + \frac{y^3}{3}\right]_0^2 = 4\tfrac{2}{3} \qquad \therefore\ I_2 = 4\tfrac{2}{3}$$

(c) CD: c_3 is $y = 2$ $\therefore$ $dy = 0$

$$\therefore\ I_3 = \int_3^1 2x\,dx = \left[x^2\right]_3^1 = -8 \qquad \therefore\ I_3 = -8$$

(d) DA: c_4 is $x = 1$ $\therefore$ $dx = 0$

$$\therefore\ I_4 = \int_2^0 (1+y^2)\,dy = \left[y + \frac{y^3}{3}\right]_2^0 = -4\tfrac{2}{3} \quad \therefore\ I_4 = -4\tfrac{2}{3}$$

Finally

$$I = I_1 + I_2 + I_3 + I_4$$
$$= 0 + 4\tfrac{2}{3} - 8 - 4\tfrac{2}{3} = -8 \qquad \therefore\ I = -8$$

Remember that, unless we are directed otherwise, we always proceed round the closed boundary in a counter-clockwise manner.

On now to the next piece of work

Line integral with respect to arc length

51

We have already established that

$$I=\int_{AB} F_t\,\mathrm{d}s=\int_{AB}\{P\,\mathrm{d}x+Q\,\mathrm{d}y\}$$

where F_t denoted the tangential force along the curve c at the sample point K (x, y).

The same kind of integral can, of course, relate to any function $f(x, y)$ which is a function of the position of a point on the stated curve, so that $I=\int_c f(x, y)\,\mathrm{d}s$.

This can readily be converted into an integral in terms of x. If δs is an element of arc length then

$$(\delta s)^2 \cong (\delta x)^2+(\delta y)^2 \text{ and so } \frac{\delta s}{\delta x}\cong\sqrt{1+\left(\frac{\delta y}{\delta x}\right)^2}.$$

In the limit as $\delta x \to 0$, $\dfrac{\mathrm{d}s}{\mathrm{d}x}=\sqrt{1+\left(\dfrac{\mathrm{d}y}{\mathrm{d}x}\right)^2}$. Therefore

$$I=\int_c f(x, y)\,\mathrm{d}s=\int_c f(x, y)\frac{\mathrm{d}s}{\mathrm{d}x}\mathrm{d}x$$

$$=\int_{x_1}^{x_2} f(x, y)\sqrt{1+\left(\frac{\mathrm{d}y}{\mathrm{d}x}\right)^2}\,\mathrm{d}x \qquad (1)$$

Example

Evaluate $I=\int_c (4x+3xy)\,\mathrm{d}s$ where c is the straight line joining O (0, 0) to A (1, 2).

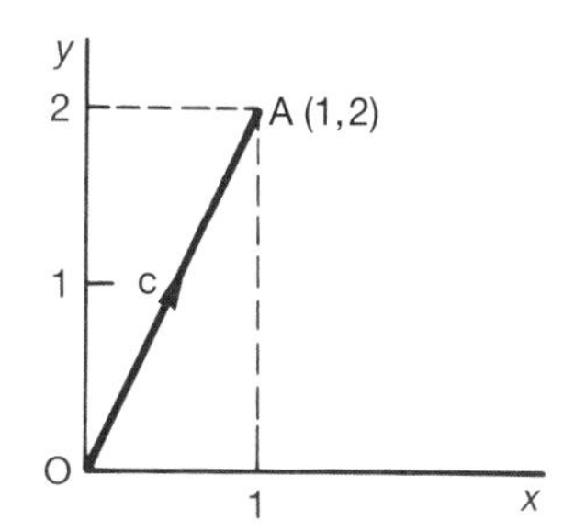

c is the line $y=2x \quad \therefore \quad \dfrac{\mathrm{d}y}{\mathrm{d}x}=2$

$$\therefore \quad \frac{\mathrm{d}s}{\mathrm{d}x}=\sqrt{1+\left(\frac{\mathrm{d}y}{\mathrm{d}x}\right)^2}=\sqrt{5}$$

$$\therefore \quad I=\int_{x=0}^{x=1}(4x+3xy)\,\mathrm{d}s=\int_0^1(4x+3xy)(\sqrt{5})\,\mathrm{d}x. \quad \text{But } y=2x$$

$$\therefore \quad I=\ldots\ldots\ldots\ldots$$

52

$$I = 4\sqrt{5}$$

Because

$$I = \int_0^1 (4x + 6x^2)(\sqrt{5})\,dx = 2\sqrt{5}\int_0^1 (2x + 3x^2)\,dx = 4\sqrt{5}$$

Try another.

The path length of the parabola defined by $y = x^2$ betwen the values $x = 0$ and $x = 2$ is given by the integral

$$I = \int_c ds = \ldots\ldots\ldots\ldots \quad \text{to 3 dp}$$

53

3.393 to 3 dp

Because

$$I = \int_c ds = \int_{x=0}^{2} \sqrt{1 + \left(\frac{dy}{dx}\right)}\,dx$$

$$= \int_{x=0}^{2} \sqrt{1 + 2x}\,dx$$

Let $u = 1 + 2x$ so that $du = 2dx$ and so

$$I = \int_{u=1}^{5} u^{1/2}\frac{du}{2}$$

$$= \frac{1}{2}\left[\frac{2}{3}u^{3/2}\right]_1^5$$

$$= \frac{1}{3}\left(125^{1/2} - 1\right)$$

$$= 3.393 \text{ to 3 dp}$$

54 Parametric equations

When x and y are expressed in parametric form, e.g. $x = f(t)$, $y = g(t)$, then

$$\frac{ds}{dt} = \sqrt{\left(\frac{dx}{dt}\right)^2 + \left(\frac{dy}{dt}\right)^2} \quad \therefore\ ds = \sqrt{\left(\frac{dx}{dt}\right)^2 + \left(\frac{dy}{dt}\right)^2}\,dt$$

and result (1) above becomes

$$I = \int_c f(x, y)\,ds = \int_{t_1}^{t_2} f(x, y)\sqrt{\left(\frac{dx}{dt}\right)^2 + \left(\frac{dy}{dt}\right)^2}\,dt \tag{2}$$

Make a note of results (1) and (2) for future use

55

Example

Evaluate $I = \oint_c 4xy\,ds$ where c is defined as the curve $x = \sin t$, $y = \cos t$ between $t = 0$ and $t = \dfrac{\pi}{4}$.

We have $x = \sin t \quad \therefore \dfrac{dx}{dt} = \cos t$

$y = \cos t \quad \therefore \dfrac{dy}{dt} = -\sin t$

$\therefore \dfrac{ds}{dt} = \ldots\ldots\ldots\ldots$

56

$$\frac{ds}{dt} = 1$$

Because

$$\frac{ds}{dt} = \sqrt{\left(\frac{dx}{dt}\right)^2 + \left(\frac{dy}{dt}\right)^2} = \sqrt{\cos^2 t + \sin^2 t} = 1$$

$$\therefore\ I = \int_{t_1}^{t_2} f(x, y)\sqrt{\left(\frac{dx}{dt}\right)^2 + \left(\frac{dy}{dt}\right)^2}\,dt = \int_0^{\pi/4} 4\sin t\,\cos t\,dt$$

$$= 2\int_0^{\pi/4} \sin 2t\,dt = -2\left[\frac{\cos 2t}{2}\right]_0^{\pi/4} = 1 \quad \therefore\ I = 1$$

Dependence of the line integral on the path of integration

We saw earlier in the Program that integration along two separate paths joining the same two end points does not necessarily give identical results.

With this in mind, let us investigate the following problem.

Example

Evaluate $I = \oint_c \{3x^2y^2\,dx + 2x^3y\,dy\}$ between O (0, 0) and A (2, 4)

(a) along c_1 i.e. $y = x^2$

(b) along c_2 i.e. $y = 2x$

(c) along c_3 i.e. $x = 0$ from (0, 0) to (0, 4) and $y = 4$ from (0, 4) to (2, 4).

Let us concentrate on section (a).

First we draw the figure and insert relevant information.

This gives …………

57

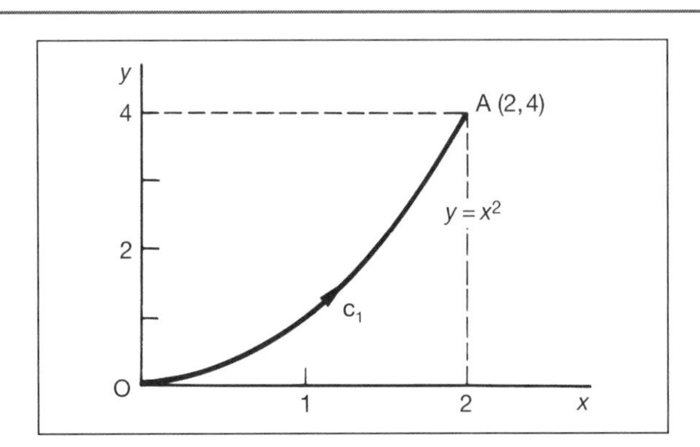

(a) $I = \int_c \{3x^2y^2\,dx + 2x^3y\,dy\}$

The path c_1 is $y = x^2$ $\quad\therefore\ dy = 2x\,dx$

$$\therefore\ I_1 = \int_0^2 \{3x^2x^4\,dx + 2x^3x^2 2x\,dx\} = \int_0^2 (3x^6 + 4x^4)\,dx$$

$$= \Big[x^7\Big]_0^2 = 128 \qquad \therefore\ I_1 = 128$$

(b) Here, the path of integration changes to c_2, i.e. $y = 2x$

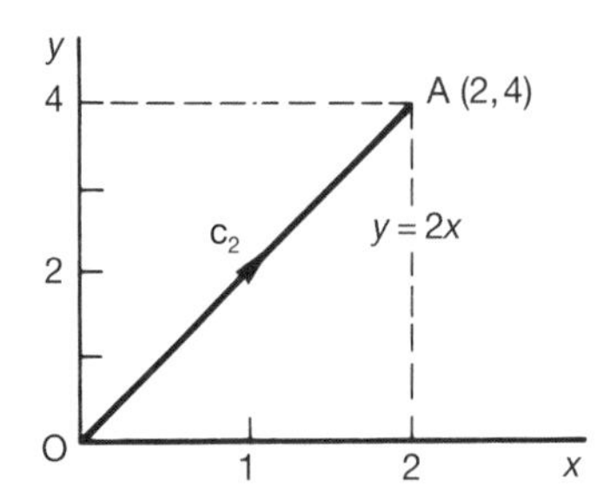

So, in this case,
$I_2 = \ldots\ldots\ldots\ldots$

58

$$\boxed{I_2 = 128}$$

Because with c_2, $y = 2x$ $\quad\therefore\ dy = 2\,dx$

$$\therefore\ I_2 = \int_0^2 \{3x^2\,4x^2\,dx + 2x^3\,2x2\,dx\} = \int_0^2 20x^4\,dx$$

$$= 4\Big[x^5\Big]_0^2 = 128 \qquad \therefore\ I_2 = 128$$

(c) In the third case, the path c_3 is split

$x = 0$ from (0, 0) to (0, 4)

$y = 4$ from (0, 4) to (2, 4)

Sketch the diagram and determine I_3.

$I_3 = \ldots\ldots\ldots\ldots$

59

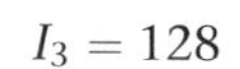

$I_3 = 128$

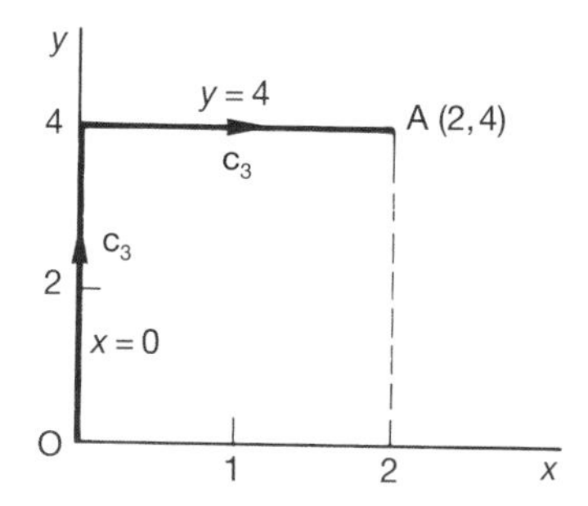

From (0, 0) to (0, 4) $\quad x = 0 \quad \therefore \; dx = 0 \quad \therefore \; I_{3a} = 0$

From (0, 4) to (2, 4) $\quad y = 4 \quad \therefore \; dy = 0 \quad \therefore \; I_{3b} = 48\int_0^2 x^2 dx = 128$

$$\therefore \; I_3 = 128$$

On to the next frame

60

In the example we have just worked through, we took three different paths and in each case, the line integral produced the same result. It appears, therefore, that in this case, the value of the integral is independent of the path of integration taken.

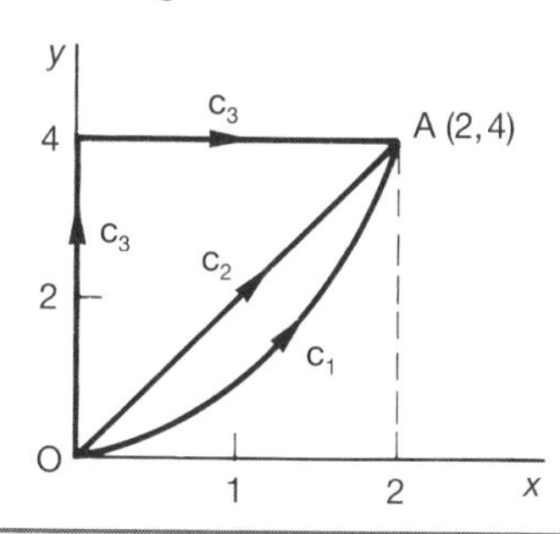

How then does this integral perhaps differ from those of previous cases?

Let us investigate

61

We have been dealing with $I = \int_c \{3x^2y^2\,dx + 2x^3y\,dy\}$

On reflection, we see that the integrand $3x^2y^2\,dx + 2x^3y\,dy$ is of the form $P\,dx + Q\,dy$ which we have met before and that it is, in fact, an *exact differential* of the function $z = x^3y^2$, because

$$\frac{\partial z}{\partial x} = 3x^2y^2 \quad \text{and} \quad \frac{\partial z}{\partial y} = 2x^3y$$

Provided P, Q and their first partial derivatives are finite and continuous at all points inside and on any closed curve, this always happens. If the integrand of the given integral is seen to be an *exact differential*, then the value of the line integral is *independent of the path taken and depends only on the coordinates of the two end points.*

Make a note of this. It is important

62

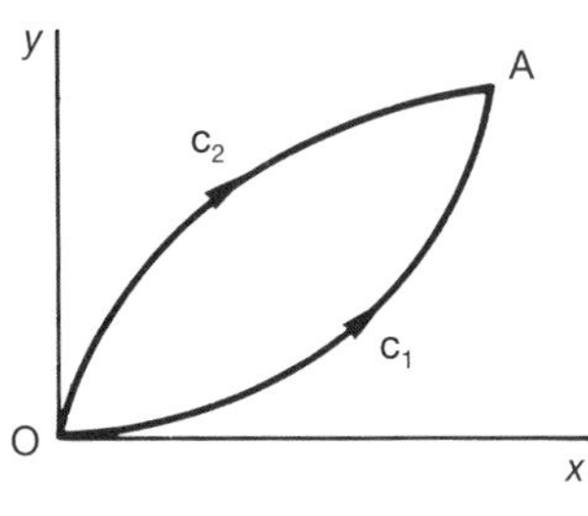

If $I = \int_c \{P\,dx + Q\,dy\}$ and $(P\,dx + Q\,dy)$ is an exact differential, then

$$I_{c_1} = I_{c_2}$$

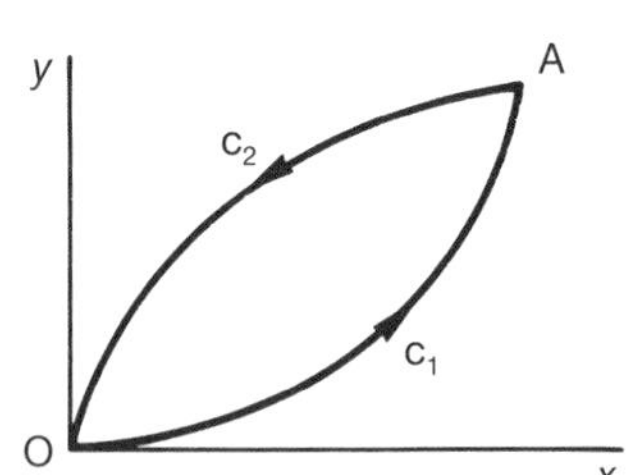

If we reverse the direction of c_2, then

$$I_{c_1} = -I_{c_2}$$

i.e. $I_{c_1} + I_{c_2} = 0$

Hence, *if $(P\,dx + Q\,dy)$ is an exact differential, then the integration taken round a closed curve is zero.*

$\therefore$ If $(P\,dx + Q\,dy)$ is an exact differential, $\oint (P\,dx + Q\,dy) = 0$

63

Example 1

Evaluate $I = \int_c \{3y\,dx + (3x + 2y)\,dy\}$ from A (1, 2) to B (3, 5).

No path is given, so the integrand is probably an exact differential of some function $z = f(x, y)$. In fact $\dfrac{\partial P}{\partial y} = 3 = \dfrac{\partial Q}{\partial x}$.

We have already dealt with the integration of exact differentials, so there is no difficulty. Compare with $I = \int_c \{P\,dx + Q\,dy\}$.

$$P = \frac{\partial z}{\partial x} = 3y \qquad \therefore\ z = \int 3y\,dx = 3xy + f(y) \qquad (1)$$

$$Q = \frac{\partial z}{\partial y} = 3x + 2y \quad \therefore\ z = \int (3x + 2y)\,dy = 3xy + y^2 + F(x) \qquad (2)$$

For (1) and (2) to agree

$f(y) = \dots\dots\dots\dots$ and $F(x) = \dots\dots\dots\dots$

64

$$f(y) = y^2; \quad F(x) = 0$$

Hence $z = 3xy + y^2$

$$\therefore \; I = \int_c \{3y\,dx + (3x + 2y)\,dy\} = \int_{(1,\,2)}^{(3,\,5)} d(3xy + y^2)$$

$$= \Big[3xy + y^2\Big]_{(1,\,2)}^{(3,\,5)}$$

$$= (45 + 25) - (6 + 4)$$

$$= 60$$

Example 2

Evaluate $I = \int_c \{(x^2 + ye^x)\,dx + (e^x + y)\,dy\}$ between A (0, 1) and B (1, 2).

As before, compare with $\int_c \{P\,dx + Q\,dy\}$.

$$P = \frac{\partial z}{\partial x} = x^2 + ye^x \quad \therefore \; z = \ldots\ldots\ldots\ldots$$

$$Q = \frac{\partial z}{\partial y} = e^x + y \quad \therefore \; z = \ldots\ldots\ldots\ldots$$

Continue the working and complete the evaluation.

When you have finished, check the result with the next frame

65

$$z = \frac{x^3}{3} + ye^x + f(y)$$

$$z = ye^x + \frac{y^2}{2} + F(x)$$

For these expressions to agree, $\quad f(y) = \dfrac{y^2}{2}; \quad F(x) = \dfrac{x^3}{3}$

$$\text{Then } I = \left[\frac{x^3}{3} + ye^x + \frac{y^2}{2}\right]_{(0,\,1)}^{(1,\,2)}$$

$$= \frac{5}{6} + 2e$$

So the main points are that, if $(P\,dx + Q\,dy)$ is an exact differential

(a) $I = \int_c (P\,dx + Q\,dy)$ is independent of the path of integration

(b) $I = \int_c (P\,dx + Q\,dy)$ is zero when c is a closed curve.

On to the next frame

66 Exact differentials in three independent variables

A line integral in space naturally involves three independent variables, but the method is very much like that for two independent variables.

$dw = P dx + Q dy + R dz$ is an exact differential of $w = f(x, y, z)$

$$\text{if} \quad \frac{\partial P}{\partial y} = \frac{\partial Q}{\partial x}; \quad \frac{\partial P}{\partial z} = \frac{\partial R}{\partial x}; \quad \frac{\partial R}{\partial y} = \frac{\partial Q}{\partial z}$$

If the test is successful, then

(a) $\int_c (P\,dx + Q\,dy + R\,dz)$ is independent of the path of integration

(b) $\oint_c (P\,dx + Q\,dy + R\,dz)$ is zero when c is a closed curve.

Example

Verify that $dw = (3x^2yz + 6x)dx + (x^3z - 8y)dy + (x^3y + 1)dz$ is an exact differential and hence evaluate $\int_c dw$ from A (1, 2, 4) to B (2, 1, 3).

First check that dw is an exact differential by finding the partial derivatives above, when $P = 3x^2yz + 6x$; $Q = x^3z - 8y$; and $R = x^3y + 1$.

We have ....

67

$$\frac{\partial P}{\partial y} = 3x^2z; \quad \frac{\partial Q}{\partial x} = 3x^2z \quad \therefore \frac{\partial P}{\partial y} = \frac{\partial Q}{\partial x}$$

$$\frac{\partial P}{\partial z} = 3x^2y; \quad \frac{\partial R}{\partial x} = 3x^2y \quad \therefore \frac{\partial P}{\partial z} = \frac{\partial R}{\partial x}$$

$$\frac{\partial R}{\partial y} = x^3; \quad \frac{\partial Q}{\partial z} = x^3 \quad \therefore \frac{\partial R}{\partial y} = \frac{\partial Q}{\partial z}$$

$\therefore$ dw is an exact differential

Now to find w. $\quad P = \frac{\partial z}{\partial x}; \quad Q = \frac{\partial z}{\partial y}; \quad R = \frac{\partial w}{\partial z}$

$$\therefore \frac{\partial w}{\partial x} = 3x^2yz + 6x \quad \therefore w = \int (3x^2yz + 6x)dx$$
$$= x^3yz + 3x^2 + f(y,z)$$

$$\frac{\partial w}{\partial y} = x^3z - 8y \quad \therefore w = \int (x^3z - 8y)\,dy$$
$$= x^3zy - 4y^2 + F(x,z)$$

$$\frac{\partial w}{\partial z} = x^3y + 1 \quad \therefore w = \int (x^3y + 1)\,dz$$
$$= x^3yz + z + g(x, y)$$

For these three expressions for z to agree

$f(y,z) = ...;\quad F(x,z) = ...;\quad g(x, y) = ...$

68

$$f(y, z) = -4y^2; \quad F(x, z) = z; \quad g(x, y) = 3x^2$$

$$\therefore \; w = x^3yz + 3x^2 - 4y^2 + z$$

$$\therefore \; I = \left[x^3yz + 3x^2 - 4y^2 + z \right]_{(1,2,4)}^{(2,1,3)}$$

$$= \dots\dots\dots\dots$$

69

$$I = 36$$

Because

$$I = \left[x^3yz + 3x^2 - 4y^2 + z \right]_{(1,2,4)}^{(2,1,3)}$$

$$= (24 + 12 - 4 + 3) - (8 + 3 - 16 + 4) = 36$$

The extension to line integrals in space is thus quite straightforward.

Finally, we have a theorem that can be very helpful on occasions and which links up with the work we have been doing.

It is important, so let us start a new section

Green's theorem

70

Let P and Q be two functions of x and y that are, along with their first partial derivatives, finite and continuous inside and on the boundary c of a region R in the x–y plane.

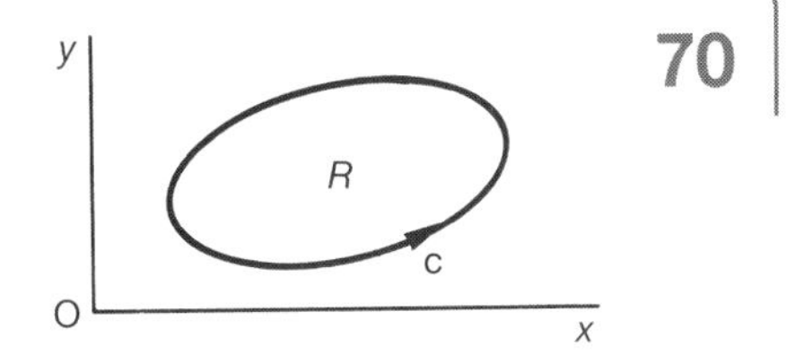

If the first partial derivatives are continuous within the region and on the boundary, then Green's theorem states that

$$\int_R \int \left(\frac{\partial P}{\partial y} - \frac{\partial Q}{\partial x} \right) \mathrm{d}x\,\mathrm{d}y = -\oint_c (P\,\mathrm{d}x + Q\,\mathrm{d}y)$$

That is, a double integral over the plane region R can be transformed into a line integral over the boundary c of the region – and the action is reversible.

Let us see how it works.

Example 1

Evaluate $I = \oint_c \{(2x - y)\,dx + (2y + x)\,dy\}$ around the boundary c of the ellipse $x^2 + 9y^2 = 16$.

The integral is of the form $I = \oint_c \{P\,dx + Q\,dy\}$ where

$$P = 2x - y \quad \therefore \quad \frac{\partial P}{\partial y} = -1$$

$$\text{and } Q = 2y + x \quad \therefore \quad \frac{\partial Q}{\partial x} = 1.$$

$$\therefore \; I = -\int_R\int \left(\frac{\partial P}{\partial y} - \frac{\partial Q}{\partial x}\right) dx\,dy$$

$$= -\int_R\int (-1 - 1)\,dx\,dy$$

$$= 2\int_R\int dx\,dy$$

But $\int_R\int dx\,dy$ over any closed region gives ....

71

the area of the figure

In this case, then, $I = 2A$ where A is the area of the ellipse

$$x^2 + 9y^2 = 16 \quad \text{i.e.} \quad \frac{x^2}{16} + \frac{9y^2}{16} = 1$$

$$\therefore \; a = 4; \; b = \frac{4}{3}$$

$$\therefore \; A = \pi ab = \frac{16\pi}{3}$$

$$\therefore \; I = 2A = \frac{32\pi}{3}$$

To demonstrate the advantage of Green's theorem, let us work through the next example (a) by the method of line integrals, and (b) by applying Green's theorem.

Example 2

Evaluate $I = \oint_c \{(2x + y)\,dx + (3x - 2y)\,dy\}$ taken in counter-clockwise manner round the triangle with vertices at O (0, 0), A (1, 0) and B (1, 2).

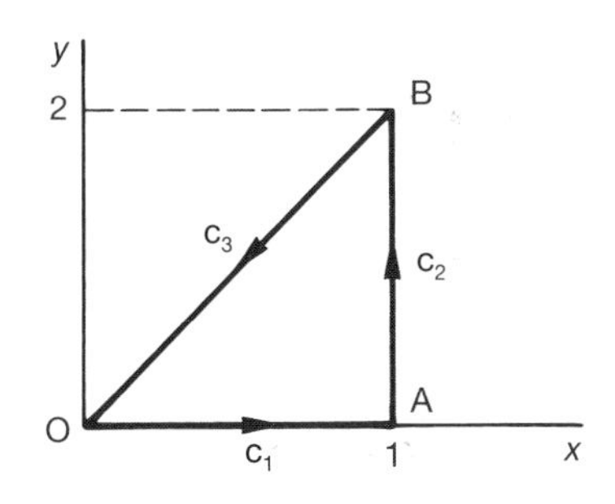

$$I = \oint_c \{(2x + y)\,dx + (3x - 2y)\,dy\}$$

(a) *By the method of line integrals*

There are clearly three stages with c_1, c_2, c_3. Work through the complete evaluation to determine the value of I. It will be good review.

When you have finished, check the result with the solution in the next frame

---

72

$$\boxed{I = 2}$$

(a) (1) c_1 is $y = 0 \quad \therefore\ dy = 0$

$$\therefore\ I_1 = \int_0^1 2x\,dx = \left[x^2\right]_0^1 = 1 \quad \therefore\ I_1 = 1$$

(2) c_2 is $x = 1 \quad \therefore\ dx = 0$

$$\therefore\ I_2 = \int_0^2 (3 - 2y)\,dy = \left[3y - y^2\right]_0^2 = 2 \quad \therefore\ I_2 = 2$$

(3) c_3 is $y = 2x \quad \therefore\ dy = 2\,dx$

$$\therefore\ I_3 = \int_1^0 \{4x\,dx + (3x - 4x)2\,dx\}$$

$$= \int_1^0 2x\,dx = \left[x^2\right]_1^0 = -1 \qquad \therefore\ I_3 = -1$$

$$I = I_1 + I_2 + I_3 = 1 + 2 + (-1) = 2 \quad \therefore\ I = 2$$

Now we will do the same problem by applying Green's theorem, so move on

73

(b) *By Green's theorem*

$$I = \oint_c \{(2x + y)\,dx + (3x - 2y)\,dy\}$$

$$P = 2x + y \quad \therefore \quad \frac{\partial P}{\partial y} = 1; \quad Q = 3x - 2y \quad \therefore \quad \frac{\partial Q}{\partial x} = 3$$

$$I = -\int_R\int\left(\frac{\partial P}{\partial y} - \frac{\partial Q}{\partial x}\right)dx\,dy$$

Finish it off. $I = \ldots\ldots\ldots\ldots$

74

$$\boxed{I = 2}$$

Because

$$I = -\int_R\int(1 - 3)\,dx\,dy$$

$$= 2\int_R\int dx\,dy = 2A$$

$$= 2 \times \text{ the area of the triangle}$$

$$= 2 \times 1 = 2 \quad \therefore \; I = 2$$

Application of Green's theorem is not always the quickest method. It is useful, however, to have both methods available. If you have not already done so, make a note of Green's theorem.

$$\int_R\int\left(\frac{\partial P}{\partial y} - \frac{\partial Q}{\partial x}\right)dx\,dy = -\oint_c (P\,dx + Q\,dy)$$

75

Example 3

Evaluate the line integral $I = \oint_c \{xy\,dx + (2x - y)\,dy\}$ round the region bounded by the curves $y = x^2$ and $x = y^2$ by the use of Green's theorem.

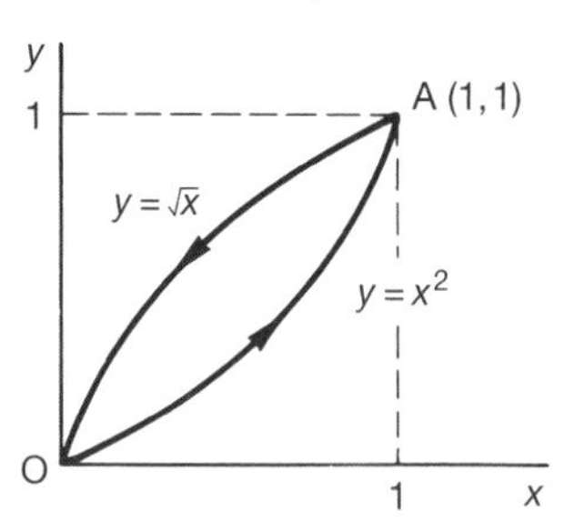

Points of intersection are O (0, 0) and A (1, 1). P and Q are known, so there is no difficulty.

Complete the working.

$I = \ldots\ldots\ldots\ldots$

76

$$I = \frac{31}{60}$$

Here is the working.

$$I = \oint_c \{xy\,dx + (2x - y)\,dy\}$$

$$\oint_c \{P\,dx + Q\,dy\} = -\int_R\int \left(\frac{\partial P}{\partial y} - \frac{\partial Q}{\partial x}\right) dx\,dy$$

$$P = xy \quad \therefore \quad \frac{\partial P}{\partial y} = x; \quad Q = 2x - y \quad \therefore \quad \frac{\partial Q}{\partial x} = 2$$

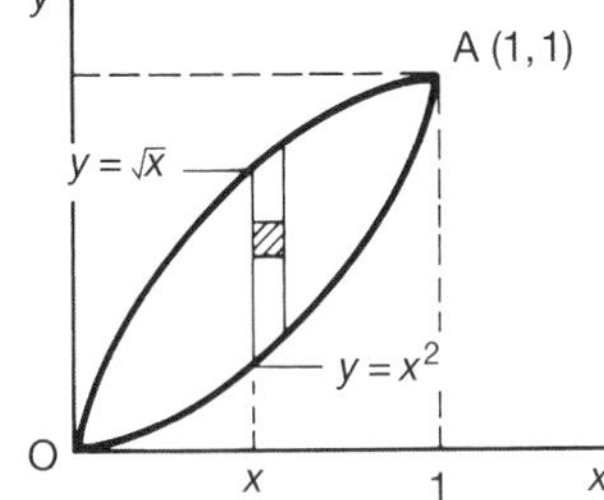

$$I = -\int_R\int (x - 2)\,dx\,dy$$

$$= -\int_0^1 \int_{y=x^2}^{y=\sqrt{x}} (x - 2)\,dy\,dx$$

$$= -\int_0^1 (x - 2)\Big[y\Big]_{x^2}^{\sqrt{x}} dx$$

$$\therefore \; I = -\int_0^1 (x - 2)(\sqrt{x} - x^2)\,dx$$

$$= -\int_0^1 (x^{3/2} - x^3 - 2x^{1/2} + 2x^2)\,dx$$

$$= -\left[\frac{2}{5}x^{5/2} - \frac{1}{4}x^4 - \frac{4}{3}x^{3/2} + \frac{2}{3}x^3\right]_0^1 = \frac{31}{60}$$

Before we finally leave this section of the work, there is one more result to note.

In the special case when $P = y$ and $Q = -x$

$$\frac{\partial P}{\partial y} = 1 \quad \text{and} \quad \frac{\partial Q}{\partial x} = -1$$

Green's theorem then states

$$\int_R\int \{1 - (-1)\}\,dx\,dy = -\oint_c (P dx + Q\,dy)$$

i.e.
$$2\int_R\int dx\,dy = -\oint_c (y\,dx - x\,dy)$$

$$= \oint_c (x\,dy - y\,dx)$$

Therefore, the area of the closed region

$$A = \int_R\int dx\,dy = \frac{1}{2}\oint_c (x\,dy - y\,dx)$$

Note this result in your record book. Then let us see an example

77

Example 1

Determine the area of the figure enclosed by $y = 3x^2$ and $y = 6x$.

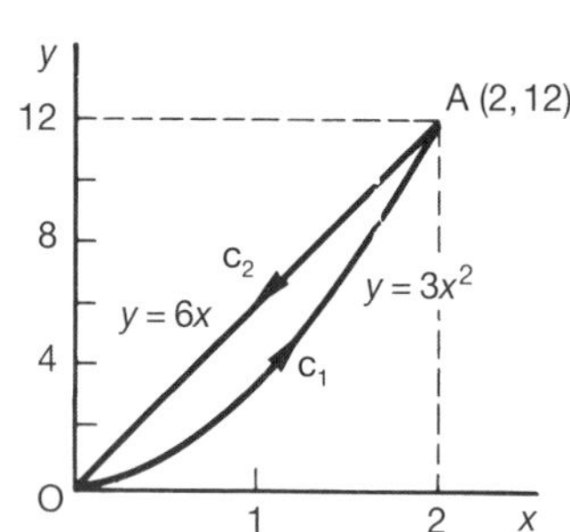

Points of intersection:

$$3x^2 = 6x \quad \therefore \; x = 0 \text{ or } 2$$

$$\text{Area } A = \frac{1}{2}\oint_c (x\,dy - y\,dx)$$

We evaluate the integral in two parts, i.e. OA along c_1
and AO along c_2

$$2A = \int_{c_1 \text{ (along OA)}} (x\,dy - y\,dx) + \int_{c_2 \text{ (along AO)}} (x\,dy - y\,dx) = I_1 + I_2$$

I_1: c_1 is $y = 3x^2 \quad \therefore \; dy = 6x\,dx$

$$\therefore \; I_1 = \int_0^2 (6x^2\,dx - 3x^2\,dx) = \int_0^2 3x^2\,dx = \left[x^3\right]_0^2 = 8$$

$$\therefore \; I_1 = 8$$

Similarly, $I_2 = \ldots\ldots\ldots\ldots$

78

$$\boxed{I_2 = 0}$$

Because

c_2 is $y = 6x \quad \therefore \; dy = 6\,dx$

$$\therefore \; I_2 = \int_2^0 (6x\,dx - 6x\,dx) = 0 \quad \therefore \; I_2 = 0$$

$$\therefore \; I = I_1 + I_2 = 8 + 0 = 8 \quad \therefore \; A = 4 \text{ square units}$$

Finally, here is one for you to do entirely on your own.

Example 2

Determine the area bounded by the curves $y = 2x^3$, $y = x^3 + 1$ and the axis $x = 0$ for $x \geq 0$.

Complete the working and see if you agree with the working in the next frame

79

Here it is.

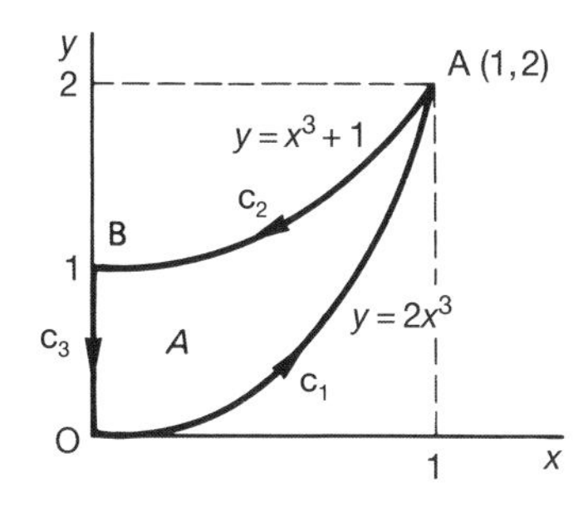

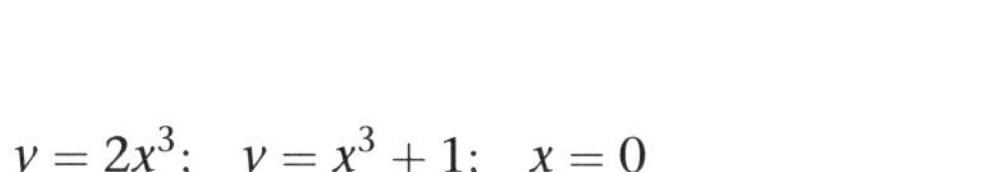

$y = 2x^3; \quad y = x^3 + 1; \quad x = 0$

Point of intersection

$$2x^3 = x^3 + 1 \quad \therefore \; x^3 = 1 \quad \therefore \; x = 1$$

$$\text{Area} \quad A = \frac{1}{2}\oint_c (x\,dy - y\,dx)$$

$$\therefore \; 2A = \oint_c (x\,dy - y\,dx)$$

(a) OA: c_1 is $y = 2x^3 \quad \therefore \; dy = 6x^2\,dx$

$$\therefore \; I_1 = \int_{c_1} (x\,dy - y\,dx) = \int_0^1 (6x^3\,dx - 2x^3\,dx)$$

$$= \int_0^1 4x^3\,dx = \Big[x^4\Big]_0^1 = 1 \qquad \therefore \; I_1 = 1$$

(b) AB: c_2 is $y = x^3 + 1 \quad \therefore \; dy = 3x^2\,dx$

$$\therefore \; I_2 = \int_1^0 \{3x^3\,dx - (x^3 + 1)\,dx\} = \int_1^0 (2x^3 - 1)\,dx$$

$$= \left[\frac{x^4}{2} - x\right]_1^0 = -(\tfrac{1}{2} - 1) = \tfrac{1}{2} \qquad \therefore \; I_2 = \tfrac{1}{2}$$

(c) BO: c_3 is $x = 0 \quad \therefore \; dx = 0$

$$I_3 = \int_{y=1}^{y=0} (x\,dy - y\,dx) = 0 \qquad \therefore \; I_3 = 0$$

$\therefore \; 2A = I = I_1 + I_2 + I_3 = 1 + \tfrac{1}{2} + 0 = 1\tfrac{1}{2}$

$\therefore \; A = \tfrac{3}{4}$ square units

And that brings this Program to an end. We have covered some important topics, so check down the **Summary** and the **Can You?** checklist that follow and review any part of the text if necessary, before working through the **Test exercise**. The **Further problems** provide an opportunity for additional practice.

Summary

80

1 *Differentials* dy and dx

(a)

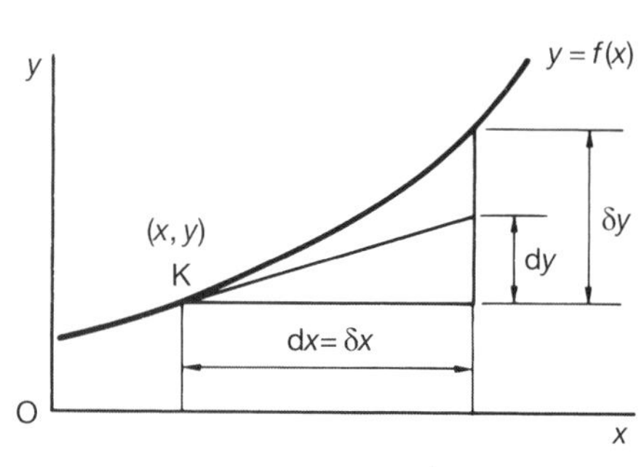

$$dy = f'(x)\,dx$$

(b) If $z = f(x, y)$, $\quad dz = \dfrac{\partial z}{\partial x}dx + \dfrac{\partial z}{\partial y}dy$

If $z = f(x, y, w)$, $\quad dz = \dfrac{\partial z}{\partial x}dx + \dfrac{\partial z}{\partial y}dy + \dfrac{\partial z}{\partial w}dw.$

(c) $dz = P\,dx + Q\,dy$, where P and Q are functions of x and y, is an *exact differential* if $\dfrac{\partial P}{\partial y} = \dfrac{\partial Q}{\partial x}$.

2 *Line integrals* – definition

$$I = \int_c f(x, y)\,ds = \int_c (P\,dx + Q\,dy)$$

3 *Properties of line integrals*

(a) Sign of line integral is reversed when the direction of integration along the path is reversed.

(b) Path of integration parallel to y-axis, $dx = 0$ $\therefore\ I_c = \int_c Q\,dy.$

Path of integration parallel to x-axis, $dy = 0$ $\therefore\ I_c = \int_c P\,dx.$

(c) The y-values on the path of integration must be continuous and single-valued.

4 *Line of integral round a closed curve* $\oint$

Positive direction $\oint$ counter-clockwise

Negative direction $\oint$ clockwise, i.e. $\oint = -\oint$.

5 *Line integral related to arc length*

$$I = \int_{AB} F\,ds = \int_{AB} (P\,dx + Q\,dy)$$

$$= \int_{x_1}^{x_2} f(x, y)\sqrt{1 + \left(\frac{dy}{dx}\right)^2}\,dx$$

With parametric equations, x and y in terms of t,

$$I = \int_c f(x, y)\,ds = \int_{t_1}^{t_2} f(x, y)\sqrt{\left(\frac{dx}{dt}\right)^2 + \left(\frac{dy}{dt}\right)^2}\,dt$$

6 *Dependence of line integral on path of integration*

In general, the value of the line integral depends on the particular path of integration.

7 *Exact differential*

If $P\,dx + Q\,dy$ is an exact differential where P, Q and their first derivatives are finite and continuous inside the simply connected region R

(a) $\dfrac{\partial P}{\partial y} = \dfrac{\partial Q}{\partial x}$

(b) $I = \int_c (P\,dx + Q\,dy)$ is independent of the path of integration where c lies entirely within R

(c) $I = \oint_c (P\,dx + Q\,dy)$ is zero when c is a closed curve lying entirely within R.

8 *Exact differentials in three variables*

If $P\,dx + Q\,dy + R\,dz$ is an exact differential where P, Q, R and their first partial derivatives are finite and continuous inside a simply connected region containing path c

(a) $\dfrac{\partial P}{\partial y} = \dfrac{\partial Q}{\partial x}$; $\dfrac{\partial P}{\partial z} = \dfrac{\partial R}{\partial x}$; $\dfrac{\partial R}{\partial y} = \dfrac{\partial Q}{\partial z}$

(b) $\int_c (P\,dx + Q\,dy + R\,dz)$ is independent of the path of integration

(c) $\oint_c (P\,dx + Q\,dy + R\,dz)$ is zero when c is a closed curve.

9 *Green's theorem*

$$\oint_c (P\,dx + Q\,dy) = -\int_R\!\!\int \left\{\frac{\partial P}{\partial y} - \frac{\partial Q}{\partial x}\right\} dx\,dy$$

and, for a simple closed curve

$$\oint_c (x\,dy - y\,dx) = 2\int_R\!\!\int dx\,dy = 2A$$

where A is the area of the enclosed figure.

Can You?

81 Checklist 7

Check this list before and after you try the end of Program test.

On a scale of 1 to 5 how confident are you that you can: **Frames**

- Evaluate double and triple integrals and apply them to the determination of the areas of plane figures and the volumes of solids? 1 to 10
  Yes ☐ ☐ ☐ ☐ ☐ *No*
- Understand the role of the differential of a function of two or more real variables? 11 to 13
  Yes ☐ ☐ ☐ ☐ ☐ *No*
- Determine exact differentials in two real variables and their integrals? 14 to 19
  Yes ☐ ☐ ☐ ☐ ☐ *No*
- Evaluate the area enclosed by a closed curve by contour integration? 20 to 26
  Yes ☐ ☐ ☐ ☐ ☐ *No*
- Evaluate line integrals and appreciate their properties? 27 to 41
  Yes ☐ ☐ ☐ ☐ ☐ *No*
- Evaluate line integrals around closed curves within a simply connected region? 42 to 50
  Yes ☐ ☐ ☐ ☐ ☐ *No*
- Link line integrals to integrals along the x-axis? 51 to 53
  Yes ☐ ☐ ☐ ☐ ☐ *No*
- Link line integrals to integrals along a contour given in parametric form? 54 to 56
  Yes ☐ ☐ ☐ ☐ ☐ *No*
- Discuss the dependence of a line integral between two points on the path of integration? 56 to 65
  Yes ☐ ☐ ☐ ☐ ☐ *No*
- Determine exact differentials in three real variables and their integrals? 66 to 69
  Yes ☐ ☐ ☐ ☐ ☐ *No*
- Demonstrate the validity and use of Green's theorem? 70 to 79
  Yes ☐ ☐ ☐ ☐ ☐ *No*

Test exercise 7

82

1 Determine the differential dz of each of the following.

(a) $z = x^4 \cos 3y$; (b) $z = e^{2y} \sin 4x$; (c) $z = x^2yw^3$.

2 Determine which of the following are exact differentials and integrate where appropriate to determine z.

(a) $\mathrm{d}z = (3x^2y^4 + 8x)\,\mathrm{d}x + (4x^3y^3 - 15y^2)\,\mathrm{d}y$

(b) $\mathrm{d}z = (2x \cos 4y - 6 \sin 3x)\mathrm{d}x - 4(x^2 \sin 4y - 2y)\,\mathrm{d}y$

(c) $\mathrm{d}z = 3e^{3x}(1 - y)\,\mathrm{d}x + (e^{3x} + 3y^2)\,\mathrm{d}y$.

3 Calculate the area of the triangle with vertices at O (0, 0), A (4, 2) and B (1, 5).

4 Evaluate the following.

(a) $I = \int_c \{(x^2 - 3y)\,\mathrm{d}x + xy^2\,\mathrm{d}y\}$ from A (1, 2) to B (2, 8) along the curve $y = 2x^2$.

(b) $I = \int_c (2x + y)\,\mathrm{d}x$ from A (0, 1) to B (0, − 1) along the semicircle $x^2 + y^2 = 1$ for $x \geq 0$.

(c) $I = \oint_c \{(1 + xy)\,\mathrm{d}x + (1 + x^2)\,\mathrm{d}y\}$ where c is the boundary of the rectangle joining A (1, 0), B (4, 0), C (4, 3) and D (1, 3).

(d) $I = \int_c 2xy\,\mathrm{d}s$ where c is defined by the parametric equations $x = 4\cos\theta$, $y = 4\sin\theta$ between $\theta = 0$ and $\theta = \dfrac{\pi}{3}$.

(e) $I = \int_c \{(8xy + y^3)\,\mathrm{d}x + (4x^2 + 3xy^2)\,\mathrm{d}y\}$ from A(1, 3) to B(2, 1).

(f) $I = \oint_c \{(3x + y)\,\mathrm{d}x + (y - 2x)\,\mathrm{d}y\}$ round the boundary of the ellipse $x^2 + 4y^2 = 36$.

5 Apply Green's theorem to determine the area of the plane figure bounded by the curves $y = x^3$ and $y = \sqrt{x}$.

6 Verify that $\mathrm{d}w = (2xyz + 2z - y^2)\mathrm{d}x + (x^2z - 2yx)\mathrm{d}y + (x^2y + 2x)\mathrm{d}z$ is an exact differential and find the value of

$$\int_c \mathrm{d}w \text{ where}$$

(a) c is the straight line joining (0, 0, 0) to (1, 1, 1)

(b) c is the curve of intersection of the unit sphere centered on the origin and the plane $x + y + z = 1$.

Further problems 7

83

1 Show that $I = \int_c \{xy^2w^2\,dx + x^2yw^2\,dy + x^2y^2w\,dw\}$ is independent of the path of integration c and evaluate the integral from A (1, 3, 2) to B (2, 4, 1).

2 Determine whether $dz = 3x^2(x^2 + y^2)\,dx + 2y(x^3 + y^4)\,dy$ is an exact differential. If so, determine z and hence evaluate $\int_c dz$ from A (1, 2) to B (2, 1).

3 Evaluate the line integral $I = \oint_c \left\{\dfrac{x\,dy - y\,dx}{x^2 + y^2 + 4}\right\}$ where c is the boundary of the segment formed by the arc of the circle $x^2 + y^2 = 4$ and the chord $y = 2 - x$ for $x \geq 0$.

4 Show that

$$I = \int_c \{(3x^2 \sin y + 2\sin 2x + y^3)\,dx + (x^3 \cos y + 3xy^2)\,dy\}$$

is independent of the path of integration and evaluate it from A (0,0) to B $\left(\dfrac{\pi}{2}, \pi\right)$.

5 Evaluate the integral $I = \int_c xy\,ds$ where c is defined by the parametric equations $x = \cos^3 t$, $y = \sin^3 t$ from $t = 0$ to $t = \dfrac{\pi}{2}$.

6 Verify that $dz = \dfrac{x\,dx}{x^2 - y^2} - \dfrac{y\,dy}{x^2 - y^2}$ for $x^2 > y^2$ is an exact differential and evaluate $z = f(x, y)$ from A (3, 1) to B (5, 3).

7 The parametric equations of a circle, center (1, 0) and radius 1, can be expressed as $x = 2\cos^2\theta$, $y = 2\cos\theta\sin\theta$.
Evaluate $I = \int_c \{(x + y)\,dx + x^2\,dy\}$ along the semicircle for which $y \geq 0$ from O (0, 0) to A (2, 0).

8 Evaluate $\oint_c \{x^3y^2\,dx + x^2y\,dy\}$ where c is the boundary of the region enclosed by the curve $y = 1 - x^2$, $x = 0$ and $y = 0$ in the first quadrant.

9 Use Green's theorem to evaluate

$$I = \oint_c \{(4x + y)\,dx + (3x - 2y)\,dy\}$$

where c is the boundary of the trapezium with vertices A (0, 1), B (5, 1), C (3, 3) and D (1, 3).

10 Evaluate $I = \int_c \{(3x^2y^2 + 2\cos 2x - 2xy)\,dx + (2x^3y + 8y - x^2)\,dy\}$

(a) along the curve $y = x^2 - x$ from A (0, 0) to B (2, 2)

(b) round the boundary of the quadrilateral joining the points (1, 0), (3, 1), (2, 3) and (0, 3)

11 Verify that $dw = \frac{y}{z}dx + \frac{x}{z}dy - \frac{xy}{z^2}dz$ is an exact differential and find the value of

$$\int_c dw$$

where c is the straight line joining (0, 0, 1) to (1, 2, 3) for either region $z > 0$ or $z < 0$.

Complex calculus

Learning outcomes

When you have completed this Program you will be able to:

- Appreciate when the derivative of a function of a complex variable exists
- Understand the notions of regular functions and singularities and be able to obtain the derivative of a regular function from first principles
- Derive the Cauchy–Riemann equations and apply them to find the derivative of a regular function
- Understand the notion of an harmonic function and derive a conjugate function
- Evaluate line and contour integrals in the complex plane
- Derive and apply Cauchy's theorem
- Apply Cauchy's theorem to contours around regions that contain singularities
- Define the essential characteristics of and conditions for a conformal mapping
- Locate critical points of a function of a complex variable
- Determine the image in the w-plane of a figure in the z-plane under a conformal transformation $w = f(z)$
- Describe and apply the Schwarz–Christoffel transformation

1

In Program 7 we introduced the ideas of mapping from one complex plane to another and considered some of the more common transformation functions. Now we pursue our consideration of the complex variable a little further.

Differentiation of a complex function

In differentiation of a function of a single real variable, $y = f(x)$, the derivative of y with respect to x can be defined as the limiting value of $\dfrac{(y+\delta y)-y}{\delta x}$ as δx tends to zero.

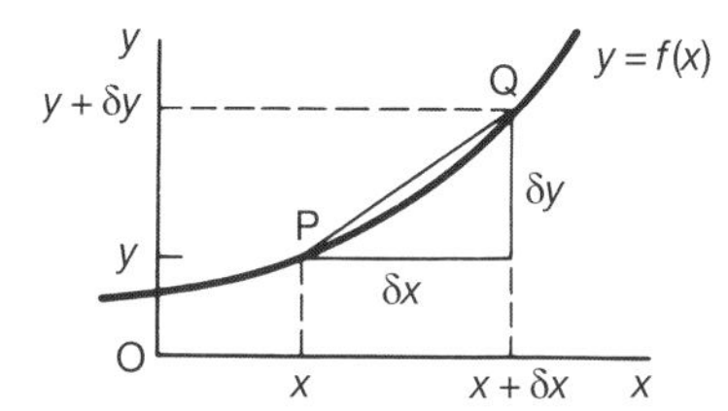

$$y = f(x) \quad \delta y = f(x+\delta x) - f(x)$$

$$\text{i.e. } \frac{dy}{dx} = \underset{\delta x \to 0}{Lim}\left\{\frac{f(x+\delta x)-f(x)}{\delta x}\right\}$$

In considering the differentiation of a function of a complex variable, $w = f(z)$, the derivative of w with respect to z can similarly be defined as the limiting value of .... as δz tends to zero.

2

$$\boxed{\frac{(w+\delta w)-w}{\delta z} \quad \text{i.e.} \quad \frac{f(z+\delta z)-f(z)}{\delta z}}$$

Now, of course, we are dealing in vectors.

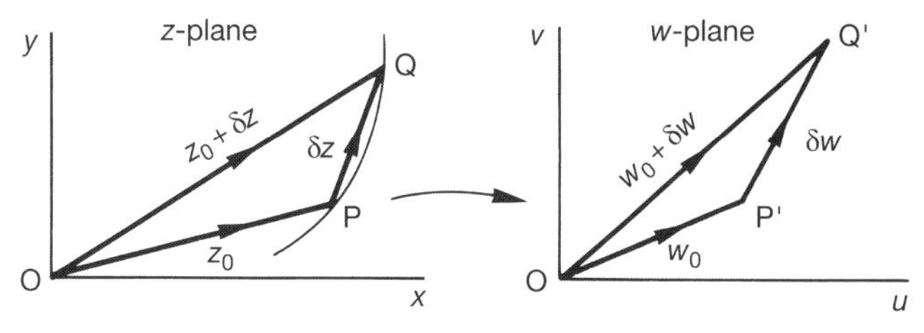

If P and Q in the z-plane map onto P′ and Q′ in the w-plane, then

$$P'Q' = \delta w = (w_0 + \delta w) - w_0 = f(z_0 + \delta z) - f(z_0)$$

Therefore, the derivative of w at P′ $(z = z_0)$ is the limiting value of $\dfrac{\delta w}{\delta z}$ as

$$\delta z \to 0, \text{ i.e. } \left[\frac{dw}{dz}\right]_{z_0} = \underset{\delta z \to 0}{Lim}\left\{\frac{f(z_0+\delta z)-f(z_0)}{\delta z}\right\} = \underset{Q \to P}{Lim}\left(\frac{P'Q'}{PQ}\right)$$

If this limiting value exists – which is not always the case as we shall see – the function $f(z)$ is said to be *differentiable at* P.

▶

Also, if $w = f(z)$ and $f'(z)$ has a limit for all points z_0 within a given region for which $w = f(z)$ is defined, then $f(z)$ is said to be differentiable in that region. From this, it follows that the limit exists whatever the path of approach from Q $(z = z_0 + \delta z)$ to P $(z = z_0)$.

Regular functions and singularities

A function $w = f(z)$ is said to be *regular* (or analytic) at a point $z = z_0$, if it is defined and single-valued, and has a derivative at every point at and around z_0. Points in a region where $f(z)$ ceases to be regular are called *singular points*, or *singularities*.

A function of a complex variable that is analytic over the entire finite complex plane is called an *entire* function. Examples of entire functions are polynomials, e^z, $\sin z$ and $\cos z$.

We have introduced quite a few new definitions, so let us pause here while you make a note of them. We shall be meeting the various terms quite often.

3

In those cases where a derivative exists, the usual rules of differentiation apply. For example, the derivative of $w = z^2$ can be found from first principles in the normal way.

$$w = z^2 \quad \therefore \quad w + \delta w = (z + \delta z)^2 = z^2 + 2z\delta z + \delta z^2$$

$$\therefore \quad \delta w = 2z\delta z + \delta z^2 \quad \therefore \quad \frac{\delta w}{\delta z} = 2z + \delta z$$

$\therefore \quad \dfrac{\mathrm{d}w}{\mathrm{d}z} = \mathop{Lim}\limits_{\delta z \to 0}(2z + \delta z) = 2z$ and does not depend on the path along which δz tends to zero.

That was elementary. Here is a rather different one.

Example

To find the derivative of $w = z\bar{z}$ where $z = x + iy$ and $\bar{z} = x - iy$.

We have $\quad w = z\bar{z} \quad \therefore \quad w + \delta w = (z + \delta z)(\bar{z} + \delta\bar{z})$ from which

$$\frac{\delta w}{\delta z} = \ldots\ldots\ldots\ldots$$

4

$$\boxed{\frac{\delta w}{\delta z} = \bar{z} + z\frac{\delta\bar{z}}{\delta z} + \delta\bar{z}}$$

Because

$$w + \delta w = (z + \delta z)(\bar{z} + \delta\bar{z}) = z\bar{z} + \bar{z}\delta z + z\delta\bar{z} + \delta z\delta\bar{z}$$

$$\therefore \quad \delta w = \bar{z}\delta z + z\delta\bar{z} + \delta z\delta\bar{z} \quad \therefore \quad \frac{\delta w}{\delta z} = \bar{z} + z\frac{\delta\bar{z}}{\delta z} + \delta\bar{z}$$

Now since $z = x + iy$ and $\bar{z} = x - iy$, we can express $\dfrac{\delta w}{\delta z}$ in terms of x and y.

$$\frac{\delta w}{\delta z} = \ldots\ldots\ldots\ldots$$

5

$$\frac{\delta w}{\delta z} = (x - iy) + (x + iy)\left\{\frac{\delta x - i\delta y}{\delta x + i\delta y}\right\} + \delta x - i\delta y$$

Because

$$\left.\begin{aligned} z &= x + iy \quad \therefore\ \delta z = \delta x + i\delta y \\ \bar{z} &= x - iy \quad \therefore\ \delta\bar{z} = \delta x - i\delta y \end{aligned}\right\} \quad \therefore\ \frac{\delta\bar{z}}{\delta z} = \frac{\delta x - i\delta y}{\delta x + i\delta y}$$

Then $\dfrac{\delta w}{\delta z} = \bar{z} + z\dfrac{\delta\bar{z}}{\delta z} + \delta\bar{z}$ gives the expression quoted above.

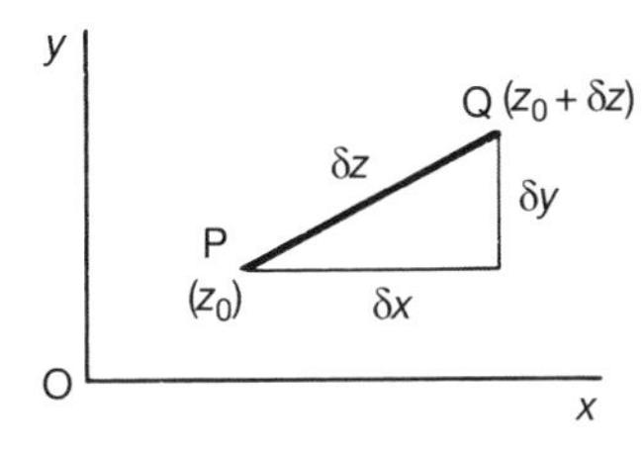

The next step is to reduce δz to zero. But δz consists of $\delta x + i\delta y$ and so reducing δz to zero can be done in one of two ways.

(1) First let $\delta y \to 0$ and afterwards let $\delta x \to 0$.

If $\delta y \to 0$, $\quad \dfrac{\delta w}{\delta z} = x - iy + (x + iy)\dfrac{\delta\bar{x}}{\delta x} + \delta\bar{x}$

Then $\dfrac{\mathrm{d}w}{\mathrm{d}z} = \lim\limits_{\delta x \to 0}\{x - iy + x + iy + \delta x\}$

$= \ldots\ldots\ldots\ldots$

6

$$\boxed{\frac{dw}{dz} = 2x}$$

On the other hand, we could have reduced δz to zero in the second way.

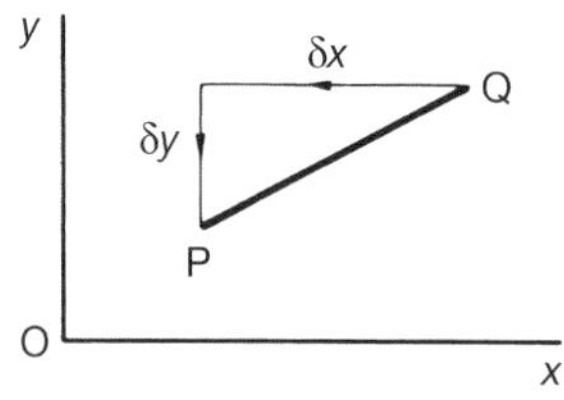

(2) First let $\delta x \to 0$ and afterwards let $\delta y \to 0$.

We have $\quad \dfrac{\delta w}{\delta z} = x - iy + (x + iy)\left\{\dfrac{\delta x - i\delta y}{\delta x + i\delta y}\right\} + \delta x - i\delta y$

If $\delta x \to 0$ $\quad \dfrac{\delta w}{\delta z} = x - iy + (x + iy)(-1) - i\delta y = -2yi - i\delta y$

Then $\quad \dfrac{dw}{dz} = \underset{\delta y \to 0}{Lim}\{-2yi - i\delta y\} = -2yi$

So, in the first case, $\dfrac{dw}{dz} = 2x$ and in the second case $\dfrac{dw}{dz} = -i2y$.
These two results are clearly not the same for all values of x and y – with one exception, i.e.

. . . .

7

$$\boxed{\text{when } x = y = 0}$$

Therefore $w = z\bar{z}$ is a function that has no specific derivative, except at $z = 0$ – and there are others. It would be convenient, therefore, to have some form of test to see whether a particular function $w = f(z)$ has a derivative $f'(z)$ at $z = z_0$. This useful tool is provided by the Cauchy–Riemann equations.

Cauchy–Riemann equations

The development is very much along the same lines as in the previous example. If $w = f(z) = u + iv$, we have to establish conditions for $w = f(z)$ to have a derivative at a given point $z = z_0$.

$w = u + iv \quad \therefore \ \delta w = \delta u + i\delta v; \qquad z = x + iy \quad \therefore \ \delta z = \delta x + i\delta y$

$$\text{Then} \quad f'(z) = \frac{dw}{dz} = \underset{\delta z \to 0}{Lim}\left\{\frac{\delta u + i\delta v}{\delta z}\right\} = \underset{\substack{\delta x \to 0 \\ \delta y \to 0}}{Lim}\left\{\frac{\delta u + i\delta v}{\delta x + i\delta y}\right\} \qquad (1)$$

(a) Let $\delta x \to 0$, followed by $\delta y \to 0$

Then from (1) above, $\quad f'(z) = \dfrac{dw}{dz} = \ldots\ldots\ldots\ldots$

8

$$\frac{dw}{dz} = \frac{\partial v}{\partial y} - i\frac{\partial u}{\partial y}$$

Because

$$f'(z) = \underset{\delta y \to 0}{Lim}\left\{\frac{\delta u + i\delta v}{i\delta y}\right\} = \underset{\delta y \to 0}{Lim}\left\{\frac{\delta v}{\delta y} - i\frac{\delta u}{\delta y}\right\} = \frac{\partial v}{\partial y} - i\frac{\partial u}{\partial y} \qquad (2)$$

We use the 'partial' notation since u and v are functions of both x and y.

Or (b) Let $\delta y \to 0$, followed by $\delta x \to 0$.

This gives ....

9

$$\frac{dw}{dz} = \frac{\partial u}{\partial x} + i\frac{\partial v}{\partial x}$$

Because

$$f'(z) = \underset{\delta x \to 0}{Lim}\left\{\frac{\delta u + i\delta v}{\delta x}\right\} = \underset{\delta x \to 0}{Lim}\left\{\frac{\delta u}{\delta x} + i\frac{\delta v}{\delta x}\right\} = \frac{\partial u}{\partial x} + i\frac{\partial v}{\partial x} \qquad (3)$$

If the results (2) and (3) are to have the same value for $f'(z)$ irrespective of the path chosen for δz to tend to zero, then

....

10

$$\frac{\partial u}{\partial x} + i\frac{\partial v}{\partial x} = \frac{\partial v}{\partial y} - i\frac{\partial u}{\partial y}$$

Equating real and imaginary parts, this gives

$$\frac{\partial u}{\partial x} = \frac{\partial v}{\partial y} \quad \text{and} \quad \frac{\partial v}{\partial x} = -\frac{\partial u}{\partial y}$$

These are the *Cauchy–Riemann equations*.

So, to sum up:

A necessary condition for $w = f(z) = u + iv$ to be regular at $z = z_0$ is that u, v and their partial derivatives are continuous and that in the neighbourhood of $z = z_0$

$$\frac{\partial u}{\partial x} = \frac{\partial v}{\partial y} \quad \text{and} \quad \frac{\partial v}{\partial x} = -\frac{\partial u}{\partial y}$$

Make a note of this important result – then move on to the next frame

11

We said earlier that where a function fails to be regular, a *singular point*, or *singularity* occurs, for example where $w = f(z)$ is not continuous or where the Cauchy–Riemann test fails.

Exercise

Determine where each of the following functions fails to be regular, i.e. where singularities occur.

1 $w = z^2 - 4$

2 $w = \dfrac{z}{z-2}$

3 $w = \dfrac{z+5}{z+1}$

4 $w = \dfrac{1}{(z-2)(z-3)}$

5 $w = z\bar{z}$

6 $w = \dfrac{x+iy}{x^2+y^2}$

Finish all six: then check with the next frame

12

Conclusions:

1 Putting $z = x + iy$, the Cauchy–Riemann conditions are satisfied everywhere. Therefore, no singularity in $w = z^2 - 4$.

2 The function becomes discontinuous at $z = 2$. Singularity at $z = 2$.

3 The function is discontinuous at $z = -1$. Singularity at $z = -1$.

4 Singularities at $z = 2$ and $z = 3$.

5 We have already seen that $w = z\bar{z}$ has no derivative for all values of z apart from $z = 0$. All points on $w = z\bar{z}$ are singularities.

6 Singularity occurs where $x^2 + y^2 = 0$, i.e. $x = 0$ and $y = 0 \;\therefore\; z = 0$. At all other points the Cauchy–Riemann equations do not hold.

Harmonic functions

13

If a function of two real variables $f(x, y)$ satisfies Laplace's equation

$$\frac{\partial^2 f(x, y)}{\partial x^2} + \frac{\partial^2 f(x, y)}{\partial y^2} = 0$$

then we say that $f(x, y)$ is an *harmonic* function. It is relatively straightforward to demonstrate that the real and imaginary parts of an analytic function are both harmonic.

Let $f(z) = u(x, y) + iv(x, y)$ be an analytic function in some region of the z-plane. Because $f(z)$ is analytic the Cauchy–Riemann equations hold true. That is

$$\frac{\partial u}{\partial x} = \frac{\partial v}{\partial y} \text{ and } \frac{\partial u}{\partial y} = -\frac{\partial v}{\partial x}$$

Differentiating the first with respect to x and the second with respect to y shows us that

$$\frac{\partial^2 u}{\partial x^2} = \frac{\partial^2 v}{\partial x \partial y} \text{ and } \frac{\partial^2 u}{\partial y^2} = -\frac{\partial^2 v}{\partial y \partial x} = -\frac{\partial^2 v}{\partial x \partial y} = -\frac{\partial^2 u}{\partial x^2}$$

$$\text{and so } \frac{\partial^2 u}{\partial x^2} + \frac{\partial^2 u}{\partial y^2} = 0$$

By a similar reasoning

$$\frac{\partial^2 \ldots}{\partial x^2} + \frac{\partial^2 \ldots}{\partial y^2} = 0$$

$$\boxed{\frac{\partial^2 v}{\partial x^2} + \frac{\partial^2 v}{\partial y^2} = 0}$$

14

Because

$$-\frac{\partial^2 v}{\partial x^2} = \frac{\partial^2 u}{\partial x \partial y} \text{ and } \frac{\partial^2 v}{\partial y^2} = \frac{\partial^2 u}{\partial y \partial x} = \frac{\partial^2 u}{\partial x \partial y} = -\frac{\partial^2 v}{\partial x^2}$$

$$\text{and so } \frac{\partial^2 v}{\partial x^2} + \frac{\partial^2 v}{\partial y^2} = 0$$

The functions $u(x, y)$ and $v(x, y)$ are called *conjugate* functions. In addition, the curves $u =$ constant, $v =$ constant are orthogonal.

Example 1

Show that the real and imaginary parts of the function defined by $f(z) = z^2$ are harmonic.

$$\begin{aligned} f(z) &= z^2 \\ &= (x + iy)^2 \\ &= (x^2 - y^2) + i2xy \end{aligned}$$

and so $u = x^2 - y^2$ and $v = 2xy$ and therefore

$$\frac{\partial^2 u}{\partial x^2} + \frac{\partial^2 u}{\partial y^2} = \ldots\ldots\ldots\ldots \text{ and } \frac{\partial^2 v}{\partial x^2} + \frac{\partial^2 v}{\partial y^2} = \ldots\ldots\ldots\ldots$$

15

$$\frac{\partial^2 u}{\partial x^2} + \frac{\partial^2 u}{\partial y^2} = 0 \text{ and } \frac{\partial^2 v}{\partial x^2} + \frac{\partial^2 v}{\partial y^2} = 0$$

Because

$$\frac{\partial u}{\partial x} = 2x \text{ so } \frac{\partial^2 u}{\partial x^2} = 2 \text{ and } \frac{\partial u}{\partial y} = -2y \text{ so } \frac{\partial^2 u}{\partial y^2} = -2$$

$$\text{therefore } \frac{\partial^2 u}{\partial x^2} + \frac{\partial^2 u}{\partial y^2} = 0$$

and

$$\frac{\partial v}{\partial x} = 2y \text{ so } \frac{\partial^2 v}{\partial x^2} = 0 \text{ and } \frac{\partial v}{\partial y} = 2x \text{ so } \frac{\partial^2 v}{\partial y^2} = 0$$

$$\text{therefore } \frac{\partial^2 v}{\partial x^2} + \frac{\partial^2 v}{\partial y^2} = 0$$

Example 2

Show that $u(x, y) = x^3y - y^3x$ is an harmonic function and find the function $v(x, y)$ that ensures that $f(z) = u(x, u) + iv(x, y)$ is analytic. That is, find the function $v(x, y)$ that is conjugate to $u(x, y)$.

$$\frac{\partial^2 u}{\partial x^2} + \frac{\partial^2 u}{\partial y^2} = \ldots\ldots\ldots\ldots$$

16

$$\frac{\partial^2 u}{\partial x^2} + \frac{\partial^2 u}{\partial y^2} = 0$$

Because

$$\frac{\partial u}{\partial x} = 3x^2y - y^3 \text{ so } \frac{\partial^2 u}{\partial x^2} = 6xy \text{ and } \frac{\partial u}{\partial y} = x^3 - 3y^2x \text{ so } \frac{\partial^2 u}{\partial y^2} = -6xy$$

$$\text{therefore } \frac{\partial^2 u}{\partial x^2} + \frac{\partial^2 u}{\partial y^2} = 0$$

This means that $u(x, y) = x^3y - y^3x$ is harmonic.

Now, if $f(z) = u(x, u) + iv(x, y)$ is analytic then $u(x, y)$ and $v(x, y)$ satisfy the .... equations.

17

Cauchy–Riemann

That is

$$\frac{\partial u}{\partial x} = 3x^2y - y^3 = \frac{\partial v}{\partial y}$$

and

$$\frac{\partial u}{\partial y} = x^3 - 3y^2x = -\frac{\partial v}{\partial x}$$

Integrating $\frac{\partial v}{\partial y} = 3x^2y - y^3$ with respect to y gives

$$v(x, y) = \ldots\ldots\ldots\ldots$$

18

$$v(x, y) = \frac{3}{2}x^2y^2 - \frac{1}{4}y^4 + a(x)$$

Because

$\frac{\partial v}{\partial y} = 3x^2y - y^3$ and so x is treated as a constant and the integral of y^n is $y^{n+1}/(n+1)$.

Did you miss the constant term in the form of $a(x)$? *Because x is treated as a constant, the integration determines y up to an expression involving x.* Differentiate the result with respect to y and you will reclaim the original form for $\frac{\partial v}{\partial y}$.

Now, differentiating this expression with respect to x gives

$$\frac{\partial v}{\partial x} = \ldots\ldots\ldots\ldots$$

19

$$\frac{\partial v}{\partial x} = 3xy^2 + a'(x)$$

Because

$v(x, y) = \frac{3}{2}x^2y^2 - \frac{1}{4}y^4 + a(x)$ and so $\frac{\partial v}{\partial x} = 3xy^2 + a'(x)$ and this is equal to $-\frac{\partial u}{\partial y}$. Now $-\frac{\partial u}{\partial y} = -x^3 + 3y^2x$ and so

$a'(x) = \ldots\ldots\ldots\ldots$ giving $a(x) = \ldots\ldots\ldots\ldots$

Therefore $v(x, y) = \ldots\ldots\ldots\ldots$

20

$$a'(x) = -x^3 \text{ giving } a(x) = -\frac{x^4}{4} + C.$$

$$\text{Therefore } v(x, y) = \frac{3x^2y^2}{2} - \frac{y^4}{4} - \frac{x^4}{4} + C$$

Because

Comparing $\frac{\partial v}{\partial x} = 3xy^2 + a'(x)$ and $-\frac{\partial u}{\partial y} = -x^3 + 3y^2x$

where $\frac{\partial v}{\partial x} = -\frac{\partial u}{\partial y}$ then it is seen that $a'(x) = -x^3$.

Therefore $a(x) = -\frac{x^4}{4} + C$ giving $v(x, y) = \frac{3x^2y^2}{2} - \frac{y^4}{4} - \frac{x^4}{4} + C$

Try one for yourself.

Example 3

Given $u(x, y) = e^{-x}\cos y$, show that $u(x, y)$ is an harmonic function and find the function $v(x, y)$ that ensures that $f(z) = u(x, y) + iv(x, y)$ is analytic. That is, find the function $v(x, y)$ that is conjugate to $u(x, y)$.

$$\frac{\partial^2 \ldots}{\partial x^2} + \frac{\partial^2 \ldots}{\partial y^2} = \ldots\ldots\ldots\ldots$$

21

$$\frac{\partial^2 u}{\partial x^2} + \frac{\partial^2 u}{\partial y^2} = 0$$

Because

$u = e^{-x}\cos y$ so $\frac{\partial u}{\partial x} = -e^{-x}\cos y$ and $\frac{\partial^2 u}{\partial x^2} = e^{-x}\cos y$.

Also $\frac{\partial u}{\partial y} = -e^{-x}\sin y$ so $\frac{\partial^2 u}{\partial y^2} = -e^{-x}\cos y$. Therefore $\frac{\partial^2 u}{\partial x^2} + \frac{\partial^2 u}{\partial y^2} = 0$, that is $u(x, y)$ is harmonic. The conjugate function $v(x, y)$ is then

$$v(x, y) = \ldots\ldots\ldots\ldots$$

22

$$v = -e^{-x}\sin y + C$$

Because

By the Cauchy–Riemann equation $\frac{\partial u}{\partial x} = \frac{\partial v}{\partial y} = -e^{-x}\cos y$. Integrating with respect to y gives $v = -e^{-x}\sin y + a(x)$. Differentiating this with respect to x gives $\frac{\partial v}{\partial x} = e^{-x}\sin y + a'(x)$.

Now, by the other Cauchy–Riemann equation $\frac{\partial v}{\partial x} = -\frac{\partial u}{\partial y} = e^{-x}\sin y$, so that $a'(x) = 0$ giving $a(x) = C$. Therefore, $v = -e^{-x}\sin y + C$.

Now we shall look at complex integration. Move to the next frame

Complex integration

23

At the beginning of this Program, we defined differentiation with respect to z in the case of a complex function, since z is a function of two independent variables x and y, i.e. $z = x + iy$. Complex integration is approached in the same way.

$z = x + iy$ and $w = f(z) = u + iv$ where u and v are also functions of x and y.

Also $\mathrm{d}z = \mathrm{d}x + i\,\mathrm{d}y$ and $\mathrm{d}w = \mathrm{d}u + i\,\mathrm{d}v$

$$\therefore \quad \int w\,\mathrm{d}z = \int f(z)\,\mathrm{d}z = \int (u + iv)(\mathrm{d}x + i\,\mathrm{d}y)$$

$$= \int \{(u\,\mathrm{d}x - v\,\mathrm{d}y) + i(v\,\mathrm{d}x + u\,\mathrm{d}y)\}$$

$$\therefore \quad \int f(z)\,\mathrm{d}z = \int (u\,\mathrm{d}x - v\,\mathrm{d}y) + i\int (v\,\mathrm{d}x + u\,\mathrm{d}y)$$

That is, the integral reduces to two real-variable integrals

$$\int (u\,\mathrm{d}x - v\,\mathrm{d}y) \quad \text{and} \quad \int (v\,\mathrm{d}x + u\,\mathrm{d}y)$$

Note that each of these two integrals is of the general form $\int (P\,\mathrm{d}x + Q\,\mathrm{d}y)$ which we met before during our work on *line integrals* and, in the complex plane, this rather neatly leads us into *contour integration*.

Let us make a fresh start

Contour integration – line integrals in the *z*-plane

24

If z moves along the curve c in the z-plane and at each position z has associated with it a function of z, i.e. $f(z)$, then summing up $f(z)$ for all such points between A and B means that we are evaluating a line integral in the z-plane between A ($z = z_1$) and B ($z = z_2$) along the curve c, i.e. we are evaluating $\int_c f(z)\,\mathrm{d}z$ where c is the particular path joining A to B.

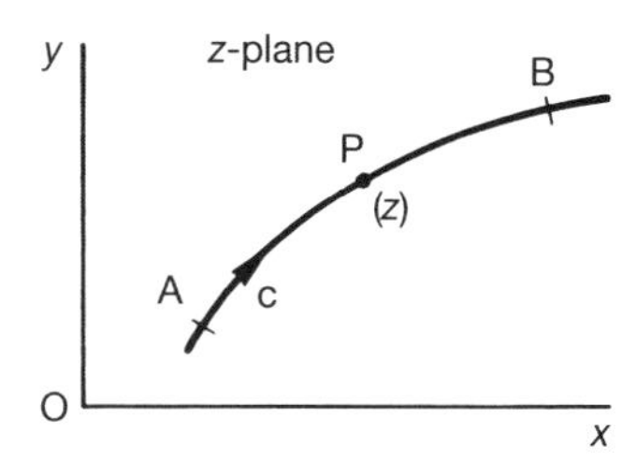

The evaluation of line integrals in the complex plane is known as *contour integration*. Let us see how it works in practice.

25

Example

Evaluate the integral $\int_c f(z)\,dz$ where $f(z) = (z - i)^2$ and c is the straight line joining A $(z = 0)$ to B $(z = 1 + 2i)$.

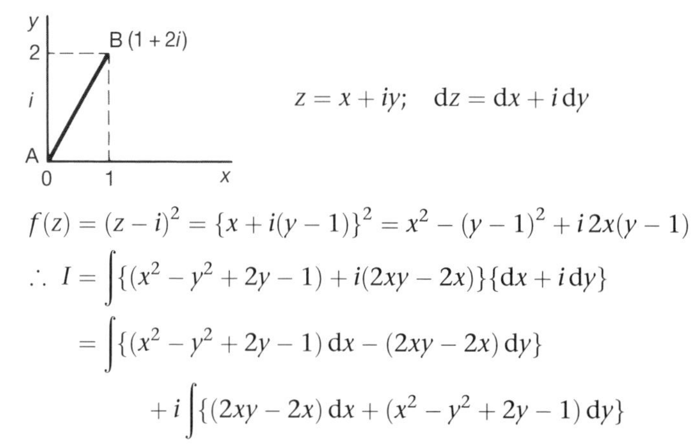

$$z = x + iy; \quad dz = dx + i\,dy$$

$$f(z) = (z - i)^2 = \{x + i(y - 1)\}^2 = x^2 - (y - 1)^2 + i\,2x(y - 1)$$

$$\therefore \; I = \int \{(x^2 - y^2 + 2y - 1) + i(2xy - 2x)\}\{dx + i\,dy\}$$

$$= \int \{(x^2 - y^2 + 2y - 1)\,dx - (2xy - 2x)\,dy\}$$

$$+ i \int \{(2xy - 2x)\,dx + (x^2 - y^2 + 2y - 1)\,dy\}$$

Now the equation of AB is $y = 2x$. $\therefore$ $dy = 2\,dx$ and substituting these in the expression for I, between the limits $x = 0$ and $x = 1$, gives

$I = \dots\dots\dots\dots$ Finish it.

26

$$\boxed{I = \frac{1}{3}(-2 + i)}$$

Because

$$I = \int_0^1 \{(x^2 - 4x^2 + 4x - 1)\,dx - (4x^2 - 2x)2\,dx\}$$

$$+ i \int_0^1 \{(4x^2 - 2x)\,dx + (2x^2 - 8x^2 + 8x - 2)\,dx\}$$

$$= \int_0^1 (-11x^2 + 8x - 1)\,dx + i \int_0^1 (-2x^2 + 6x - 2)\,dx$$

and this, by elementary integration, gives $I = \frac{1}{3}(-2 + i)$.

Now you will remember that, in general, the value of a line integral depends on the path of integration between the end points, but that the line integral $\int (P\,dx + Q\,dy)$ is independent of the path of integration in a simply connected region if $\dfrac{\partial P}{\partial y} = \dfrac{\partial Q}{\partial x}$ throughout the region.

In our example

$$I = \int \{(x^2 - y^2 + 2y - 1)\,dx - (2xy - 2x)\,dy\}$$

$$+ i \int \{(2xy - 2x)\,dx + (x^2 - y^2 + 2y - 1)\,dy\} \equiv I_1 + i I_2$$

If we apply the test to I_1, we get $\dots\dots\dots\dots$

27

$$\boxed{\frac{\partial P}{\partial y} = \frac{\partial Q}{\partial x}}$$

Because

$$\text{for } I_1 = \int\{(x^2 - y^2 + 2y - 1)\,dx - (2xy - 2x)\,dx\} \equiv \int(P\,dx + Q\,dy)$$

$$\left.\begin{aligned} \therefore\ P &= x^2 - y^2 + 2y - 1 \quad \therefore\ \frac{\partial P}{\partial y} = -2y + 2 \\ Q &= -2xy + 2x \quad\quad\quad \therefore\ \frac{\partial Q}{\partial x} = -2y + 2 \end{aligned}\right\} \therefore\ \frac{\partial P}{\partial y} = \frac{\partial Q}{\partial x}$$

Similarly

$$\text{for } I_2 = \int\{(2xy - 2x)\,dx + (x^2 - y^2 + 2y - 1)dy\} \equiv \int(Pdx + Qdy)$$

$$\left.\begin{aligned} \therefore\ P &= 2xy - 2x \quad\quad\quad \therefore\ \frac{\partial P}{\partial y} = 2x \\ Q &= x^2 - y^2 + 2y - 1 \ \therefore\ \frac{\partial Q}{\partial x} = 2x \end{aligned}\right\} \therefore\ \frac{\partial P}{\partial y} = \frac{\partial Q}{\partial x}$$

Therefore, in this example, the value of the line integral is independent of the path of integration.

Just to satisfy our conscience, determine the value of the line integral between the same two end points, but along the parabola $y = 2x^2$.

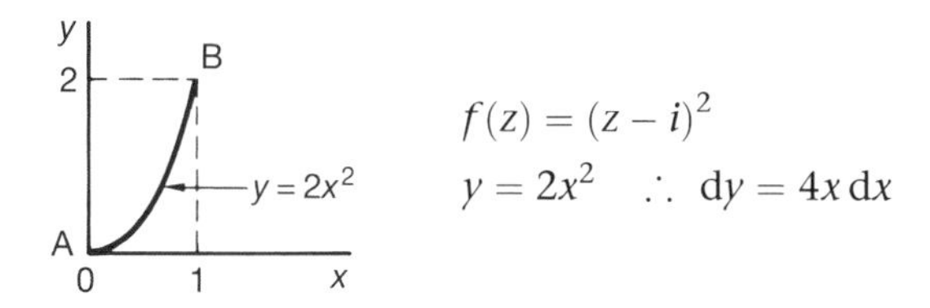

$$f(z) = (z - i)^2$$

$$y = 2x^2 \quad \therefore\ dy = 4x\,dx$$

As before we have

$$I = \int\{(x^2 - y^2 + 2y - 1)\,dx - (2xy - 2x)\,dy\}$$
$$+\, i\int(2xy - 2x)\,dx + (x^2 - y^2 + 2y - 1)\,dy\}$$

Substituting $y = 2x^2$ and $dy = 4x\,dx$, the evaluation gives

$$I = \ldots\ldots\ldots\ldots$$

28

$$I = \frac{1}{3}(-2 + i)$$

We have

$$I = \int_0^1 \{(x^2 - 4x^4 + 4x^2 - 1)\,dx - (4x^3 - 2x)4x\,dx\}$$
$$+ i\int_0^1 \{(4x^3 - 2x)\,dx + (x^2 - 4x^4 + 4x^2 - 1)4x\,dx\}$$
$$= \int_0^1 (-20x^4 + 13x^2 - 1)\,dx + i\int_0^1 (-16x^5 + 24x^3 - 6x)\,dx$$

The rest is easy enough, giving $I = \frac{1}{3}(-2 + i)$ which is, of course, the same result as before. Note that all results in Frames 25–28 can be obtained very easily by integrating the function of z with respect to z.

For example, the integral $\int_c f(z)\,dz$ where $f(z) = (z - i)^2$ and c is the straight line joining A ($z = 0$) to B ($z = 1 + 2i$) can be evaluated as

$$\int_c f(z)\,dz = \int_{z=0}^{1+2i} (z - i)^2\,dz$$
$$= \left[\frac{(z - i)^3}{3}\right]_0^{1+2i}$$
$$= \left(\frac{(1 + 2i - i)^3}{3} - \frac{(-i)^3}{3}\right)$$
$$= \frac{1}{3}(-2 + i)$$

Now on to the next frame

29 Cauchy's theorem

We have already seen that if $w = f(z)$ where, as usual, $w = u + iv$ and $z = x + iy$, then $dz = dx + i\,dy$ and

$$\int f(z)\,dz = \int (u + iv)(dx + i\,dy)$$
$$= \int (u\,dx - v\,dy) + i\int (v\,dx + u\,dy)$$

If c is a closed curve as the path of integration, then

$$\oint_c f(z)\,dz = \oint_c (u\,dx - u\,dy) + i\oint_c (v\,dx + u\,dy)$$

▶

Applying Green's theorem to each of the two integrals on the right-hand side in turn, we have

(a) $\displaystyle\oint_c (u\,dx - v\,dy) = \iint_S \left(-\frac{\partial v}{\partial x} - \frac{\partial u}{\partial y}\right) dx\,dy$

where S is the region enclosed by the curve c.

Also, if $f(z)$ is regular at every point within and on c, then the Cauchy–Riemann equations give

$$\frac{\partial u}{\partial y} = -\frac{\partial v}{\partial x} \text{ and therefore } -\frac{\partial v}{\partial x} - \frac{\partial u}{\partial y} = 0$$

$$\therefore \oint_c (u\,dx - v\,dy) = 0 \qquad (1)$$

(b) Similarly, with the second integral, we have

....

30

$$\boxed{\oint_c (v\,dx + u\,dy) = 0}$$

Because

$$\oint_c (v\,dx + u\,dy) = \iint_S \left(\frac{\partial u}{\partial x} - \frac{\partial v}{\partial y}\right) dx\,dy$$

Again, if $f(z)$ is regular at every point within and on c, then the Cauchy–Riemann equations give

$$\frac{\partial u}{\partial x} = \frac{\partial v}{\partial y} \text{ and therefore } \frac{\partial u}{\partial x} - \frac{\partial v}{\partial y} = 0$$

$$\therefore \oint_c (v\,dx + u\,dy) = 0 \qquad (2)$$

Combining the two results (1) and (2) we have the following result.

If $f(z)$ is regular at every point within and on a closed curve c, then $\displaystyle\oint_c f(z)\,dz = 0$

This is Cauchy's theorem. Make a note of the result; then we can see an example

31

Example 1

Verify Cauchy's theorem by evaluating the integral $\oint_C f(z)\,dz$ where $f(z) = z^2$ around the square formed by joining the points $z = 1$, $z = 2$, $z = 2 + i$, $z = 1 + i$.

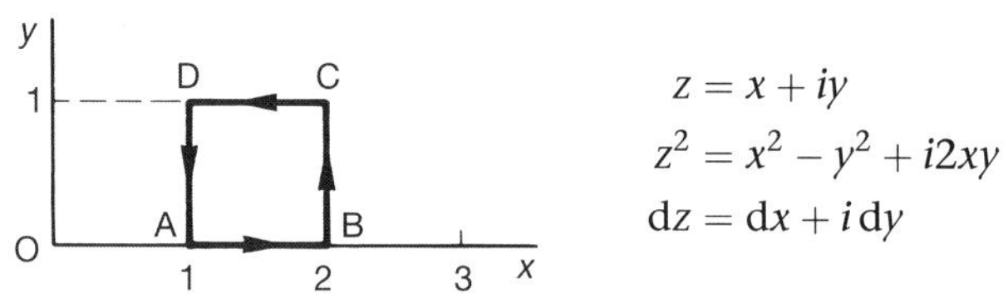

$$z = x + iy$$
$$z^2 = x^2 - y^2 + i2xy$$
$$dz = dx + i\,dy$$

$$\oint_C f(z)\,dz = \oint_C z^2\,dz = \oint_C \{x^2 - y^2 + i2xy\}\{dx + i\,dy\}$$
$$= \oint_C \{(x^2 - y^2)\,dx - 2xy\,dy\} + i\oint_C \{2xy\,dx + (x^2 - y^2)\,dy\}$$

We now take each of the sides in turn.

(a) AB: $y = 0 \quad \therefore \; dy = 0$

$$\therefore \quad \int_{AB} f(z)\,dz = \int_1^2 x^2 dx = \left[\frac{x^3}{3}\right]_1^2 = \frac{8}{3} - \frac{1}{3} = \frac{7}{3}$$

(b) BC: $x = 2 \quad \therefore \; dx = 0$

$$\therefore \quad \int_{BC} f(z)\,dz = \int_0^1 (-4y dy) + i\int_0^1 (4 - y^2)\,dy$$
$$= \left[-2y^2\right]_0^1 + i\left[4y - \frac{y^3}{3}\right]_0^1$$
$$= -2 + i\left(4 - \frac{1}{3}\right) = -2 + \frac{11}{3}i$$

Continuing in the same way, the results for the remaining two sides are

.... and ....

32

$$\text{CD: } -\frac{4}{3} - 3i; \quad \text{DA: } 1 - \frac{2}{3}i$$

Because

(c) CD: $y = 1 \quad \therefore \; dy = 0$

$$\therefore \quad \int_{CD} f(z)\,dz = \int_2^1 (x^2 - 1)\,dx + i\int_2^1 2x\,dx$$
$$= \left[\frac{x^3}{3} - x\right]_2^1 + i\left[x^2\right]_2^1 = -\tfrac{4}{3} - 3i$$

▶

(d) DA: $x = 1 \quad \therefore \; dx = 0$

$$\therefore \int_{DA} f(z)\,dz = \int_1^0 (-2y\,dy) + i\int_1^0 (1 - y^2)\,dy$$

$$= \left[-y^2\right]_1^0 + i\left[y - \frac{y^3}{3}\right]_1^0 = 1 - \tfrac{2}{3}i$$

So, collecting the four results, $\oint_c f(z)\,dz = \dots\dots\dots\dots$

33

$$\boxed{\oint_c f(z)\,dz = 0}$$

Because

$$\oint_c f(z)\,dz = \frac{7}{3} + \left(-2 + \frac{11}{3}i\right) + \left(-\frac{4}{3} - 3i\right) + \left(1 - \frac{2}{3}i\right) = 0$$

Example 2

A region in the z-plane has a boundary c consisting of

(a) OA joining $z = 0$ to $z = 2$

(b) AB a quadrant of the circle $|z| = 2$ from $z = 2$ to $z = 2i$

(c) BO joining $z = 2i$ to $z = 0$.

Verify Cauchy's theorem by evaluating the integral $\int_c (z^2 + 1)\,dz$

(1) along the arc from A to B
(2) along BO and OA.

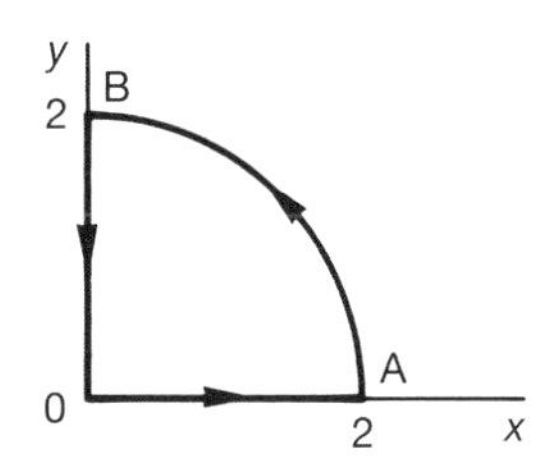

$$f(z) = z^2 + 1 = (x + iy)^2 + 1$$
$$= (x^2 - y^2 + 1) + i2xy$$
$$z = x + iy \quad \therefore \; dz = dx + i\,dy$$

So the general expression for $\int f(z)\,dz = \dots\dots\dots\dots$

34

$$\int\{(x^2-y^2+1)+i2xy\}\{dx+i\,dy\}$$
$$=\int\{(x^2-y^2+1)\,dx-2xy\,dy\}+i\int\{2xy\,dx+(x^2-y^2+1)\,dy\}$$

(1) Arc AB: $x^2+y^2=4 \quad \therefore\ y^2=4-x^2 \quad \therefore\ y=\sqrt{4-x^2}$

$$dy=\frac{1}{2}(4-x^2)^{-1/2}(-2x)\,dx \qquad \therefore\ dy=\frac{-x}{\sqrt{4-x^2}}dx$$

$$\therefore\ \int_{AB} f(z)\,dz$$
$$=\int_2^0\left\{(x^2-4+x^2+1)\,dx-2x\sqrt{4-x^2}\left(\frac{(-x)}{\sqrt{4-x^2}}\right)dx\right\}$$
$$+\,i\int_2^0\left\{2x\sqrt{4-x^2}\,dx+(x^2-4+x^2+1)\left(\frac{(1-x)}{\sqrt{4-x^2}}\right)dx\right\}$$
$$=\int_2^0(4x^2-3)\,dx+i\int_2^0\frac{11x-4x^3}{\sqrt{4-x^2}}\,dx=-\frac{14}{3}+iI_1$$

Now we must attend to $I_1=\displaystyle\int_2^0\frac{11x-4x^3}{\sqrt{4-x^2}}\,dx.$

Substituting $x=2\sin\theta$ and $dx=2\cos\theta\,d\theta$ with appropriate limits we have

....

35

$$I_1=-\frac{2}{3}$$

Because

$$I_1=\int_{\pi/2}^0\left(\frac{22\sin\theta-32\sin^3\theta}{2\cos\theta}\right)2\cos\theta\,d\theta$$
$$=\int_0^{\pi/2}(32\sin^3\theta-22\sin\theta)\,d\theta$$
$$=32\frac{2}{(3)(1)}+\Big[22\cos\theta\Big]_0^{\pi/2}=\frac{64}{3}-22=-\frac{2}{3}$$
$$\therefore\ \int_{AB} f(z)\,dz=-4\frac{2}{3}-\frac{2}{3}i=-\frac{2}{3}(7+i)$$

(2) Along BO and OA. Complete this section on your own in the same way.

$$\int_{BO} f(z)\,dz=\ldots\ldots\ldots\ldots; \quad \int_{OA} f(z)\,dz=\ldots\ldots\ldots\ldots$$

36

$$\int_{BO} f(z)\,dz = \frac{2}{3}i; \quad \int_{OA} f(z)\,dz = 4\frac{2}{3}$$

Because we have

BO: $x = 0 \quad \therefore \; dx = 0$

$$\therefore \quad \int_{BO} f(z)\,dz = i\int_2^1 (1 - y^2)\,dy = i\left[y - \frac{y^3}{3}\right]_2^0 = \frac{2}{3}i$$

OA: $y = 0 \quad \therefore \; dy = 0$

$$\therefore \quad \int_{OA} f(z)\,dz = \int_0^2 (x^2 + 1)\,dx = \left[\frac{x^3}{3} + x\right]_0^2 = 4\frac{2}{3}$$

Collecting the results together, therefore

$$\int_{AB} f(z)\,dz = -\frac{14}{3} - \frac{2}{3}i$$

$$\int_{BO+OA} f(z)\,dz = \frac{2}{3}i + 4\frac{2}{3} = \frac{14}{3} + \frac{2}{3}i$$

$$\therefore \quad \oint_c f(z)\,dz = \int_{AB} f(z)\,dz + \int_{BO+OA} f(z)\,dz = 0$$

which, once again, verifies Cauchy's theorem.

Just by way of review, Cauchy's theorem actually states that

....

37

If $f(z)$ is *regular* at every point within and on a closed curve c, then $\oint_c f(z)\,dz = 0$

In our examples so far, $f(z)$ has been regular and no problems have arisen. Let us now consider a case where one or more singularities occur within the region enclosed by the curve c.

Deformation of contours at singularities

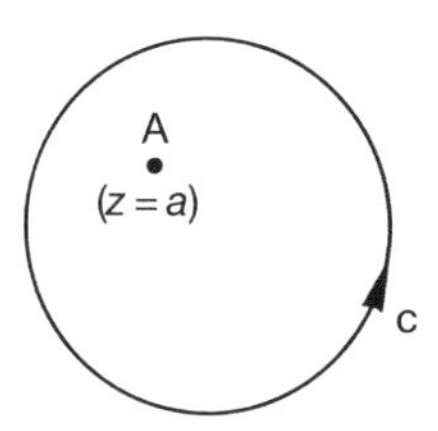

If c is the boundary curve (or *contour*) of a region and $f(z)$ is regular for all points within and on the contour, then the evaluation of $\oint_c f(z)\,dz$ around the contour is straightforward.

However, if $f(z) = \dfrac{1}{z - a}$, where a is a complex constant, and point A corresponds to $z = a$, then at A, $f(z)$ ceases to be regular and a singularity occurs at that point.

▶

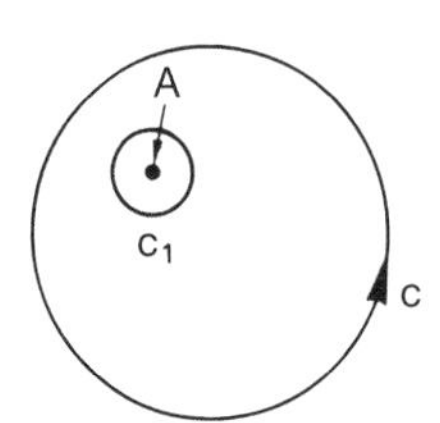

We can isolate A in a very small region within a contour c_1 and then $f(z)$ will be regular at all points within the region c and outside c_1. But the original region is now no longer simply connected (it now has a 'hole' in it) and this was one of our initial conditions.

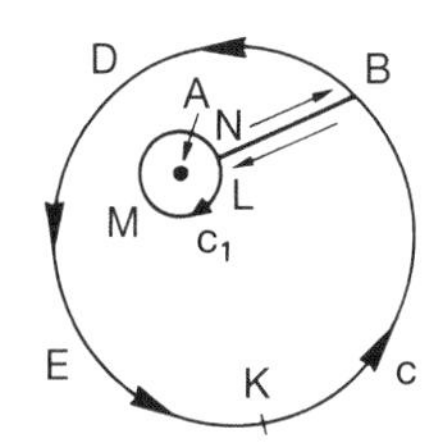

However, all is not lost! We select a suitable point B on the contour c and join it to the inner contour c_1. If we now consider the integration $\int f(z)\,dz$ starting from a point K and proceeding counter-clockwise, the path of integration can be taken as K B L M N B D E K.

Therefore

$$\int f(z)\,dz = I = I_{KB} + I_{BL} + I_{LMN} + I_{NB} + I_{BDEK} = \ldots\ldots\ldots\ldots$$

38

0

The function $f(z)$ is now regular at all points within and on the deformed contour. Remember that the inner contour c_1 can be made as small as we wish.

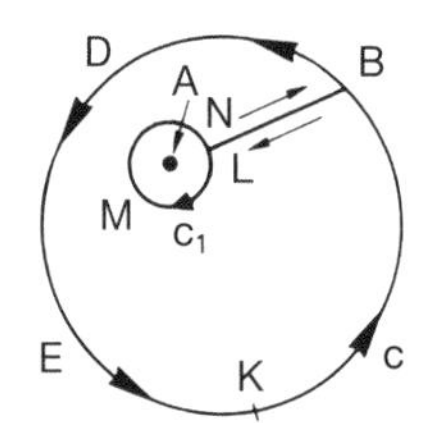

Note that $I_{NB} = -I_{BL}$, being in opposite directions, and these therefore cancel out.

The previous result then becomes

$$I_{KB} + I_{LMN} + I_{BDEK} = 0 \qquad \text{i.e.} \qquad I_{KB} + I_{BDEK} + I_{LMN} = 0$$

But $I_{KB} + I_{BDEK} = \oint_c f(z)\,dz$ and $I_{LMN} = \circlearrowright\!\!\oint_{c_1} f(z)\,dz$

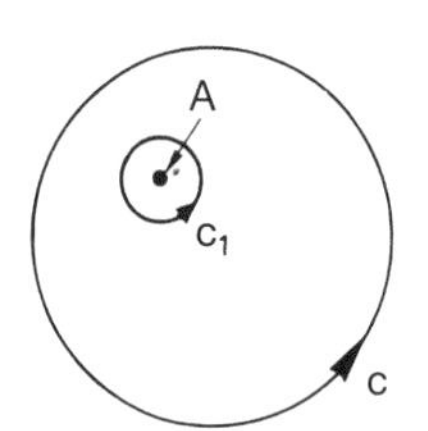

$$\therefore \quad \oint_c f(z)\,dz + \circlearrowright\!\!\oint_{c_1} f(z)\,dz = 0$$

$$\therefore \quad \oint_c f(z)\,dz - \oint_{c_1} f(z)\,dz = 0$$

$$\therefore \quad \oint_c f(z)\,dz = \oint_{c_1} f(z)\,dz$$

The process can, of course, be extended to cases with more than one such singularity.

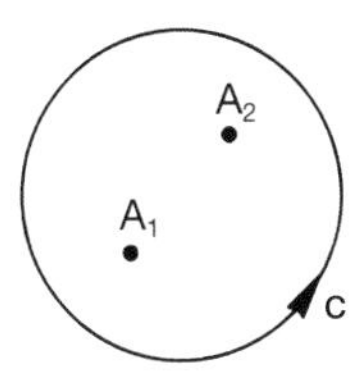

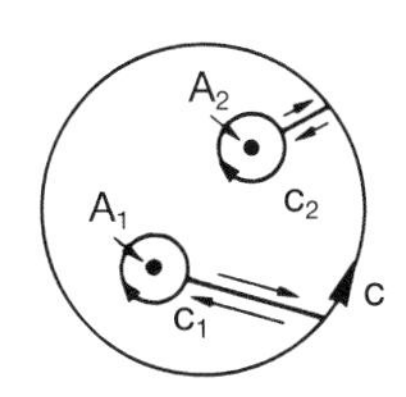

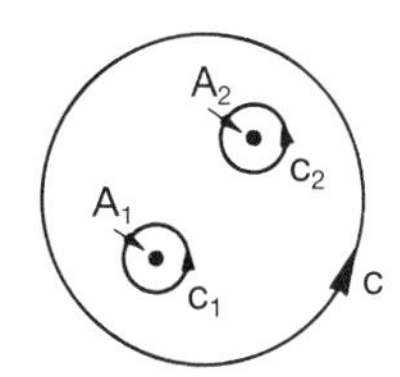

The corresponding result then becomes

$$\oint_C f(z)\,dz = \oint_{C_1} f(z)\,dz + \oint_{C_2} f(z)\,dz \ldots \text{etc.}$$

Now let us apply these ideas to an example.

Example 1

Consider the integral $\oint_C f(z)\,dz$ where $f(z) = \dfrac{1}{z}$, evaluated round a closed contour in the z-plane.

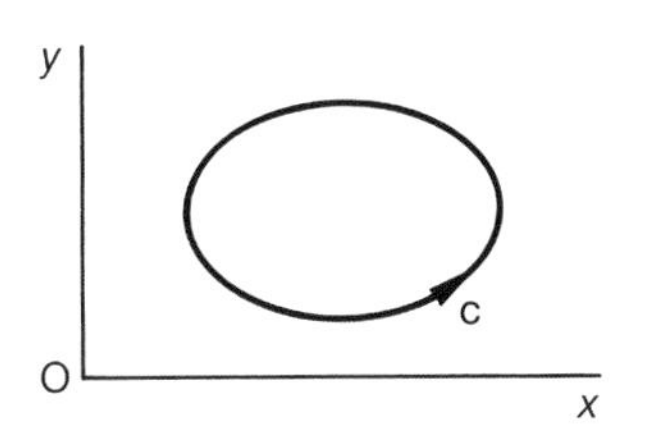

We first check the function $f(z) = \dfrac{1}{z}$ for singularities and find at once that

....

39

At $z = 0$, $f(z) = \dfrac{1}{z}$ ceases to be regular and a singularity occurs at that point

The actual position of the closed contour is not specified in the problem, so there are two possibilities: either the contour does enclose the origin, or it does not.

Let us consider them in turn.

(a) The contour does not enclose the origin.

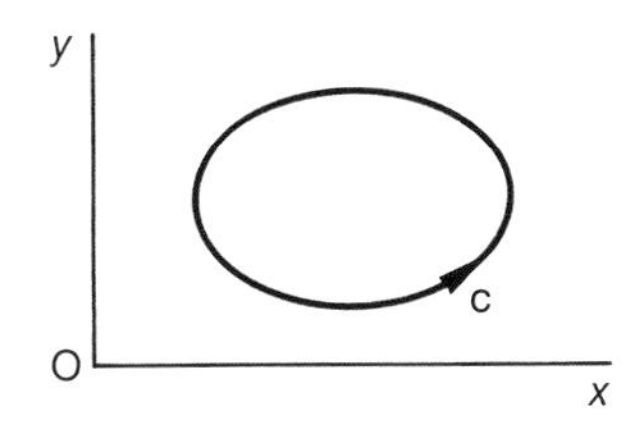

No difficulty arises here and by Cauchy's theorem

....

40

$$\oint_c f(z)\,dz = 0$$

(b) If the contour *does* enclose the origin, the singularity must be taken into account. Then

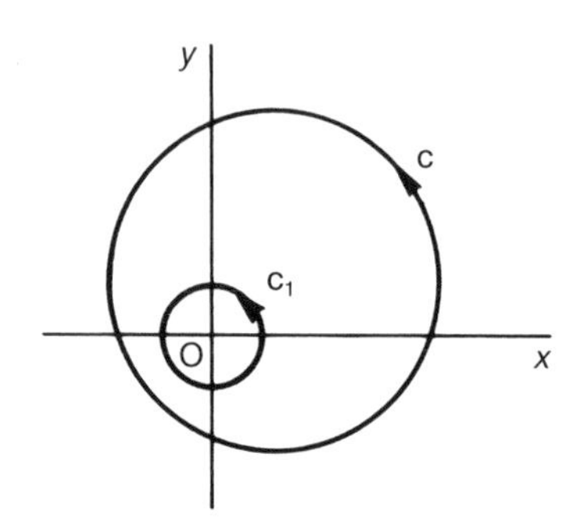

$$\oint_c f(z)\,dz = \oint_{c_1} f(z)\,dz = \oint_{c_1} \frac{1}{z}dz$$

and we attend to evaluating $\oint_{c_1} \frac{1}{z}dz$

where c_1 is a small circle of radius r entirely within the region bounded by c.

If we take an enlarged view of the small circle c_1, we have $z = x + iy$ which can be expressed in polar form .... and in exponential form ....

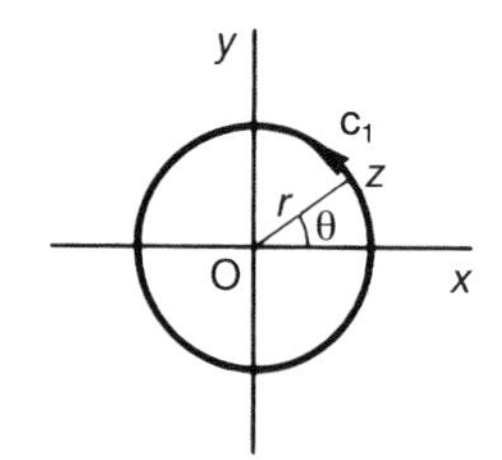

41

$$z = r(\cos\theta + i\sin\theta)$$
$$z = re^{i\theta}$$

Using $z = re^{i\theta}$ then $dz = ire^{i\theta}d\theta$ and $\oint_{c_1} \frac{1}{z}dz = \ldots\ldots\ldots\ldots$

Complete it

42

$$2\pi i$$

Because

$$\oint_{c_1} \frac{1}{z}\,dz = \int_0^{2\pi} \frac{1}{re^{i\theta}}\{ire^{i\theta}\}d\theta = \int_0^{2\pi} i\,d\theta = 2\pi i$$

$$\therefore \oint_c \frac{1}{z}dz = \oint_{c_1} \frac{1}{z}dz = 2\pi i$$

So we have:

(a) $\oint_c \frac{1}{z}dz = 0$ if the contour c does not enclose the origin

(b) $\oint_c \frac{1}{z}dz = 2\pi i$ if the contour c does enclose the origin.

These two constitute an important result, so note them well

43

Example 2

Consider the integral $\oint_c f(z)\,dz$ where $f(z) = \frac{1}{z^n}$ $(n = 2, 3, 4, \ldots)$.

Again, a singularity clearly occurs at $z = 0$ and again also we have two possible cases.

(a) If the contour c does not enclose the origin, then by Cauchy's theorem $\oint_c f(z)\,dz = 0$.

(b) If the contour c does enclose the origin, then we proceed very much as before.

Using $z = re^{i\theta}$, $dz = ire^{i\theta}d\theta$ and $z^n = r^n e^{in\theta}$

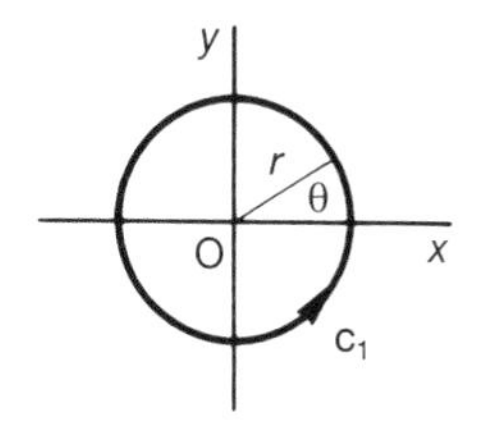

Then $\oint_c f(z)\,dz = \oint_{c_1} f(z)\,dz$

$$= \int_0^{2\pi} \frac{1}{r^n e^{in\theta}} \{ire^{i\theta}\}\,d\theta$$

$$= \frac{i}{r^{n-1}} \int_0^{2\pi} e^{-i(n-1)\theta}d\theta$$

$$= \frac{-1}{(n-1)r^{n-1}} \left[e^{-i(n-1)\theta}\right]_0^{2\pi}$$

$$= \ldots\ldots\ldots\ldots$$

Finish it off

44

$$\boxed{0}$$

Because

$$\oint_c \frac{1}{z^n}dz = \frac{-1}{(n-1)r^{n-1}}\{e^{-i(n-1)2\pi} - 1\}$$

$$= \frac{-1}{(n-1)r^{n-1}}\{\cos(n-1)2\pi - i\sin(n-1)2\pi - 1\}$$

$$= 0 \quad \text{since} \quad \left.\begin{array}{r}\cos(n-1)2\pi = 1 \\ \sin(n-1)2\pi = 0\end{array}\right\} \quad n = 2, 3, 4, \ldots$$

So $\oint_c \frac{1}{z^n}dz = 0$ for all positive integer values of n other than $n = 1$, where c is any closed contour.

The particular case when $n = 1$ we have seen in Example 1.

Now we can easily cope with this next example.

▶

Example 3

Consider $\oint_c f(z)\,dz$ where $f(z) = \dfrac{1}{(z-a)^n}$ for $n = 1, 2, 3, \ldots$

This is a simple extension of the previous piece of work. Here we see that a singularity occurs at $z = a$ and yet again we have two cases to consider.

(a)

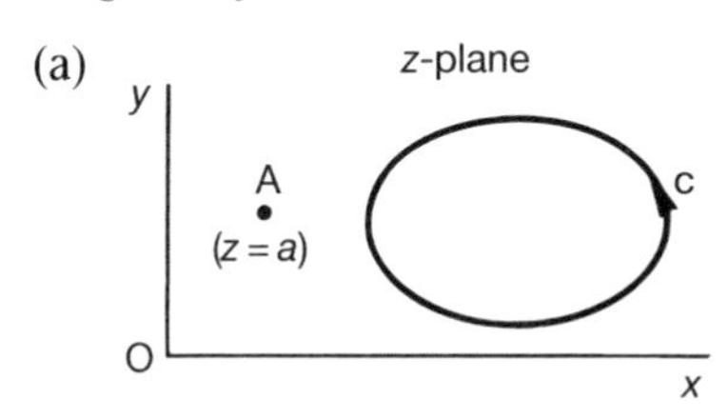

If the contour c does not enclose $z = a$, then by Cauchy's theorem

$$\oint_c f(z)\,dz = 0$$

(b)

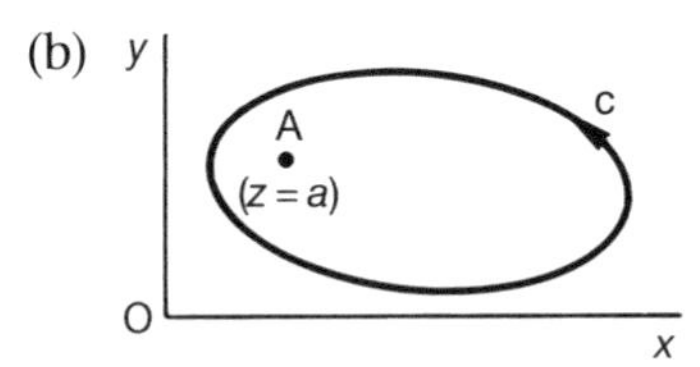

If c encloses A $(z = a)$ we consider separately the cases when

(1) $n = 1$ and (2) $n > 1$.

(1) If $n = 1$, $\displaystyle\oint_c f(z)\,dz = \oint_c \frac{1}{z-a}\,dz$

Putting $z - a = w \quad \therefore \; dz = dw \quad \therefore \; \displaystyle\oint_c \frac{1}{z-a}\,dz = \oint_c \frac{1}{w}\,dw$

and this we have already established has a value . . . .

45

$$\boxed{2\pi i}$$

(2) If $n > 1$, $\displaystyle\oint_c f(z)\,dz = \oint_c \frac{1}{(z-a)^n}\,dz = \oint_c \frac{1}{w^n}\,dw = 0$ for $n \neq 1$.

So collecting our results together, we have the following.

For $\oint_c f(z)\,dz$, where $f(z) = \dfrac{1}{(z-a)^n}$, $n = 1, 2, 3, \ldots$ and c is a closed contour

$$\begin{aligned} \oint_c \frac{1}{(z-a)^n}\,dz &= 0 && n \neq 1 \\ &= 0 && n = 1 \text{ and c does not enclose } z = a \\ &= 2\pi i && n = 1 \text{ and c does enclose } z = a. \end{aligned}$$

You will notice that this is a more general result and includes the results obtained from Examples 1 and 2. *Make a note of it, therefore: it is quite important.*

Then on to Example 4

46

Example 4

Finally, we can go one stage further and consider the contour integral of functions such as $f(z) = \dfrac{z - i - 4}{(z + i)(z - 2)}$.

First we express $f(z)$ in partial fractions

$$\frac{z - i - 4}{(z + i)(z - 2)} = \frac{A}{z + i} + \frac{B}{z - 2}$$

One quick way of finding A and B is by the 'cover up' method.

(a) *To find A*, temporarily cover up the denominator $(z + i)$ in the partial fraction $\dfrac{A}{[z + i]}$ and in the function $\dfrac{z - i - 4}{[z + i](z - 2)}$ and substitute $z + i = 0$, i.e. $z = -i$ in the remainder of the function.

$$A = \frac{-i - i - 4}{-i - 2} = \frac{4 + 2i}{2 + i} = 2 \ \therefore \ A = 2$$

(b) *To find B*, cover up the denominator $(z - 2)$ in the partial fraction $\dfrac{B}{[z - 2]}$ and in the function $\dfrac{z - i - 4}{(z + i)[z - 2]}$ and substitute $z - 2 = 0$, i.e. $z = 2$ in the remainder of the function.

$B = \ldots\ldots\ldots\ldots$

47

$$\boxed{B = -1}$$

Because

$$B = \frac{2 - i - 4}{2 + i}$$
$$= \frac{-2 - i}{2 + i}$$
$$= -1$$

Therefore the function $f(z)$ becomes

$$f(z) = \frac{z - i - 4}{(z + i)(z - 2)} \equiv \frac{2}{z + i} - \frac{1}{z - 2}$$

Now we can see that there are singularities at $\ldots\ldots\ldots\ldots$

48

$$z = -i \quad \text{and} \quad z = 2$$

Denote the singularities by L and M.

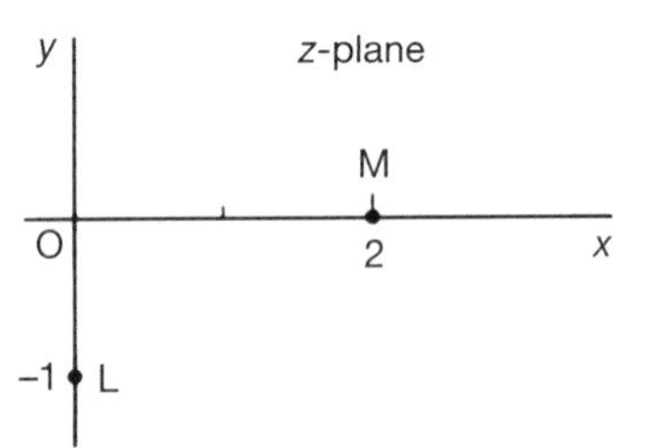

$$\therefore \oint_c \frac{z-i-4}{(z+i)(z-2)}\,dz = \oint_c \left\{\frac{2}{z+i} - \frac{1}{z-2}\right\}dz$$
$$= \oint_c \left\{2\left(\frac{1}{z+i}\right) - \frac{1}{z-2}\right\}dz$$

So we now have *four* cases to consider, depending on whether L, M, neither, or both, are enclosed within the contour c.

(a) *Neither L nor M enclosed*

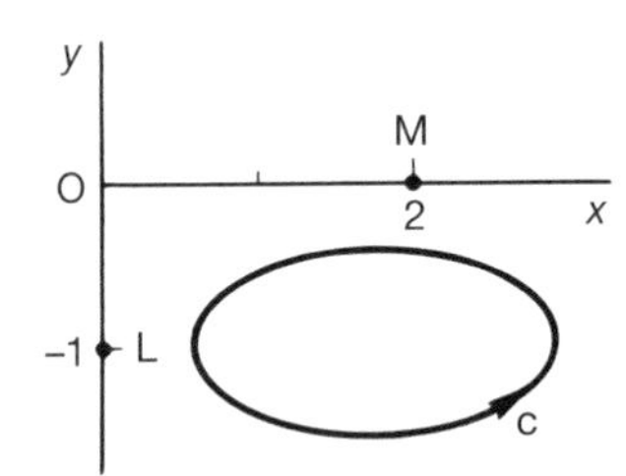

Then, once again, by Cauchy's theorem

$$\oint_c f(z)\,dz = 0$$

(b) *L enclosed but not M*

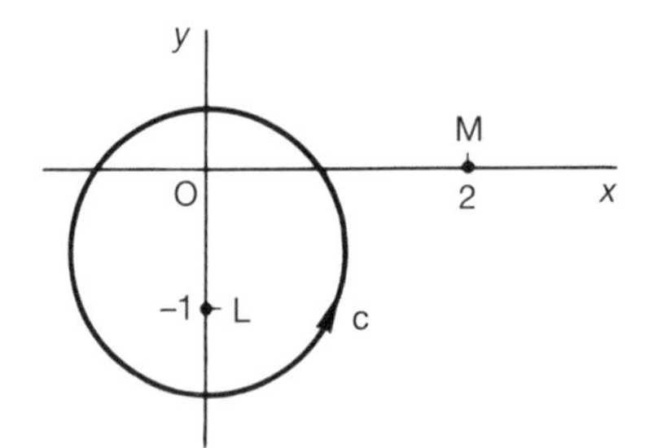

Then, in this case

$$\oint_c f(z)\,dz = 2(2\pi i) - 0 = 4\pi i$$

(c) *M enclosed but not L*

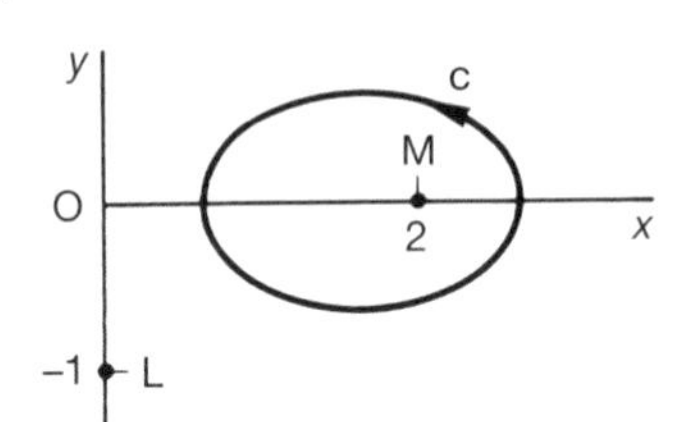

Here

$$\oint_c f(z)\,dz = 0 - (2\pi i) = -2\pi i$$

(d) *Both L and M enclosed*

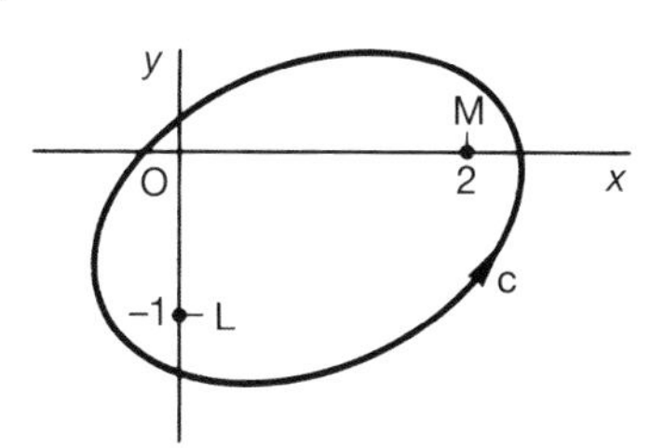

In this case

$$\oint_c f(z)\,dz = \ldots\ldots\ldots\ldots$$

49

$$2\pi i$$

Because, when both L and M are enclosed

$$\oint_c f(z)\,dz = \oint_c \left\{2\left(\frac{1}{z+i}\right) - \frac{1}{z-2}\right\}dz$$
$$= 2(2\pi i) - 2\pi i$$
$$= 2\pi i$$

The key is provided by the results we established earlier.

$$\oint_c \frac{1}{(z-a)^n}\,dz = \ldots\ldots\ldots\ldots \text{ if } \ldots\ldots\ldots\ldots$$
$$= \ldots\ldots\ldots\ldots \text{ if } \ldots\ldots\ldots\ldots$$
$$= \ldots\ldots\ldots\ldots \text{ if } \ldots\ldots\ldots\ldots$$

50

$$\oint_c \frac{1}{(z-a)^n}\,dz = 0 \quad \text{if } n \neq 1$$
$$= 0 \quad \text{if } n = 1 \text{ and c does not enclose } z = a$$
$$= 2\pi i \quad \text{if } n = 1 \text{ and c does enclose } z = a.$$

Now for something somewhat different.

Conformal transformation (conformal mapping)

A mapping from the z-plane onto the w-plane is said to be *conformal* if the angles between lines in the z-plane are preserved both in magnitude and in sense of rotation when transformed onto the corresponding lines in the w-plane.

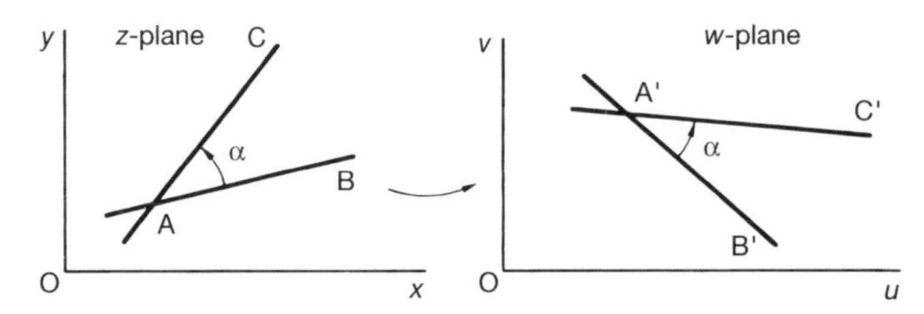

The angle between two intersecting curves in the z-plane is defined by the angle α $(0 \leq \alpha \leq \pi)$ between their two tangents at the point of intersection, and this is preserved.

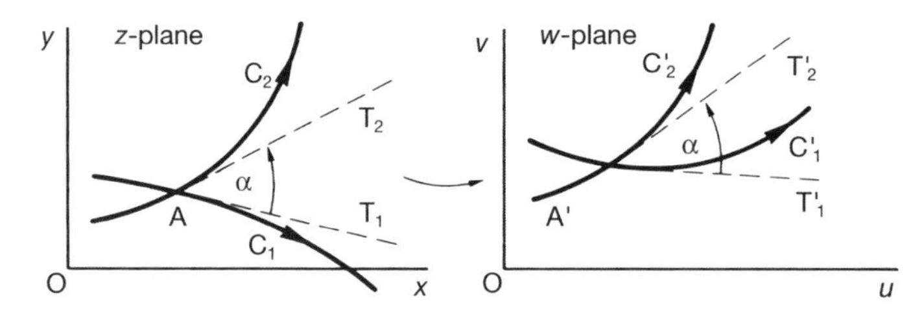

The essential characteristic of a conformal mapping is that

....

51

angles are preserved both in magnitude and in sense of rotation

Conditions for conformal transformation

The conditions necessary in order that a transformation shall be conformal are as follows.

1 The transformation function $w = f(z)$ must be a regular function of z. That is, it must be defined and single-valued, have a continuous derivative at every point in the region and satisfy the Cauchy–Riemann equations.

2 The derivative $\dfrac{dw}{dz}$ must not be zero, i.e. $f'(z) \neq 0$ at a point of intersection.

▶

Critical points

A point at which $f'(z) = 0$ is called a *critical point* and, at such a point, the transformation is not conformal.

So, if $w = f(z)$ is a regular function, then, except for points at which $f'(z) = 0$, the transformation function will preserve both the magnitude of the angle and its sense of rotation.

Now for a short exercise by way of practice.

Exercise

Determine critical points (if any) which occur in the following transformations $w = f(z)$.

1 $f(z) = (z-1)^2$

2 $f(z) = e^z$

3 $f(z) = \dfrac{1}{z^2}$

4 $f(z) = z + \dfrac{1}{z}$

5 $f(z) = (2z+3)^3$

6 $f(z) = z^3 + 6z + 9$

7 $f(z) = \dfrac{z-i}{z+i}$

8 $f(z) = (z+3)(z-i)$.

Finish the whole set before checking with the results in the next frame.

52

1	$z = 1$	**5**	$z = -\frac{3}{2}$
2	none	**6**	$z = \pm\sqrt{2}i$
3	none	**7**	none
4	$z = \pm 1$	**8**	$z = \frac{1}{2}(i-3)$

All that is required is to differentiate each function and to find for which values of z, $f'(z) = 0$.

Now one or two simple examples on conformal mapping.

Example 1

Linear transformation $w = az + b$, $a \neq 0$, a and b complex.

(1) Cauchy–Riemann conditions satisfied.

(2) $f'(z) = a$ i.e. not zero $\therefore$ no critical points.

Therefore, the transformation $w = az + b$ provides conformal mapping throughout the entire z-plane.

Example 2

Non-linear transformation $w = z^2$.

First check for singularities and critical points. These, if any, occur at

....

53

no singularities; critical point at $z = 0$

Because

$$f'(z) = 2z \quad \therefore \; f'(z) = 0 \text{ at } z = 0.$$

Therefore, the transformation is not conformal at the origin.

If we choose to express z in exponential form $z = x + iy = re^{i\theta}$, then $w = z^2 = r^2 e^{i2\theta}$, i.e. r is squared and the angle doubled.

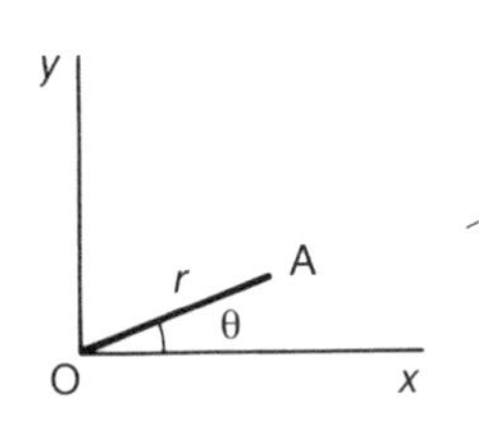

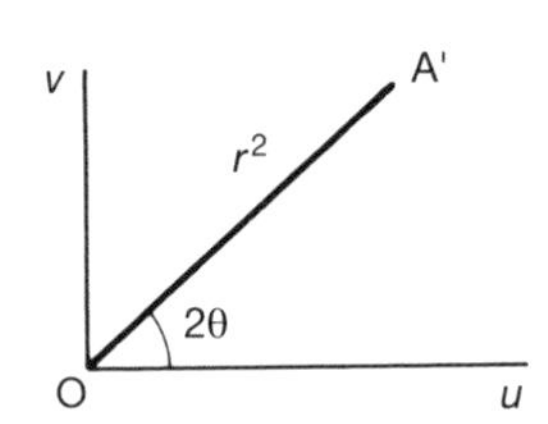

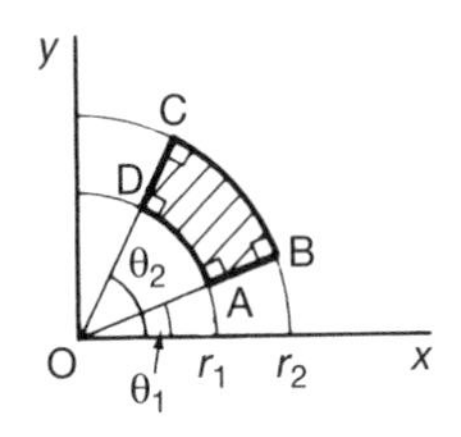

So ABCD, a section of an annulus of inner and outer radii r_1 and r_2 respectively, will be mapped onto

....

54

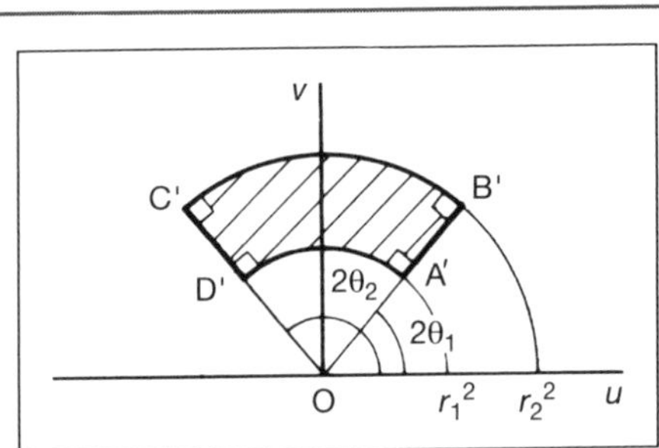

The angles at the origin are doubled, but notice that the right angles at A, B, C, D are preserved at A′, B′, C′, D′, i.e. the transformation there is conformal.

Example 3

Consider the mapping of the circle $|z| = 1$ under the transformation $w = z + \dfrac{4}{z}$ onto the w-plane.

First, as always, check for singularities and critical points. We find

....

55

singularity at $z = 0$; critical points at $z = \pm 2$

A singularity occurs at $z = 0$, i.e. $f'(z)$ does not exist at $z = 0$. Also $f(z) = z + \frac{4}{z}$ $\therefore$ $f'(z) = 1 - \frac{4}{z^2}$ $\therefore$ $f'(z) = 0$ at $z = \pm 2$.
Therefore the transformation is not conformal at $z = 0$ and at $z = \pm 2$.

In fact, if we carry out the transformation $w = z + \frac{4}{z}$ on the unit circle $|z| = 1$, we get ....

Complete it: it is good review

56

the ellipse $\frac{u^2}{5^2} + \frac{v^2}{3^2} = 1$

Because

$$w = u + iv = z + \frac{4}{z}$$

$$= x + iy + \frac{4}{x + iy}$$

$$= x + iy + \frac{4(x - iy)}{x^2 + y^2}$$

$$\therefore \quad u = x + \frac{4x}{x^2 + y^2}; \qquad v = y - \frac{4y}{x^2 + y^2}$$

$$|z| = 1 \quad \therefore \quad x^2 + y^2 = 1 \quad \therefore \quad u = x(1 + 4) = 5x; \quad v = y(1 - 4) = -3y$$

$$\therefore \quad x = \frac{u}{5} \quad \text{and} \quad y = -\frac{v}{3}$$

Then $x^2 + y^2 = 1$ gives $\frac{u^2}{5^2} + \frac{v^2}{3^2} = 1$

The image of the unit circle is therefore an ellipse with center at the origin; semi major axis 5; semi minor axis 3.

Now let us move on to a new section

57 Schwarz–Christoffel transformation

Example 1

Consider a semi-infinite strip on BC as base, the arrows at A and D indicating that the ordinate boundaries extend to infinity in the positive y-direction and that progression round the boundary is to be taken in the direction indicated.

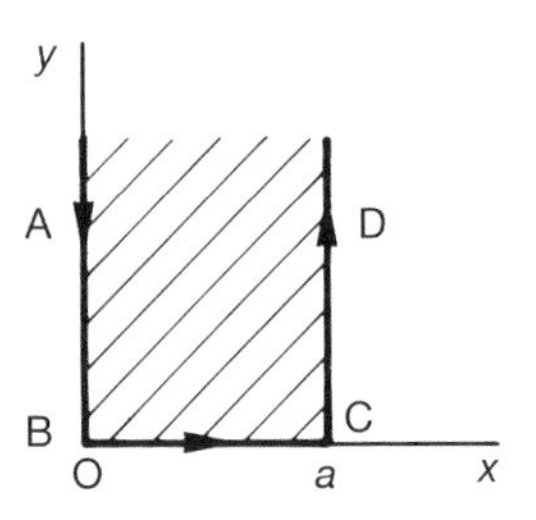

Let us apply the transformation $w = -\cos\dfrac{\pi z}{a}$ to the shaded region.

$$\text{Then } w = u + iv = -\cos\frac{\pi z}{a}$$

$$= -\cos\frac{\pi(x+iy)}{a}$$

$$= -\left\{\cos\frac{\pi x}{a}\cos\frac{i\pi y}{a} - \sin\frac{\pi x}{a}\sin\frac{i\pi y}{a}\right\}$$

Now $\cos i\theta = \cosh\theta$ and $\sin i\theta = i\sinh\theta$.

$$\therefore\ w = u + iv$$

$$= -\cos\frac{\pi x}{a}\cosh\frac{\pi y}{a} + i\sin\frac{\pi x}{a}\sinh\frac{\pi y}{a}$$

$$\therefore\ u = -\cos\frac{\pi x}{a}\cosh\frac{\pi y}{a}; \qquad v = \sin\frac{\pi x}{a}\sinh\frac{\pi y}{a}$$

So B and C map onto B′ and C′ where

B′ = ...; C′ = ...

58

B′: $u = -1,\ v = 0$; C′: $u = 1,\ v = 0$

Because

(1) at B, $x = 0,\ y = 0$ $\therefore\ u = -(1)(1) = -1;\ v = (0)(0) = 0$
and (2) at C, $x = a,\ y = 0$ $\therefore\ u = -(-1)(1) = 1;\ v = (0)(0) = 0$

So we have

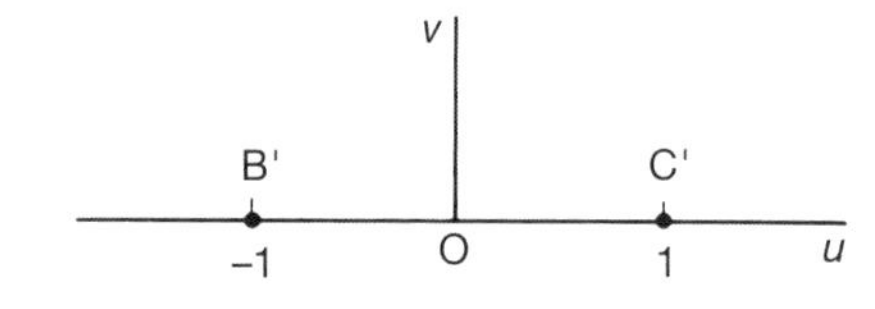

Now we map AB, BC, CD onto the w-plane giving $A'B', B'C', C'D'$.

(a) AB: $x = 0 \quad \therefore \quad A'B'$: $u = -\cosh\dfrac{\pi y}{a}$; $v = 0$

$\therefore$ As y decreases from ∞ to 0, u increases from $-\infty$ to -1.

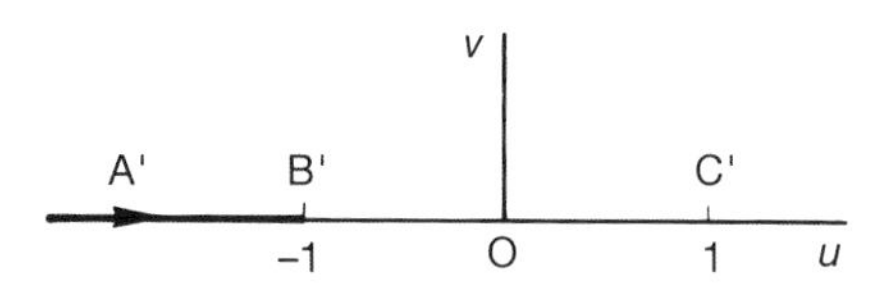

(b) BC: $y = 0 \quad \therefore \quad B'C'$: $u = -\cos\dfrac{\pi x}{a}$; $v = 0$

$\therefore$ As x increases from 0 to a, u increases from -1 to 1.

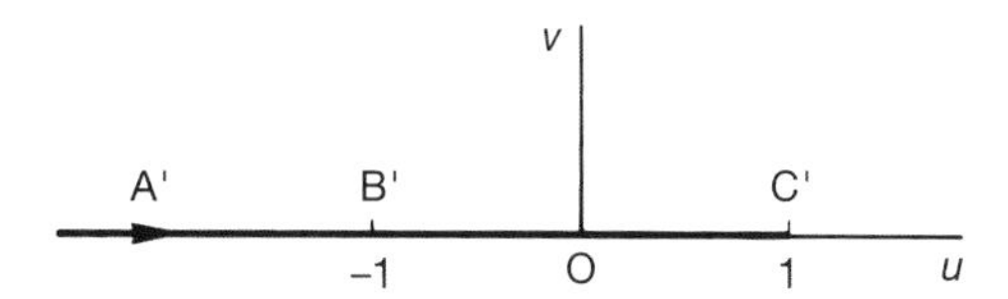

(c) CD: In the same way you can map CD and $C'D'$ in the w-plane and the mapping then becomes ....

59

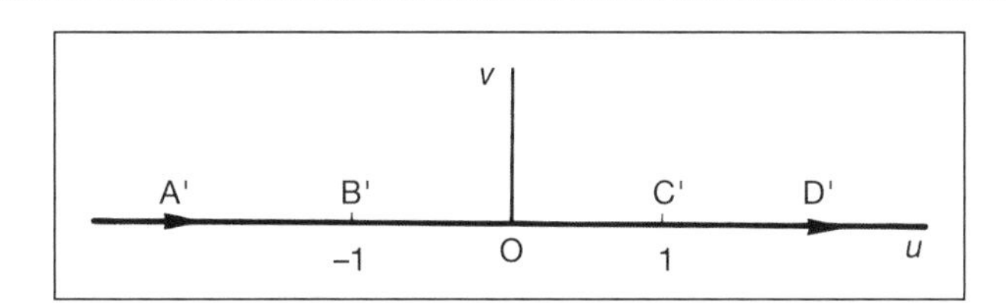

Because

CD: $x = a \quad \therefore \quad C'D'$: $u = \cosh\dfrac{\pi y}{a}$; $v = 0$.

Therefore, as y increases from 0 to ∞, u increases from 1 to ∞.

Notice the direction of the arrows. These correspond to the directed travel round the boundary shown in the z-plane.

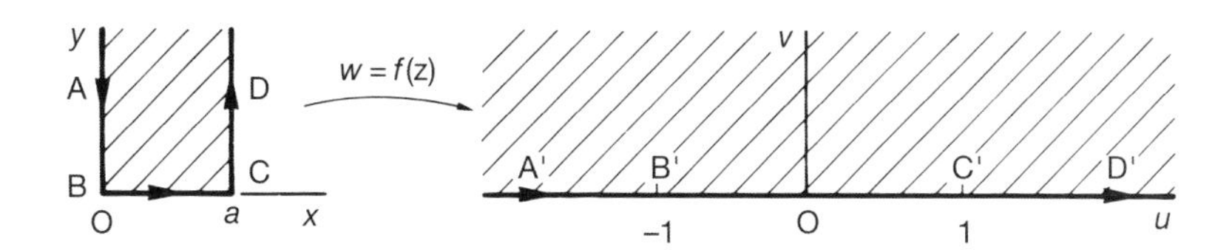

The shaded region in the z-plane is on the left-hand side of the boundary as traversed. This maps onto the left-hand side of the image on the w-plane, i.e. the entire upper half of the plane.

Note that $\dfrac{dw}{dz} = \dfrac{\pi}{a}\sin\dfrac{\pi z}{a} \quad \therefore$ at B $(z = 0)$ and C $(z = a)$, $\dfrac{dw}{dz} = 0$.

Therefore, the conformal property does not hold at these points. The internal angle at B and at C is $\dfrac{\pi}{2}$, while at B' and C' it is π.

Example 2

Consider an infinite strip in the z-plane bounded by the real axis and $z = ai$

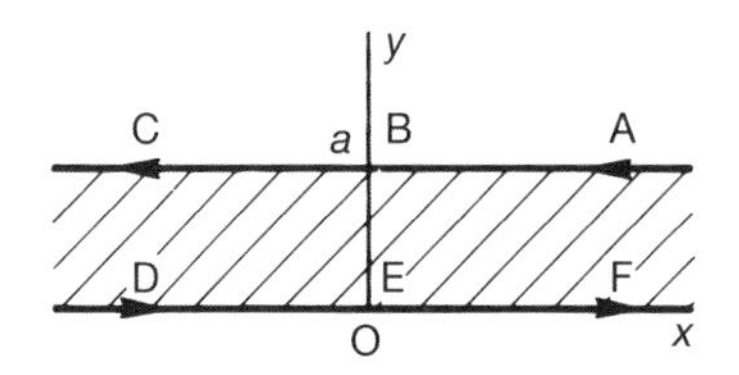

Note the arrows. The boundary comes from $+\infty$ (A) and continues to $-\infty$ (C); then returns from $-\infty$ (D) to $+\infty$ (F).

The strip can be considered as a closed figure with the left- and right- hand vertices at infinity.

We now map the infinite strip onto the w-plane by the transformation $w = e^{\pi z/a}$.

$$\therefore \; w = u + iv = e^{\pi z/a}, \text{ from which}$$

$$u = \dots\dots\dots\dots; \; v = \dots\dots\dots\dots$$

60

$$\boxed{u = e^{\pi x/a}\cos\frac{\pi y}{a}; \quad v = e^{\pi x/a}\sin\frac{\pi y}{a}}$$

Because

$$\begin{aligned} u + iv &= e^{\pi z/a} \\ &= e^{\pi(x+iy)/a} \\ &= e^{\pi x/a}e^{i\pi y/a} \\ &= e^{\pi x/a}\left(\cos\frac{\pi y}{a} + i\sin\frac{\pi y}{a}\right) \end{aligned}$$

$$\therefore \; u = e^{\pi x/a}\cos\frac{\pi y}{a}; \qquad v = e^{\pi x/a}\sin\frac{\pi y}{a}$$

Now we map points B and E onto B′ and E′.

(1) B: $x = 0, \; y = a \quad \therefore$ B′: $\; u = -1, \; v = 0$

(2) E: $x = 0, \; y = 0 \quad \therefore$ E′: $\; u = 1, \; v = 0$

i.e.

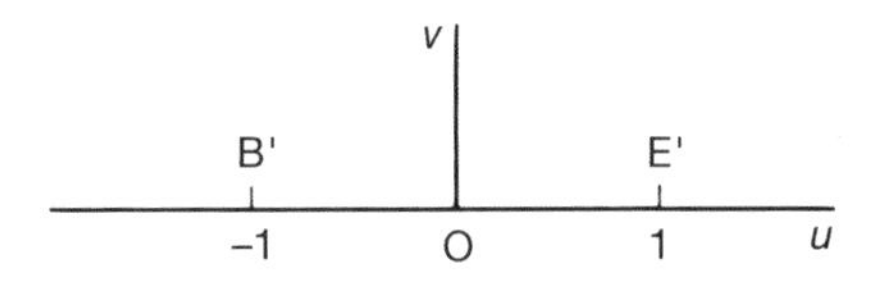

Now we map the lines AB, BC, DE, EF onto the w-plane.

(a) AB: $y = a \quad \therefore \quad u = -e^{\pi x/a}, \quad v = 0$

$\therefore$ As x decreases from $+\infty$ to 0, u increases from $-\infty$ to -1.

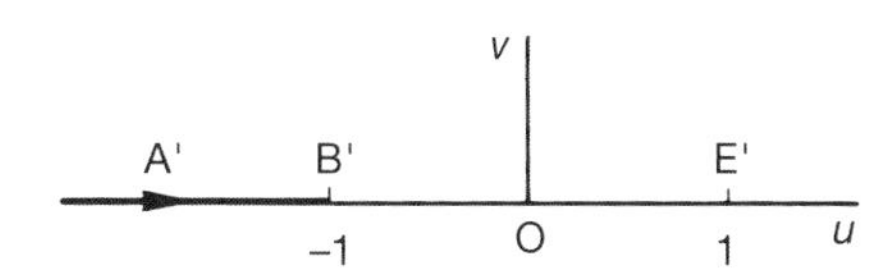

(b) BC: $y = a \quad \therefore \quad u = -e^{\pi x/a}, \quad v = 0$ (as for AB)

$\therefore$ As x decreases from 0 to $-\infty$, u increases from -1 to 0.

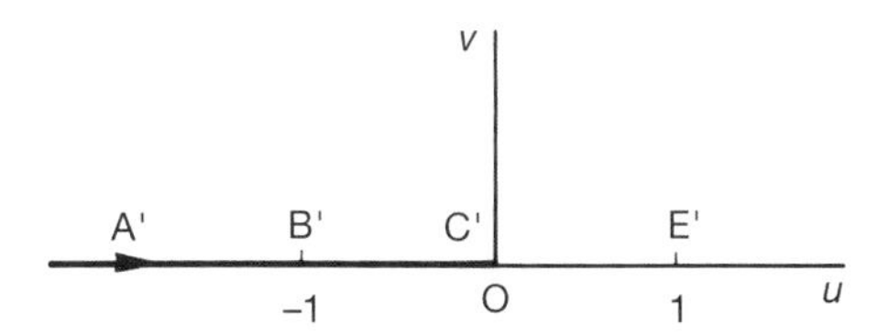

(c) Now there is DE which maps onto

. . . .

61

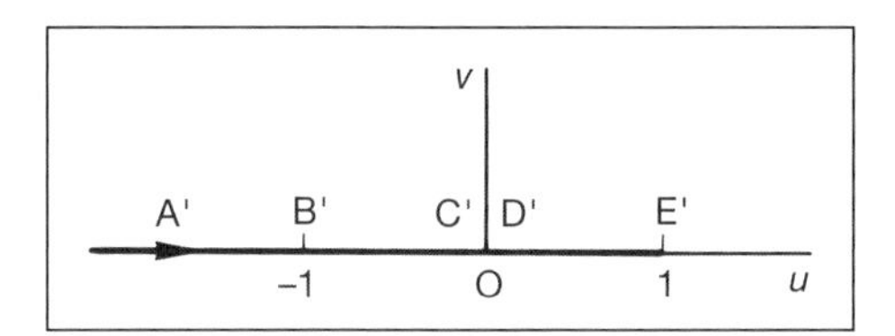

Because

(c) DE: $y = 0 \quad \therefore \quad u = e^{\pi x/a}, \quad v = 0$

$\therefore$ As x increases from $-\infty$ to 0, u increases from 0 to 1.

(d) EF: $y = 0 \quad \therefore \quad u = e^{\pi x/a}, \quad v = 0$ (as for DE)

$\therefore$ As x increases from 0 to $+\infty$, u increases from 1 to $+\infty$.

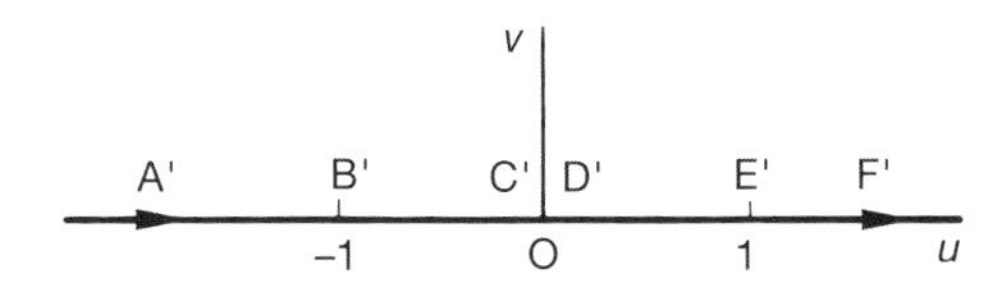

Notice that C and D map to the same point, namely $u = v = 0$.

Finally, what about the shaded region in the z-plane? This maps onto

. . . .

62

the upper half of the w-plane

because it is on the left-hand side of the directed boundary in the z-plane.

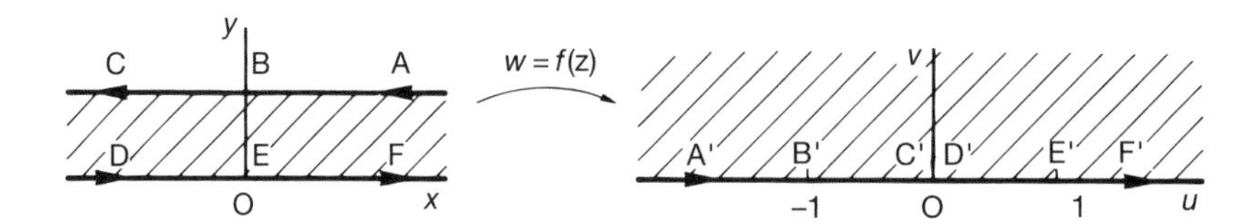

The previous two examples have been simple cases of the application of the Schwarz–Christoffel transformation under which any polygon in the z-plane can be made to map onto the entire *upper half* of the w-plane and the boundary of the polygon onto the *real axis* of the w-plane.

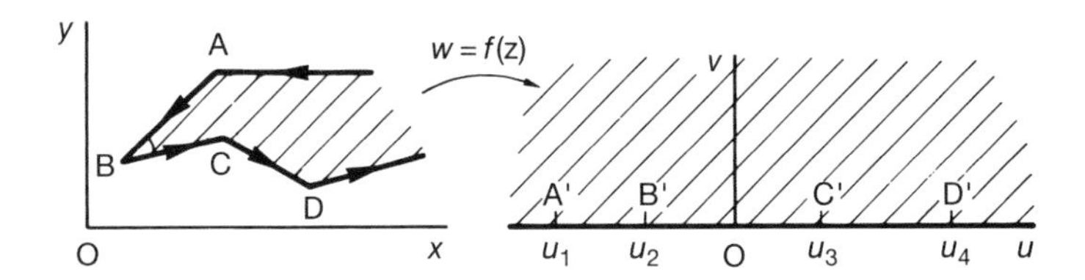

The process depends, of course, on the right choice of transformation function for any particular polygon, which can be defined by its vertices and the internal angle at each vertex.

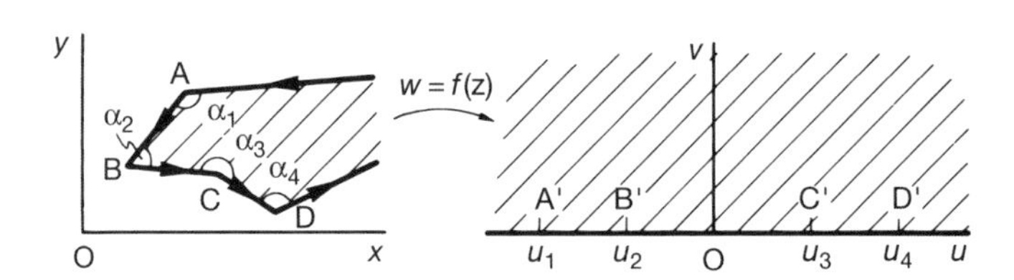

The Schwarz–Christoffel transformation function is given by

$$\frac{\mathrm{d}z}{\mathrm{d}w} = A(w - u_1)^{\alpha_1/\pi - 1}(w - u_2)^{\alpha_2/\pi - 1}(w - u_3)^{\alpha_3/\pi - 1}\dots$$

$$\therefore\ z = A\int (w - u_1)^{\alpha_1/\pi - 1}(w - u_2)^{\alpha_2/\pi - 1}\dots(w - u_n)^{\alpha_n/\pi - 1}\mathrm{d}w + B$$

where A and B are complex constants, determined by the physical properties of the polygon.

This is not as bad as it looks!

Make a careful note of it: then we will apply it

63

Here it is again.

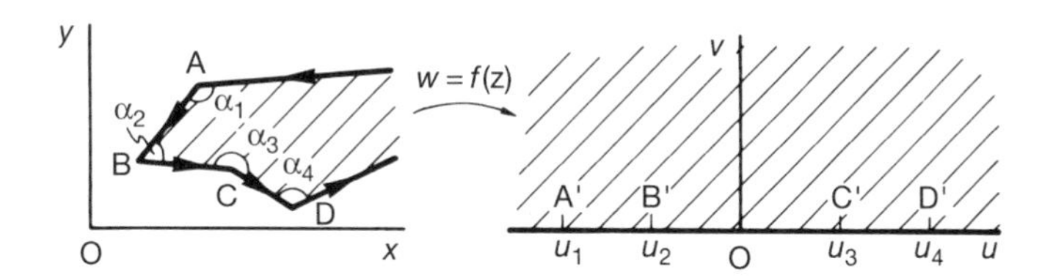

$$\frac{dz}{dw} = A(w - u_1)^{\alpha_1/\pi - 1}(w - u_2)^{\alpha_2/\pi - 1}(w - u_3)^{\alpha_3/\pi - 1}\dots$$

$$\therefore\ z = A\int (w - u_1)^{\alpha_1/\pi - 1}(w - u_2)^{\alpha_2/\pi - 1}\dots(w - u_n)^{\alpha_n/\pi - 1}dw + B$$

where A and B are complex constants.

Three other points also have to be noted.

1. Any three points u_1, u_2, u_3 on the u-axis can be selected as required.
2. It is convenient to choose one such point, u_n, at infinity, in which case the relevant factor in the integral above does not occur.
3. Infinite open polygons are regarded as limiting cases of closed polygons where one (or more) vertex is taken to infinity.

Open polygons

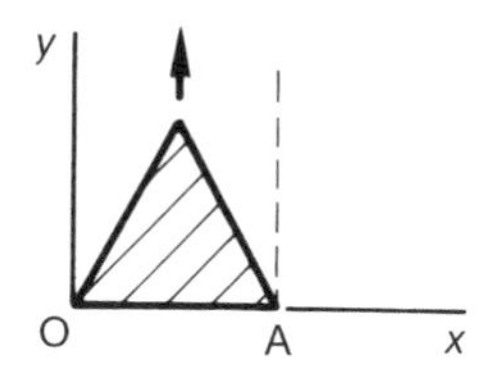

We have already introduced these in Examples 1 and 2 of this section.

In Example 1, the semi-infinite strip is a case of a triangle with one vertex that is

. . . .

64

taken to infinity in the y-direction

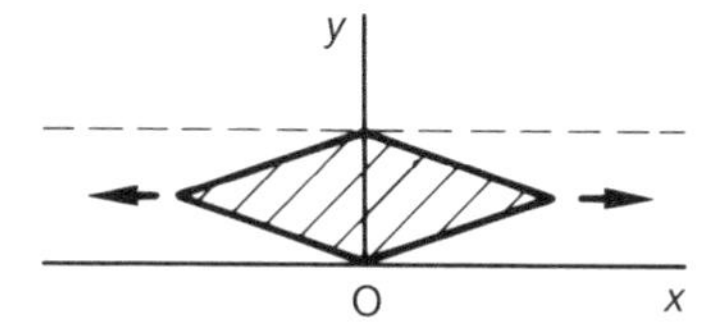

In Example 2, the infinite strip is a case of a double triangle, or quadrilateral, with two vertices taken to infinity.

An open polygon with n sides with one vertex at infinity will have $(n - 1)$ internal angles.

An open polygon with n sides with two vertices at infinity will have $(n - 2)$ internal angles.

Now for an example to see how all this works.

65

Example 3

To determine the transformation that will map the semi-infinite strip ABCD onto the w-plane so that the images of B and C occur at $u = -1$ and $u = 1$, respectively, and the shaded region maps onto the upper half of the w-plane.

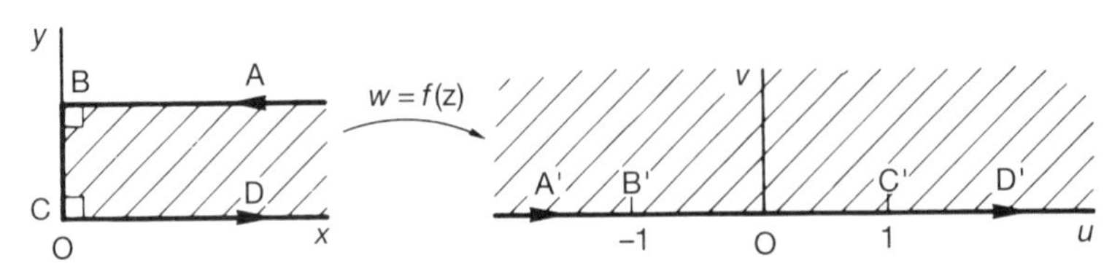

In this case, B′ is $u_1 = -1$ and C′ is $u_2 = 1$.

The corresponding internal angles are:

at B $(z = ia)$, $\alpha_1 = \dfrac{\pi}{2}$ and at C $(z = 0)$, $\alpha_2 = \dfrac{\pi}{2}$.

So we have

$$\frac{dz}{dw} = A(w+1)^{(\pi/2)/\pi - 1}(w-1)^{(\pi/2)/\pi - 1} \quad \text{where } A \text{ is a complex constant}$$
$$= A(w+1)^{-1/2}(w-1)^{-1/2}$$
$$= A(w^2-1)^{-1/2}$$
$$= K(1-w^2)^{-1/2} = \frac{K}{\sqrt{1-w^2}}$$
$$\therefore\; z = \int \frac{K}{\sqrt{1-w^2}}\,dw = \ldots\ldots\ldots\ldots$$

66

$$\boxed{z = K \arcsin w + \overline{B}}$$

$$\therefore\; \arcsin w = \frac{z - \overline{B}}{K} \qquad \therefore\; w = \sin\frac{z - \overline{B}}{K}$$

Now we have to find $\overline{B}$ and K.

(a) We require B $(z = ia)$ to map onto B′ $(w = -1)$

$$\therefore\; -1 = \sin\frac{ia - \overline{B}}{K}$$
$$\therefore\; \frac{ia - \overline{B}}{K} = -\frac{\pi}{2} \qquad \therefore\; i2a - 2\overline{B} = -K\pi \tag{1}$$

(b) We also require C $(z = 0)$ to map onto C′ $(w = 1)$ $\therefore\; 1 = \sin\dfrac{0 - \overline{B}}{K}$

$$\therefore\; -\frac{\overline{B}}{K} = \frac{\pi}{2} \qquad \therefore\; -2\overline{B} = K\pi \tag{2}$$

Then, from (1) and (2), $\overline{B} = \ldots\ldots\ldots\ldots$; $K = \ldots\ldots\ldots\ldots$

67

$$\overline{B} = \frac{ia}{2}; \quad K = -\frac{ia}{\pi}$$

$$\therefore \quad w = \sin\left\{\frac{z - (ia)/2}{-ia/\pi}\right\} = \sin\left\{iz\frac{\pi}{a} + \frac{\pi}{2}\right\} = \cos\frac{iz\pi}{a}$$

But $\cos i\theta = \cosh\theta \quad \therefore \; w = \cosh\dfrac{\pi z}{a}$

To verify that this is the required transformation, let us apply it to the figure given in the z-plane.

We will do that in the next frame

68

We have

$$w = u + iv = \cosh\frac{\pi z}{a} = \cosh\frac{(x + iy)\pi}{a}$$

$$\therefore \; u + iv = \cosh\frac{x\pi}{a}\cosh\frac{iy\pi}{a} + \sinh\frac{x\pi}{a}\sinh\frac{iy\pi}{a}$$

But $\cosh i\theta = \cosh\theta$ and $\sinh i\theta = i\sin\theta$

$$\therefore \; u + iv = \cosh\frac{x\pi}{a}\cos\frac{y\pi}{a} + i\sinh\frac{x\pi}{a}\sin\frac{y\pi}{a}$$

$$\therefore \; u = \cosh\frac{x\pi}{a}\cos\frac{y\pi}{a}; \quad v = \sinh\frac{x\pi}{a}\sin\frac{y\pi}{a}$$

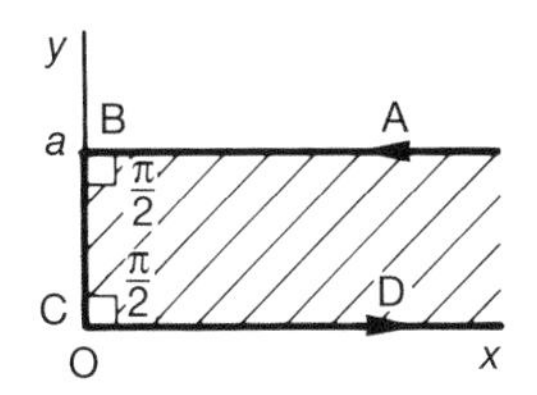

First map the points B and C onto B′ and C′ in the w-plane.

B′: ....; C′: ....

69

B′: $u = -1, v = 0$; C′: $u = 1, v = 0$

Because

B: $x = 0, y = a \quad \therefore$ B′: $u = \cos\pi = -1, v = 0 \quad \therefore$ B′: $u = -1, v = 0$

C: $x = 0, y = 0 \quad \therefore$ C′: $u = 1, v = 0 \quad \therefore$ C′: $u = 1, v = 0$.

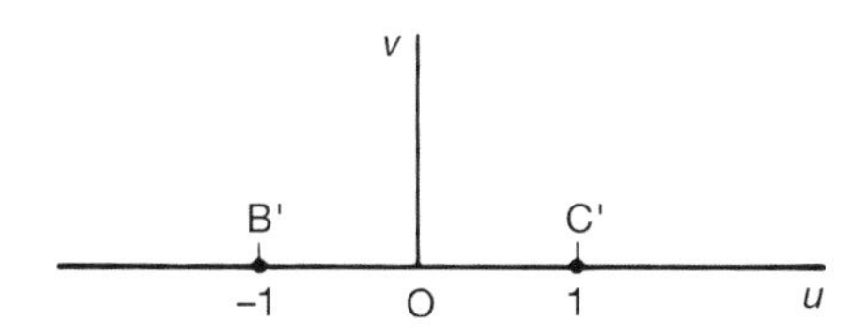

Now we map AB, BC, CD in turn.

(a) AB: $y = a \quad \therefore \ u = -\cosh\dfrac{x\pi}{a}, \quad v = 0$

$\therefore$ As x decreases from ∞ to 0, u increases from $-\infty$ to -1.

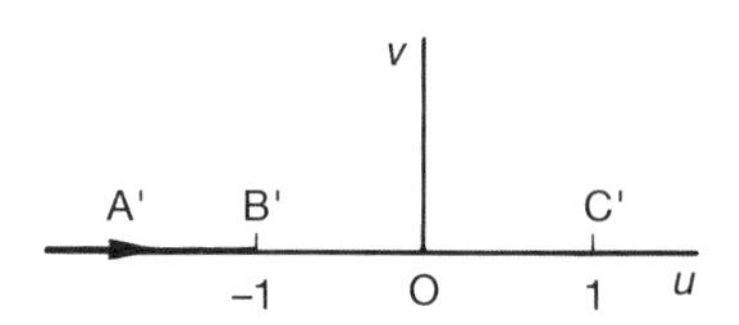

(b) BC: }
(c) CD: } Complete the working and show the mapped region

which is ....

70

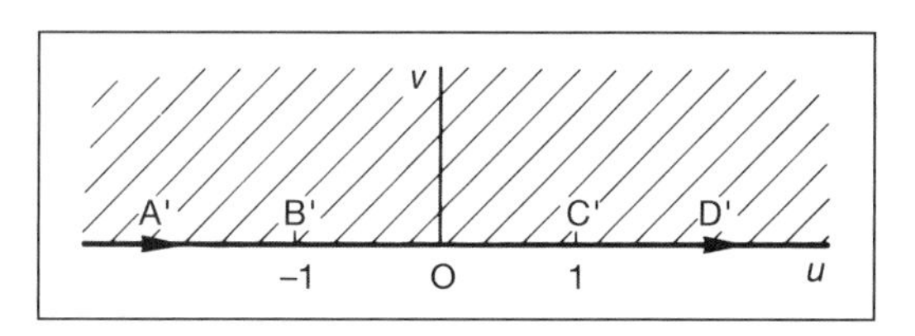

Because we have

(b) BC: $x = 0 \ \therefore \ u = \cos\dfrac{y\pi}{a}, \quad v = 0$

$\therefore$ As y decreases from a to 0, u increases from -1 to 1.

CD: $y = 0 \quad \therefore \quad u = \cosh\dfrac{x\pi}{a}, \quad v = 0$

$\therefore$ As x increases from 0 to ∞, u increases from 1 to ∞.

In each plane, the shaded region is on the left-hand side of the boundary.

We will now finish with one further example.

So move on

71

Example 4

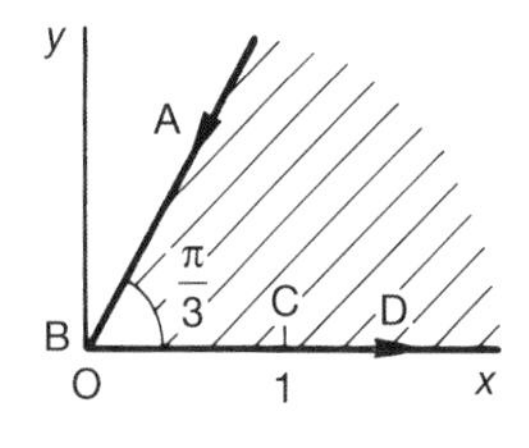

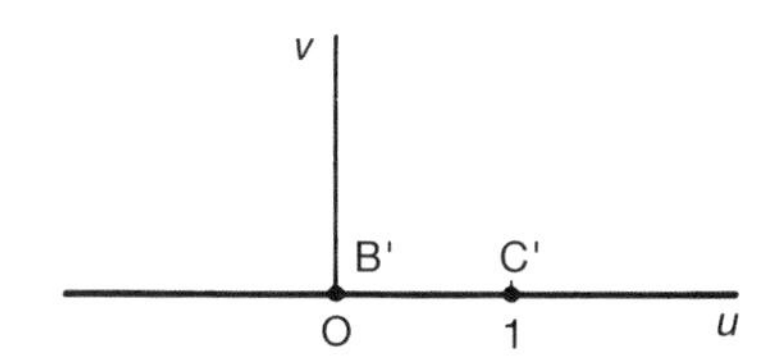

Determine the transformation function $w = f(z)$ that maps the infinite sector in the z-plane onto the upper half of the w-plane with points B and C mapping onto B′ and C′ as shown.

The transformation function $w = f(z)$ is given by ....

72

$$\frac{dz}{dw} = A(w - u_1)^{\alpha_1/\pi - 1}(w - u_2)^{\alpha_2/\pi - 1} \dots (w - u_n)^{\alpha_n/\pi - 1}$$

At B, $\alpha_1 = \frac{\pi}{3}$. At C, $\alpha_2 = \pi$.

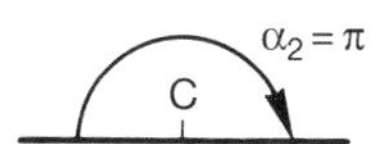

With that reminder, you can now work through on your own, just as we did before, finally obtaining

$$w = \dots\dots\dots\dots$$

73

$$w = z^3$$

Check with the working.

$$\frac{dz}{dw} = A(w - 0)^{(\pi/3)/\pi - 1}(w - 1)^{\pi/\pi - 1}$$
$$= Aw^{-2/3}(w - 1)^0$$
$$= Aw^{-2/3}$$
$$\therefore\ z = 3Aw^{1/3} + \overline{B}$$
$$= Kw^{1/3} + \overline{B}$$
$$\therefore\ w = \left(\frac{z - \overline{B}}{K}\right)^3$$

To find $\overline{B}$ and K

(a) At B: $z = 0$ At B′: $w = 0$ $\therefore\ 0 = \left(\frac{-\overline{B}}{K}\right)^3$ $\therefore\ \overline{B} = 0$ $\therefore\ w = \left(\frac{z}{K}\right)^3$

(b) At C: $z = 1$ At C′: $w = 1$ $\therefore\ 1 = \left(\frac{1}{K}\right)^3$ $\therefore\ K = 1$ $\therefore\ w = z^3$

$\therefore$ the transformation function is $w = z^3$

Finally, as a check – and a little more valuable practice – apply the function $w = z^3$ to the region shaded in the z-plane.

$$w = u + iv = (x + iy)^3 = x^3 + 3x^2(iy) + 3x(iy)^2 + (iy)^3$$

$$\therefore\ u = \dots\dots\dots\dots;\quad v = \dots\dots\dots\dots$$

74

$$u = x^3 - 3xy^2; \quad v = 3x^2y - y^3$$

At B: $x = 0,\ y = 0 \quad \therefore\ u = 0,\ v = 0 \quad \therefore$ B′: $u = 0,\ v = 0$
At C: $x = 1,\ y = 0 \quad \therefore\ u = 1,\ v = 0 \quad \therefore$ C′: $u = 1,\ v = 0$

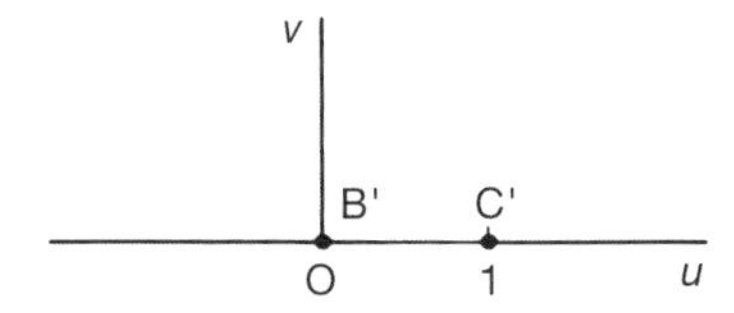

Now we map AB, BC, CD onto A′B′, B′C′, C′D′.

AB: $y = \sqrt{3}x \quad \therefore\ u = x^3 - 9x^3 = -8x^3, \quad v = 0$

$\therefore$ As x decreases from ∞ to 0, u increases from $-\infty$ to 0.

You can now deal with BC and CD in the same way and finally show the transformed region.

So we get ....

75

Here is the remaining working.

BC: $y = 0 \ \therefore\ u = x^3, \quad v = 0$

$\therefore$ As x increases from 0 to 1, u increases from 0 to 1.

CD: $y = 0 \ \therefore\ u = x^3, \quad v = 0$

$\therefore$ As x increases from 1 to ∞, u increases from 1 to ∞.

So we have

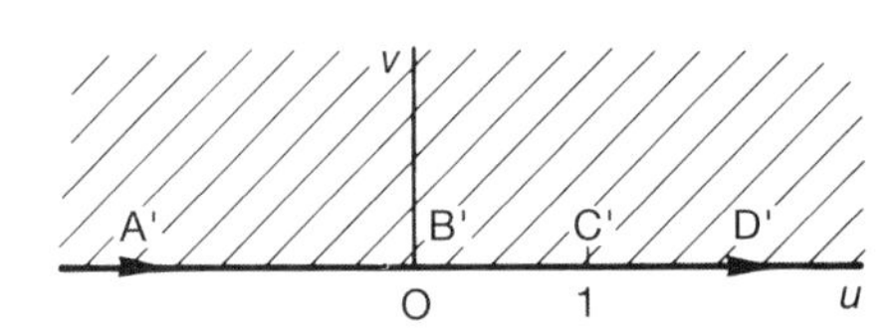

The shaded region is to the left of the directed boundary in the z-plane. This therefore maps onto the region to the left of the directed real axis in the w-plane, i.e. the upper half of the plane.

We have just touched on the fringe of the work on Schwarz– Christoffel transformation. The whole topic of mapping between planes has applications in fluid mechanics, heat conduction, electromagnetic theory, etc. and it is at times convenient to solve a problem relating to the z-plane by transforming to the upper half of the w-plane and later to transform back to the z-plane. The transformation function can be operated in either direction.

And that is it. The **Summary** follows and the **Can You?** checklist. Then on to the **Test exercise** and the **Further problems** for additional practice.

Summary

76

1 *Differentiation of a complex function*

$$w = f(z) \qquad \frac{dw}{dz} = f'(z) = \lim_{\delta z \to 0}\left\{\frac{f(z_0 + \delta z) - f(z_0)}{\delta z}\right\}$$

2 *Regular (or analytic) function*
$w = f(z)$ is *regular* at z_0 if it is defined, single-valued and has a derivative at every point at and around $z = z_0$.

3 *Singularities* or singular points – points at which $f(z)$ ceases to be regular.

4 *Cauchy–Riemann equations* test whether $w = f(z)$ has a derivative $f'(z)$ at $z = z_0$. $\quad w = u + iv = f(z)$ where $z = x + iy$.

Then $\dfrac{\partial u}{\partial x} = \dfrac{\partial v}{\partial y}$ and $\dfrac{\partial v}{\partial x} = -\dfrac{\partial u}{\partial y}$

5 If a function of two real variables $f(x, y)$ satisfies Laplace's equation

$$\frac{\partial^2 f(x, y)}{\partial x^2} + \frac{\partial^2 f(x, y)}{\partial y^2} = 0$$

then $f(x, y)$ is an harmonic function. The real and imaginary parts of an analytic function are both harmonic and form a conjugate pair of functions.

6 *Complex integration*

$$\int w\,dz = \int f(z)\,dz = \int (u\,dx - v\,dy) + i\int (v\,dx + u\,dy)$$

7 *Contour integration* – evaluation of line integrals in the z-plane.

8 *Cauchy's theorem* If $f(z)$ is regular at every point within and on closed curve c, then $\oint_c f(z)\,dz = 0$.

9 *Deformation of contours*

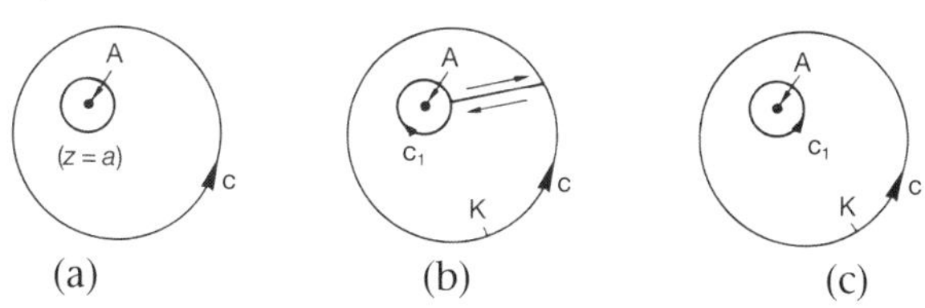

(a) Singularity at A

(b) Restored to a closed curve

(c) $\oint_c f(z)\,dz = \oint_{c_1} f(z)\,dz.$

For $\oint_c f(z)\,dz$ where $f(z) = \dfrac{1}{(z-a)^n} \quad n = 1, 2, 3, \ldots$

$$\oint_c \frac{1}{(z-a)^n}\,dz = 0 \quad \text{if } n \neq 1$$
$$= 0 \quad \text{if } n = 1 \text{ and c does not enclose } z = a$$
$$= 2\pi i \quad \text{if } n = 1 \text{ and c does enclose } z = a.$$

10 *Conformal transformation* – mapping in which angles are preserved in size and sense of rotation.

Conditions

1 $w = f(z)$ must be a regular function of z.

2 $f'(z)$, i.e. $\dfrac{dw}{dz}$, $\neq 0$ at the point of intersection.

If $f'(z) = 0$ at $z = z_0$, then z_0 is a *critical point.*

11 *Schwarz–Christoffel transformation* maps any polygon in the z-plane onto the entire *upper half* of the w-plane and the boundary of the polygon onto the *real axis* of the w-plane.

$$\frac{dz}{dw} = A(w - u_1)^{\alpha_1/\pi - 1}(w - u_2)^{\alpha_2/\pi - 1} \dots (w - u_n)^{\alpha_n/\pi - 1}$$

1 Any three points u_1, u_2, u_3 can be selected on the u-axis.

2 One such point can be chosen at infinity.

3 Infinite open polygons are regarded as limiting cases of closed polygons.

Can You?

77

Checklist 8

Check this list before and after you try the end of Program test.

On a scale of 1 to 5 how confident are you that you can: **Frames**

- Appreciate when the derivative of a function of a complex variable exists? 1 to 3
  Yes ☐ ☐ ☐ ☐ ☐ *No*
- Understand the notions of regular functions and singularities and be able to obtain the derivative of a regular function from first principles? 3 to 6
  Yes ☐ ☐ ☐ ☐ ☐ *No*
- Derive the Cauchy–Riemann equations and apply them to find the derivative of a regular function? 7 to 12
  Yes ☐ ☐ ☐ ☐ ☐ *No*
- Understand the notion of an harmonic function and derive a conjugate function? 13 to 22
  Yes ☐ ☐ ☐ ☐ ☐ *No*
- Evaluate line and contour integrals in the complex plane? 23 to 28
  Yes ☐ ☐ ☐ ☐ ☐ *No*
- Derive and apply Cauchy's theorem? 29 to 36
  Yes ☐ ☐ ☐ ☐ ☐ *No*

- Apply Cauchy's theorem to contours around regions that contain singularities?
  Yes ☐ ☐ ☐ ☐ ☐ *No* 37 to 49
- Define the essential characteristics of and conditions for a conformal mapping?
  Yes ☐ ☐ ☐ ☐ ☐ *No* 50 and 51
- Locate critical points of a function of a complex variable?
  Yes ☐ ☐ ☐ ☐ ☐ *No* 51 and 52
- Determine the image in the w-plane of a figure in the z-plane under a conformal transformation $w = f(z)$?
  Yes ☐ ☐ ☐ ☐ ☐ *No* 52 to 56
- Describe and apply the Schwarz–Christoffel transformation?
  Yes ☐ ☐ ☐ ☐ ☐ *No* 57 to 75

Text exercise 8

78

1 Determine where each of the following functions fails to be regular.

(a) $w = z^3 + 4$ (d) $w = \dfrac{z-2}{(z-4)(z+1)}$

(b) $w = \dfrac{z}{z+5}$ (e) $w = \dfrac{x - iy}{x^2 + y^2}$.

(c) $w = e^{2z+4}$

2 Demonstrate that each of the following is harmonic and obtain the conjugate function.

(a) $u(x, y) = \sinh x \cos y$

(b) $u(x, y) = 4y(1 + 3x)$.

3 Verify Cauchy's theorem by evaluating $\oint_c f(z)\,dz$ where $f(z) = z^2$ round the rectangle formed by joining the points $z = 2 + i$, $z = 2 + 4i$, $z = 4i$, $z = i$.

4 Evaluate the integral $\oint_c f(z)dz$ where $f(z) = \dfrac{3z - 6 - i}{(z-i)(z-3)}$ round the contour $|z| = 2$.

5 Determine critical points, if any, at which the following transformation functions $w = f(z)$ fail to be conformal.

(a) $w = z^4$ (d) $w = z + \dfrac{2}{z}$

(b) $w = z^3 - 3z$ (e) $w = e^{(z^2)}$

(c) $w = e^{1-z}$ (f) $w = \dfrac{z+i}{z-i}$.

▶

6 Determine the Schwarz–Christoffel transformation function $w = f(z)$ that will map the semi-infinite strip shaded in the z-plane onto the upper half of the w-plane, so that the image of B is B′ ($w = -1$) and that of C is C′ ($w = 0$). Obtain the image of the point D.

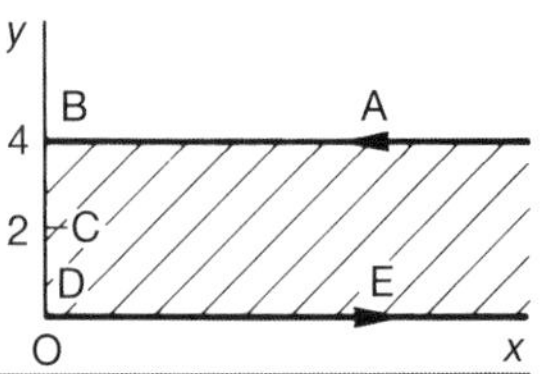

Further problems 8

79

1 Verify Cauchy's theorem for the closed path c consisting of three straight lines joining A $(1 + i)$, B $(3 + 3i)$, C $(-1 + 3i)$ where $f(z) = z - 1 + i$.

2 If $z = 2 + iy$ is mapped onto the w-plane under the transformation $w = f(z) = \dfrac{1}{z}$, show that the locus of w is a circle with center $w = 0.25$ and radius 0.25.

3 Determine the image in the w-plane of the circle $|z - 2| = 1$ in the z-plane under the transformation $w = (1 - i)z + 3$.

4 The unit circle $|z| = 1$ in the z-plane is generated in a counter-clockwise manner from the point A $(z = 1)$ and is transformed onto the w-plane by $w = \dfrac{z}{z - 2}$.
Determine the locus of w and the direction in which it is generated.

5 Find the conjugate function of each of the following.

(a) $u(x, y) = x^2 - 2x - y^2$

(b) $u(x, y) = x^3 - 3xy^2 - x^2 + y^2 + x$

(c) $u(x, y) = 2y(x - 1)$

(d) $u(x, y) = e^{x^2 - y^2} \cos 2xy$.

6 Evaluate $\oint_c f(z)\,dz$ where $f(z) = \dfrac{5z - 2 - 3i}{(z - i)(z - 1)}$ around the closed contour c for the two cases when

(a) c is the path $|z| = 2$

(b) c is the path $|z - 1| = 1$.

7 If $f(z) = \dfrac{5z + i}{(z - i)(z + 2i)}$, evaluate $\oint_c f(z)dz$ along the contours

(a) $|z - 1| = 1$; (b) $|z| = \dfrac{3}{2}$; (c) $|z| = 3$.

8 If $z = x + iy$ and $w = f(z)$, show that, if $\dfrac{i(w + z)}{w - z}$ is entirely real, then $|w| = |z|$.

9 Evaluate $\oint_c f(z)\,dz$, where $f(z) = \dfrac{3z - 5i}{(z + 1 - 2i)(z - 2 - i)}$, around the perimeter of the rectangle formed by the lines $z = 1$, $z = 3i$, $z = -2$, $z = -i$.

10 If $f(z) = \dfrac{8z^2 - 2}{z(z - 1)(z + 1)}$, evaluate $\oint_c f(z)\,dz$ along the contour c where c is the triangle joining the points $z = 2$, $z = i$, $z = -1 - i$.

11 (a) For the transformation $w = z + \dfrac{1}{z}$, state (1) singularities, (2) critical points.

(b) Apply $w = z + \dfrac{1}{z}$ to map the circle $|z| = 2$ onto the w-plane.

12 Find the images in the w-plane of (a) the line $y = 0$ and (b) the line $y = x$ that result from the mapping $w = \dfrac{z - i}{z + i}$. Show that the curves intersect at the points $(\pm 1, 0)$ in the w-plane and determine the angle at which they intersect.

13 Use the transformation $w = \dfrac{i(1 + z)}{1 - z}$ to map the unit circle $|z| = 1$ in the z-plane onto the w-plane. Determine also the image in the w-plane of the region bounded by $|z| = 1$ and inside the circle.

14 Determine the transformation that will map the semi-infinite strip shown, onto the upper half of the w-plane, where the image of B is B′ ($w = -1$) and that of C is C′ ($w = 1$).

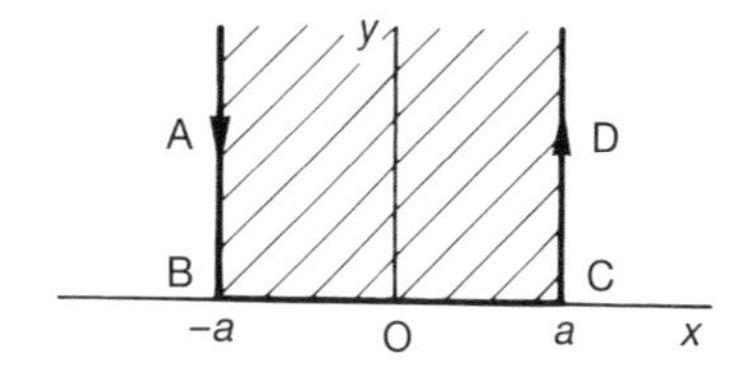

Residues

Learning outcomes

When you have completed this Program you will be able to:

- Expand a function of a complex variable about the origin in a Maclaurin series
- Determine the circle and radius of convergence of a Maclaurin series expansion
- Recognize singular points in the form of poles of order n, removable and essential singularities
- Expand a function of a complex variable about a point in the complex plane in a Taylor series, transforming the coordinates with a shift of origin
- Expand a function of a complex variable about a singular point in a Laurent series
- Recognize the principal and analytic parts of the Laurent series and link the form of the principal part to the type of singularity
- Recognize the residue of a Laurent series and state the Residue theorem
- Calculate the residues at the poles of an expression without resort to deriving the Laurent series
- Evaluate certain types of real integrals using the Residue theorem

Maclaurin series

1

You will recall that the Maclaurin series expansion of the function of a real variable x with output $f(x)$ is given as

$$f(x) = f(0) + xf'(0) + x^2\frac{f''(0)}{2!} + x^3\frac{f'''(0)}{3!} + \cdots + x^n\frac{f^{(n)}(0)}{n!} + \cdots$$

This is an infinite series expansion of $f(x)$ about the point $x = 0$. Because the series on the right-hand side of this equation contains an infinite number of terms, the right-hand side may only converge for a restricted set of values of x. Consequently, this expansion is only valid for that restricted set of values. For example, the expression $f(x) = (1 - x)^{-1}$ has the Maclaurin series expansion ....

2

$$\boxed{f(x) = 1 + x + x^2 + x^3 + \cdots + x^n + \cdots}$$

Because

$$\begin{aligned}
f(x) &= (1 - x)^{-1} \text{ and so } f(0) = (1 - 0)^{-1} = 1\\
f'(x) &= (1 - x)^{-2} \text{ and so } f'(0) = (1 - 0)^{-2} = 1\\
f''(x) &= 2(1 - x)^{-3} \text{ and so } f''(0) = 2(1 - 0)^{-3} = 2\\
f'''(x) &= 3!(1 - x)^{-4} \text{ and so } f'''(0) = 3!(1 - 0)^{-4} = 3!\\
&\vdots \qquad\qquad\qquad\qquad \vdots\\
f^{(n)}(x) &= n!(1 - x)^{-(n+1)} \text{ and so } f^{(n)}(0) = n!(1 - 0)^{-(n+1)} = n!
\end{aligned}$$

Therefore, substituting into the Maclaurin series expansion, we find

$$\begin{aligned}
f(x) &= f(0) + xf'(0) + x^2\frac{f''(0)}{2!} + x^3\frac{f'''(0)}{3!} + \cdots + x^n\frac{f^{(n)}(0)}{n!} + \cdots\\
&= 1 + x \times 1 + x^2 \times \frac{2!}{2!} + x^3 \times \frac{3!}{3!} + \cdots + x^n \times \frac{n!}{n!} + \cdots\\
&= 1 + x + x^2 + x^3 + \cdots + x^n + \cdots
\end{aligned}$$

Ratio test for convergence

This same result could also be derived by using the binomial theorem or even by performing the long division of 1 by $1 - x$. However, performing the algorithmic procedure is one thing, but knowing that the result of the procedure is valid is another. To determine the validity of the expansion we resort to convergence tests, and in this case we use the ratio test. To refresh your memory, the ratio test for the infinite series

$$f(x) = a_0(x) + a_1(x) + a_2(x) + a_3(x) + \cdots + a_n(x) + \cdots$$

is that given

$$\underset{n\to\infty}{Lim}\left|\frac{a_{n+1}(x)}{a_n(x)}\right| = L \text{ then if } \begin{array}{l} L < 1 \text{ the series converges}\\ L > 1 \text{ the series diverges}\\ L = 1 \text{ the test fails and an alternative}\\ \qquad\text{convergence test is required.}\end{array}$$

▶

Applying the ratio test to the Maclaurin series expansion

$$f(x) = 1 + x + x^2 + x^3 + \cdots + x^n + \cdots$$

tells us that

The series converges for ....
The series diverges for ....
The test fails for ....

3

The series converges for $-1 < x < 1$
The series diverges for $x < -1$ or $x > 1$
The test fails for $x = \pm 1$

Because

$$\underset{n\to\infty}{Lim}\left|\frac{a_{n+1}(x)}{a_n(x)}\right| = \underset{n\to\infty}{Lim}\left|\frac{x^{n+1}}{x^n}\right| = \underset{n\to\infty}{Lim}|x| = |x|, \text{ so}$$

if $|x| < 1$, that is $-1 < x < 1$, the series converges and so the expansion is valid
$|x| > 1$, that is $x < -1$ or $x > 1$, the series diverges and so the expansion is invalid
$|x| = 1$, that is $x = \pm 1$, the ratio test fails to give a conclusion.

By inspection, when $x = 1$ the series clearly diverges and when $x = -1$ the sum of terms alternates between 1 and 0 as each successive term is added. Clearly the series does not converge and so, therefore, it must diverge when $x = -1$.

Everything that has been said about the Maclaurin series expansion of an expression involving a real variable x can equally be said about an expression involving a complex variable z. That is, if $f(z)$ is a function in the complex variable z, analytic at $z = 0$, then the Maclaurin series expansion is

$$f(z) = f(0) + zf'(0) + z^2\frac{f''(0)}{2!} + z^3\frac{f'''(0)}{3!} + \cdots$$

So, the Maclaurin series expansion of $f(z) = \sin z$ is ....

4

$$f(z) = z - \frac{z^3}{3!} + \frac{z^5}{5!} - \cdots + \frac{(-1)^n z^{2n+1}}{(2n+1)!} + \cdots$$

Because

$f(z) = \sin z$ and so $f(0) = \sin 0 = 0$

$f'(z) = \cos z$ and so $f'(0) = \cos 0 = 1$

$f''(z) = -\sin z$ and so $f''(0) = -\sin 0 = 0$

$f'''(z) = -\cos z$ and so $f'''(0) = -\cos 0 = -1$

$\vdots \qquad\qquad \vdots$

Therefore

$$\begin{aligned} f(z) &= f(0) + zf'(0) + z^2\frac{f''(0)}{2!} + z^3\frac{f'''(0)}{3!} + \cdots \\ &= 0 + z \times 1 + z^2 \times \frac{0}{2!} + z^3 \times \frac{(-1)}{3!} + \cdots \\ &= z - \frac{z^3}{3!} + \frac{z^5}{5!} - \cdots + \frac{(-1)^n z^{2n+1}}{(2n+1)!} + \cdots \end{aligned}$$

Furthermore, applying the ratio test tells us that this series expansion is valid for ....

5

all finite values of z

Because

$$\begin{aligned} \underset{n\to\infty}{Lim}\left|\frac{a_{n+1}(z)}{a_n(z)}\right| &= \underset{n\to\infty}{Lim}\left|\frac{(-1)^{n+1}z^{2(n+1)+1}/[2(n+1)+1]!}{(-1)^n z^{2n+1}/[2n+1]!}\right| \\ &= \underset{n\to\infty}{Lim}\left|\frac{z^2}{(2n+3)(2n+2)}\right| = 0 < 1 \end{aligned}$$

So the expansion is valid for all finite values of z.

Try this one. The Maclaurin series expansion of $f(z) = \ln(1+z)$ is

$$\ln(1+z) = \ldots\ldots\ldots\ldots$$

6

$$\ln(1+z) = z - \frac{z^2}{2} + \frac{z^3}{3} - \cdots + \frac{(-1)^{n+1}z^n}{n} + \cdots \quad n = 1, 2, \ldots$$

Because

$$f(z) = \ln(1+z) \text{ and so } f(0) = (1+0) = 0$$

$$f'(z) = (1+z)^{-1} \text{ and so } f'(0) = (1+0)^{-1} = 1$$

$$f''(z) = -(1+z)^{-2} \text{ and so } f''(0) = -(1+0)^{-2} = -1$$

$$f'''(z) = 2(1+z)^{-3} \text{ and so } f'''(0) = 2(1+0)^{-3} = 2$$

$$f^{(iv)}(z) = -3!(1+z)^{-4} \text{ and so } f^{(iv)}(0) = -3!(1+0)^{-4} = -3!$$

$$\vdots \qquad \vdots$$

$$f^{(n)}(z) = (-1)^{n+1}n!(1+z)^{-n} \text{ and so } f^{(n)}(0) = (-1)^{n+1}n!(1+0)^{-n} = (-1)^{n+1}n!$$

Therefore

$$\ln(1+z) = z - \frac{z^2}{2} + \frac{z^3}{3} - \cdots + \frac{(-1)^{n+1}z^n}{n} + \cdots$$

This series is valid for

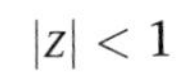

....

7

$$|z| < 1$$

Because

$$\underset{n\to\infty}{Lim}\left|\frac{a_{n+1}(z)}{a_n(z)}\right| = \underset{n\to\infty}{Lim}\left|\frac{(-1)^{n+2}z^{n+1}/[n+1]}{(-1)^{n+1}z^n/[n]}\right| = \underset{n\to\infty}{Lim}\left|\frac{nz}{n+1}\right| = |z|$$

So if $|z| < 1$ the series converges and so the expansion is valid

$|z| > 1$ the series diverges and so the expansion is invalid

$|z| = 1$ the ratio test fails

We shall look at the case $|z| = 1$ a little later.

Move to the next frame

Radius of convergence

8

We have just seen that the Maclaurin expansion of $\ln(1+z)$ is valid for $|z|<1$. This inequality defines the interior of a circle of radius 1 centered on the origin, namely $z = 1e^{i\theta}$.

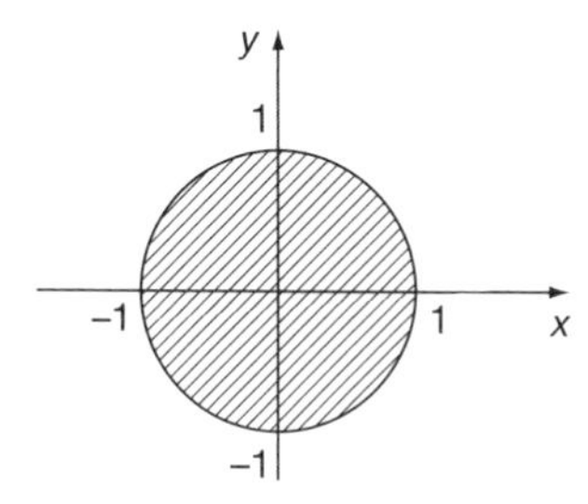

This means that the expansion is valid for all z-values lying within this circle. The radius of the circle within which a series expansion is valid is called the *radius of convergence* of the series and the circle is called the *circle of convergence*.

Example

To find the infinite series expansion and radius of convergence of the expression $f(z) = \dfrac{z}{(1-3z)^2}$, we progress in stages, noting that

$$\frac{z}{(1-3z)^2} = z(1-3z)^{-2}.$$

We expand $(1-3z)^{-2}$ first.

By the binomial theorem, the expansion of $(1-3z)^{-2}$ is

$$(1-3z)^{-2} = \ldots\ldots\ldots\ldots$$

9

$$\boxed{(1-3z)^{-2} = 1 + 6z + 27z^2 + 108z^3 + 405z^4 + \cdots}$$

Because

$$\begin{aligned}(1-3z)^{-2} &= \left(1 + (-2)\times(-3z) + \frac{(-2)(-3)\times(-3z)^2}{2!} + \frac{(-2)(-3)(-4)\times(-3z)^3}{3!} + \ldots\right)\\ &= \left(1 + 6z + 3(-3z)^2 - 4(-3z)^3 + 5(-3z)^4 + \ldots + (-1)^n(n+1)(-3)^n z^n + \ldots\right)\\ &= 1 + 6z + 27z^2 + 108z^3 + 405z^4 + \ldots + (n+1)3^n z^n + \ldots\end{aligned}$$

and so

$$z(1-3z)^{-2} = z + 6z^2 + 27z^3 + 108z^4 + 405z^5 + \ldots + (n+1)3^n z^{n+1} + \ldots$$

The radius of convergence is then $\ldots\ldots\ldots\ldots$

10

1/3

Because

The general term of the expansion is $a_n(z) = (n+1)3^n z^{n+1}$ and so the ratio test tells us that

$$\underset{n\to\infty}{Lim}\left|\frac{a_{n+1}(z)}{a_n(z)}\right| = \underset{n\to\infty}{Lim}\left|\frac{(n+2)3^{n+1}z^{n+2}}{(n+1)3^n z^{n+1}}\right| = \underset{n\to\infty}{Lim}\left|\frac{3(n+2)z}{(n+1)}\right| = |3z|$$

So, if $|3z| < 1$, that is $|z| < 1/3$, then the series converges and the expansion is valid. The radius of convergence is therefore $1/3$.

Move to the next frame

Singular points

11

Any point at which $f(z)$ fails to be analytic, that is where the derivative does not exist, is called a *singular point* (also called a singularity). For example

$$f(z) = \frac{1}{z-1}$$

is analytic everywhere in the finite complex plane except at the point $z = 1$ where not only is the derivative $f'(z)$ not defined but neither is $f(z)$. Accordingly, the point $z = 1$ is a singular point. There are different types of singular points, for now we shall look at just two of them.

Poles

If $f(z)$ has a singular point at z_0 and for some natural number n, $\underset{z\to z_0}{Lim}\{(z-z_0)^n f(z)\} = L \neq 0$ then the singular point is called a *pole of order n*. For example

$$f(z) = \frac{2z}{(z+4)^2}$$

has a singular point at $z = -4$ and because

$$\underset{z\to -4}{Lim}\left\{(z+4)^2 f(z)\right\} = \underset{z\to -4}{Lim}\{2z\} = -8 \neq 0$$

the singularity is a *pole of order* 2 (also called a *double pole*).

▶

Removable singularities

If $f(z)$ has a singular point at z_0 and $\underset{z \to z_0}{Lim}\{f(z)\}$ exists then the singular point is called a *removable singularity*. For example

$$f(z) = \frac{\sin z}{z}$$

has a singular point at $z = 0$. However, $\underset{z \to 0}{Lim}\left\{\frac{\sin z}{z}\right\} = 1$ and so the singularity at $z = 0$ is a removable singularity. We can see this from the Maclaurin series expansion of $f(z)$ where

$$f(z) = \frac{\sin z}{z} = \frac{1}{z}\left(z - \frac{z^3}{3!} + \frac{z^5}{5!} - \dots\right) = 1 - \frac{z^2}{3!} + \frac{z^4}{5!} - \dots$$

While we cannot substitute $z = 0$ into $f(z) = \frac{\sin z}{z}$, we can define $f(0) = 1$ in complete consistency with the series expansion. In this sense the singularity at $z = 0$ is removable by virtue of the fact that we can assign a value to $f(z)$ at the singularity which is consistent with the series expansion.

Move to the next frame

Circle of convergence

12

When an expression is expanded in a Maclaurin series, the circle of convergence is always centered on the origin and the radius of convergence is determined by the location of the first singular point met as $|z|$ increases from $|z| = 0$. For example, the Maclaurin series expansion of $f(z) = \ln(1 + z)$ is

$$\ln(1 + z) = z - \frac{z^2}{2} + \frac{z^3}{3} - \dots + \frac{(-1)^{n+1} z^n}{n} + \dots$$

which is valid inside the circle of convergence $|z| = 1$. The first singular point met by this function as $|z|$ increases from zero is at $z = -1$, for at that point $\ln(1 + z)$ is not defined and the series

$$-1 - \frac{1}{2} - \frac{1}{3} - \dots - \frac{1}{n} - \dots$$

diverges – it is the negative of the harmonic series. Hence the radius of convergence is 1. When $z = 1$, substitution into the series expansion gives

$$\ln 2 = 1 - \frac{1}{2} + \frac{1}{3} - \dots + \frac{(-1)^{n+1}}{n} + \dots$$

▶

The right-hand side is the alternating harmonic series which we know converges by the *alternating sign test* which states that if the magnitude of the terms decreases and the signs alternate then the series converges. Now we know that it converges to $\ln 2$. Notice that the circle of convergence is identified by the location of the *first* singularity as $|z|$ increases from $|z| = 0$. This does not mean that the function is singular at all points on the circle of convergence.

There are times when it is desirable to have a series expansion of an expression that is singular at the origin. Because the Maclaurin expansion requires the function to be analytic everywhere within the circle of convergence which is centered on the origin, we cannot use that method. Fortunately, we do have a method of expanding a function about *any point* in the complex plane – this is Taylor's expansion.

Move to the next frame

Taylor's series

13

Provided $f(z)$ is analytic inside and on a simple closed curve c, the Taylor series expansion of $f(z)$ about the point z_0 which is interior to c is given as

$$f(z) = f(z_0) + (z - z_0)f'(z_0) + \frac{(z - z_0)^2 f''(z_0)}{2!} + \cdots + \frac{(z - z_0)^n f^{(n)}(z_0)}{n!} + \cdots$$

where here, the point z_0 is the center of the circle of convergence. The circle of convergence is given as $|z - z_0| = R$. That is $z - z_o = Re^{i\theta}$ or $z = z_0 + Re^{i\theta}$ where R is the radius of convergence.

Notice that Maclaurin's series is a special case of Taylor's series where $z_0 = 0$.

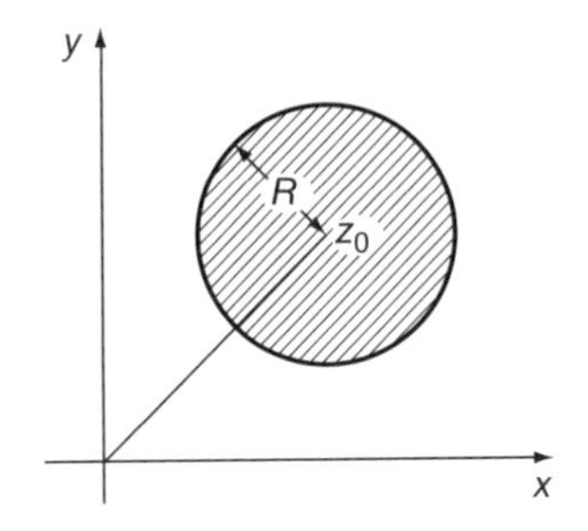

Example

Expand $f(z) = \dfrac{1}{z+1}$ in a Taylor series about the point $z = 1$ and find the values of z for which the expansion is valid.

The simplest way of doing this is to perform a coordinate transformation that moves the origin of the new coordinate to the point $z = 1$ and then derive the series about the new origin. To do this we define a new complex variable $u = z - 1$ so that $z = u + 1$ and so

$$\frac{1}{z+1} \text{ becomes } \frac{1}{u+2} = (2+u)^{-1} = \frac{1}{2}\left(1 + \frac{u}{2}\right)^{-1}.$$

▶

The expansion of this expression can now be derived using either Maclaurin or, as here, the binomial theorem to obtain

$$\begin{aligned}\frac{1}{u+2}&=\frac{1}{2}\left(1+(-1)\frac{u}{2}+\frac{(-1)(-2)}{2!}\left(\frac{u}{2}\right)^2+\cdots\right)\\&=\frac{1}{2}-\frac{u}{4}+\frac{u^2}{8}-\frac{u^3}{16}+\cdots\end{aligned}$$

Transforming back to the original variable z gives

$$\frac{1}{z+1}=\frac{1}{2}-\frac{z-1}{4}+\frac{(z-1)^2}{8}-\frac{(z-1)^3}{16}+\cdots$$

The circle of convergence is given by $\left|\frac{u}{2}\right|=1$, that is $\left|\frac{z-1}{2}\right|=1$ or $|z-1|=2$. Consequently, this series expansion is valid provided z is inside the circle defined by

$$z-1=2e^{i\theta} \text{ that is } z=1+2e^{i\theta}$$

By the same reasoning, the Taylor series expansion of $f(z)=\cos z$ about the point $z=\pi/3$ is

....

14

$$\boxed{\frac{1}{2}\left(1-\sqrt{3}(z-\pi/3)-\frac{(z-\pi/3)^2}{2!}+\sqrt{3}\frac{(z-\pi/3)^3}{3!}+\frac{(z-\pi/3)^4}{4!}-\cdots\right)}$$

Because

If $u=z-\pi/3$ then

$$\begin{aligned}\cos z&=\cos(u+\pi/3)\\&=\cos u\cos\pi/3-\sin u\sin\pi/3\\&=\frac{1}{2}\left(\cos u-\sqrt{3}\sin u\right)\\&=\frac{1}{2}\left(\left[1-\frac{u^2}{2!}+\frac{u^4}{4!}-\cdots\right]-\sqrt{3}\left[u-\frac{u^3}{3!}+\frac{u^5}{5!}-\cdots\right]\right)\\&=\frac{1}{2}\left(1-\frac{u^2}{2!}+\frac{u^4}{4!}-\cdots-\sqrt{3}u+\sqrt{3}\frac{u^3}{3!}-\sqrt{3}\frac{u^5}{5!}-\cdots\right)\\&=\frac{1}{2}\left(1-\sqrt{3}u-\frac{u^2}{2!}+\sqrt{3}\frac{u^3}{3!}+\frac{u^4}{4!}-\sqrt{3}\frac{u^5}{5!}-\cdots\right)\\&=\frac{1}{2}\left(1-\sqrt{3}(z-\pi/3)-\frac{(z-\pi/3)^2}{2!}+\sqrt{3}\frac{(z-\pi/3)^3}{3!}\right.\\&\qquad\left.+\frac{(z-\pi/3)^4}{4!}-\cdots\right)\text{ for } z<\infty\end{aligned}$$

Laurent's series

15

Sometimes a valid series expansion of a function is required within a specific region of the complex plane that contains a singular point. In this case we cannot avoid the singular point as we did with Taylor's series by expanding about an alternative non-singular point, because then we move away from part of the specified region. To accommodate this case we can use the *Laurent series expansion* which provides a series expansion valid within an annular region *centered on the singular point.*

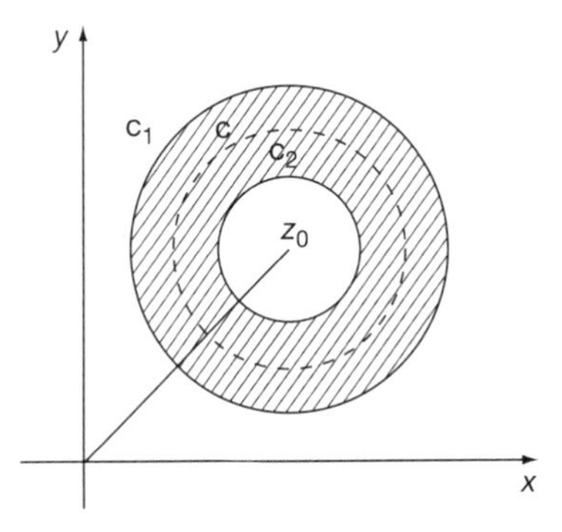

Let $f(z)$ be singular at $z = z_0$ and let c_1 and c_2 be two concentric circles centered on z_0. Then if $f(z)$ is analytic in the annular region between c_1 and c_2 and if c is any concentric circle lying within the annular region between c_1 and c_2 we can expand $f(z)$ as a Laurent series in the form

$$\begin{aligned} f(z) &= \cdots + \frac{a_{-2}}{(z-z_0)^2} + \frac{a_{-1}}{(z-z_0)} + a_0 + a_1(z-z_0) + a_2(z-z_0)^2 + \cdots \\ &= \sum_{n=-\infty}^{\infty} a_n(z-z_0)^n \end{aligned}$$

where $a_n = \dfrac{1}{2\pi i}\oint_c \dfrac{f(z)}{(z-z_0)^{n+1}}\,dz$

Example

Expand $\dfrac{e^{3z}}{(z-2)^4}$ in a Laurent series about the point $z = 2$ and determine the nature of the singularity at $z = 2$.

$f(z) = \dfrac{e^{3z}}{(z-2)^4}$ and $f'(z) = \dfrac{e^{3z}(3z-10)}{(z-2)^5}$ so $f(z)$ is analytic everywhere except at $z = 2$. The first thing we must do is to transform the coordinate system by shifting the origin to the point $z = 2$ by defining $u = z - 2$ so that $z = u + 2$. Then

$$\frac{e^{3z}}{(z-2)^4} = \frac{e^{3(u+2)}}{u^4} = e^6\frac{e^{3u}}{u^4}.$$

▶

Now we can expand using the Maclaurin series expansion

$$= \frac{e^6}{u^4}\left\{1 + 3u + \frac{(3u)^2}{2!} + \frac{(3u)^3}{3!} + \frac{(3u)^4}{4!} + \frac{(3u)^5}{5!} + \cdots\right\}$$

$$= e^6\left\{\frac{1}{u^4} + \frac{3u}{u^4} + \frac{(3u)^2}{2!u^4} + \frac{(3u)^3}{3!u^4} + \frac{(3u)^4}{4!u^4} + \frac{(3u)^5}{5!u^4} + \cdots\right\}$$

$$= e^6\left\{\frac{1}{u^4} + \frac{3}{u^3} + \frac{9}{2u^2} + \frac{27}{6u} + \frac{81}{24} + \frac{243u}{120} + \cdots\right\}$$

$$= e^6\left\{\frac{1}{(z-2)^4} + \frac{3}{(z-2)^3} + \frac{9}{2(z-2)^2} + \frac{9}{2(z-2)} + \frac{27}{8} + \frac{81(z-2)}{40} + \cdots\right\}$$

This series converges for all finite z except $z = 2$ at which point there is a pole of order 4.

Essential singularity

The part of the Laurent series that contains negative powers of the variable is called the *principal part* of the series and the remaining terms constitute what is called the *analytic part* of the series. *If, in the principal part the highest power of $1/z$ is n, then the function possesses a pole of order n; and if the principal part contains an infinite number of terms, the function possesses an* **essential singularity**.

Now you try one

16

The Laurent series expansion of $z^2 \cos\frac{1}{z}$ about the point $z = 0$ is .... valid for .... at which point there is ....

17

$$z^2 - \frac{1}{2!} + \frac{1}{4!z^2} - \frac{1}{6!z^4} + \cdots \text{ valid for all } z \neq 0$$

at which point there is an essential singularity

Because

$f(z) = z^2 \cos\frac{1}{z}$ and $f'(z) = 2z\cos\frac{1}{z} + \sin\frac{1}{z}$ and so $f(z)$ is analytic everywhere except at $z = 0$. Expanding about $z = 0$ gives

$$z^2 \cos\frac{1}{z} = z^2\left(1 - \frac{(1/z)^2}{2!} + \frac{(1/z)^4}{4!} - \frac{(1/z)^6}{6!} + \cdots\right)$$

$$= z^2 - \frac{1}{2!} + \frac{1}{4!z^2} - \frac{1}{6!z^4} + \cdots$$

valid for all $z \neq 0$, at which point there is an essential singularity because there is an infinity of terms in the principal part of the series.

Try another. The Laurent series expansion of $\frac{z}{(z+2)(z+4)}$ valid for

$2 < |z| < 4$ is ....

18

$$\cdots+\frac{8}{z^4}-\frac{4}{z^3}+\frac{2}{z^2}-\frac{1}{z}+\frac{1}{2}-\frac{z}{8}+\frac{z^2}{32}-\frac{z^3}{128}+\cdots$$

Because

$$\frac{z}{(z+2)(z+4)}=\frac{2}{z+4}-\frac{1}{z+2}\quad\text{(separating into partial fractions)}$$

If $|z|>2$ then we can write $\dfrac{1}{z+2}=\dfrac{1}{z(1+2/z)}=\dfrac{(1+2/z)^{-1}}{z}$

and because $|z|>2$, that is, $|2/z|<1$, we can now use the binomial theorem

$$\frac{1}{z+2}=\frac{1}{z(1+2/z)}=\frac{1}{z}\left\{1-\frac{2}{z}+\frac{4}{z^2}-\frac{8}{z^3}+\cdots\right\}=\frac{1}{z}-\frac{2}{z^2}+\frac{4}{z^3}-\frac{8}{z^4}+\cdots$$

and if $|z|<4$ then

$$\frac{2}{z+4}=\frac{1}{2(1+z/4)}=\frac{1}{2}\left\{1-\frac{z}{4}+\frac{z^2}{16}-\frac{z^3}{64}+\cdots\right\}$$
$$=\frac{1}{2}-\frac{z}{8}+\frac{z^2}{32}-\frac{z^3}{128}+\cdots$$

Note the expansion of $(1+z/4)^{-1}$ which is valid for $|z/4|<1$, that is $|z|<4$.

The first expansion for $|z|>2$ is still valid for $|z|<4$ since $4>2$ and the second expansion for $|z|<4$ is still valid for $|z|>2$ since $2<4$. Consequently, if $2<|z|<4$, then, by subtracting the first series from the second

$$\frac{z}{(z+2)(z+4)}=\frac{2}{z+4}-\frac{1}{z+2}$$
$$=\left\{\frac{1}{2}-\frac{z}{8}+\frac{z^2}{32}-\frac{z^3}{128}+\cdots\right\}-\left\{\frac{1}{z}-\frac{2}{z^2}+\frac{4}{z^3}-\frac{8}{z^4}+\cdots\right\}$$
$$=\cdots+\frac{8}{z^4}-\frac{4}{z^3}+\frac{2}{z^2}-\frac{1}{z}+\frac{1}{2}-\frac{z}{8}+\frac{z^2}{32}-\frac{z^3}{128}+\cdots$$

Take care here! You may be tempted to think that this displays an essential singularity at $z=0$. This is not the case because the expansion is only valid inside the annular region $2<|z|<4$ centered on the origin. Consequently, the point $z=0$ is outside this region and the series expansion is invalid at that point.

The series expansion of the same function valid for $|z|<2$ is

....

19

$$\frac{z}{8}-\frac{3z^2}{32}+\frac{7z^3}{128}+\cdots$$

Because

If $|z|<2$ then $\dfrac{1}{z+2}=\dfrac{1}{2(1+z/2)}=\dfrac{1}{2}\left\{1-\dfrac{z}{2}+\dfrac{z^2}{4}-\dfrac{z^3}{8}+\cdots\right\}$

$$=\frac{1}{2}-\frac{z}{4}+\frac{z^2}{8}-\frac{z^3}{16}+\cdots$$

We have already seen that if $|z|<4$ then

$$\frac{2}{z+4}=\frac{1}{2}-\frac{z}{8}+\frac{z^2}{32}-\frac{z^3}{128}+\cdots$$

This is still valid for $|z|<2$ since $2<4$. Conseqently, if $|z|<2$, then, by subtracting the first series from the second

$$\begin{aligned}\frac{z}{(z+2)(z+4)}&=\frac{2}{z+4}-\frac{1}{z+2}\\&=\left\{\frac{1}{2}-\frac{z}{8}+\frac{z^2}{32}-\frac{z^3}{128}+\cdots\right\}-\left\{\frac{1}{2}-\frac{z}{4}+\frac{z^2}{8}-\frac{z^3}{16}+\cdots\right\}\\&=\frac{z}{8}-\frac{3z^2}{32}+\frac{7z^3}{128}-\cdots\end{aligned}$$

Notice that for different regions of convergence we obtain different series expansions. Furthermore, each series expansion is unique within its own particular radius of convergence.

Try one more just to make sure that you can derive these expansions.

The Laurent series of $\dfrac{1-\cos(z-6)}{(z-6)^2}$ about the point $z=6$ is

.... valid for .... at which point there is ....

20

$\dfrac{1}{2!}-\dfrac{(z-6)^2}{4!}+\dfrac{(z-6)^4}{6!}-\cdots$ valid for all $z\neq 6$
at which point there is a removable singularity

Because

If we let $u=z-6$ then

$$\begin{aligned}\frac{1-\cos(z-6)}{(z-6)^2}&=\frac{1-\cos u}{u^2}\\&=\frac{1}{u^2}\left\{1-\left(1-\frac{u^2}{2!}+\frac{u^4}{4!}-\frac{u^6}{6!}+\cdots\right)\right\}\\&=\frac{1}{2!}-\frac{u^2}{4!}+\frac{u^4}{6!}-\cdots\\&=\frac{1}{2!}-\frac{(z-6)^2}{4!}+\frac{(z-6)^4}{6!}-\cdots\end{aligned}$$

▶

This is valid for all finite values of $z \neq 6$ at which point there is a removable singularity which can be removed by defining $\dfrac{1-\cos(z-6)}{(z-6)^2}$ at $z=6$ as $\dfrac{1}{2!}$. Notice that here the principal part has no terms, so that the Laurent series is identical to the Taylor series.

Next frame

Residues

21

In the Laurent series

$$f(z) = \cdots + \frac{a_{-2}}{(z-z_0)^2} + \frac{a_{-1}}{(z-z_0)} + a_0 + a_1(z-z_0) + a_2(z-z_0)^2 + \cdots$$

the coefficient a_{-1} is referred to as the *residue* of $f(z)$ for reasons that will soon become apparent. Recall the integral in Frame 45 of Program 8 which states that if the simple closed contour c has z_0 as an interior point, then

$$\oint_c \frac{dz}{(z-z_0)^n} = 2\pi i\delta_{n1}$$

where the Kronecker delta $\delta_{n1} = \begin{cases} 1 & \text{if } n=1 \\ 0 & \text{if } n \neq 1 \end{cases}$. Applying this fact to the Laurent series of $f(z)$ yields

$$\begin{aligned}\oint_c f(z)\,dz &= \oint_c \left[\cdots + \frac{a_{-2}}{(z-z_0)^2} + \frac{a_{-1}}{(z-z_0)} + a_0 + a_1(z-z_0) \right.\\ &\qquad\qquad \left. + a_2(z-z_0)^2 + \cdots\right] dz \\ &= \cdots + \oint_c \frac{a_{-2}\,dz}{(z-z_0)^2} + \oint_c \frac{a_{-1}\,dz}{(z-z_0)} + \oint_c a_0\,dz \\ &\qquad + \oint_c a_1(z-z_0)\,dz + \oint_c a_2(z-z_0)^2 dz + \cdots \\ &= \cdots + 0 + 2\pi i a_{-1} + 0 + 0 + 0 + \cdots \\ &= 2\pi i a_{-1}\end{aligned}$$

That is, provided $f(z)$ is analytic at all points inside and on the simple closed contour c, apart from the single isolated singularity at z_0 which is interior to c, then

$$\oint_c f(z)\,dz = 2\pi i a_{-1}$$

Hence the name *residue* for a_{-1} because it is all that remains when the Laurent series is integrated term by term. This statement is called the **Residue theorem** and it has many far reaching consequences – we shall see some of these later. For now, just try an example.

▶

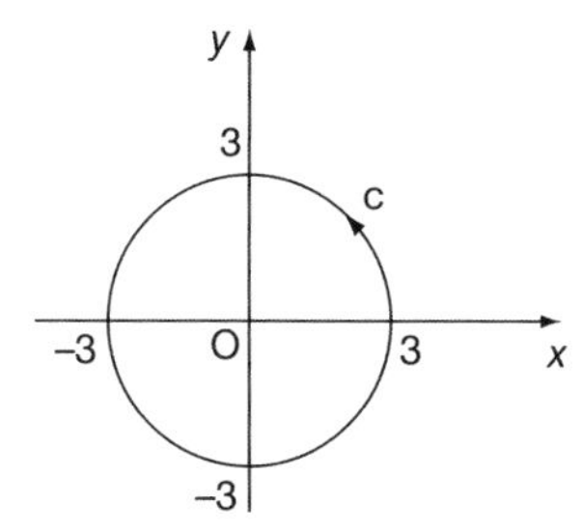

If c is a circle, centered on the origin and of radius 3, then

$$\oint_c \frac{z\,dz}{(z+2)(z+4)} = \ldots\ldots\ldots\ldots$$

22

$$\oint_c \frac{z\,dz}{(z+2)(z+4)} = -2\pi i$$

Because

The circle $|z| = 3$ lies within the annular region $2 < |z| < 4$ and we have already found the Laurent series for the integrand valid for $2 < |z| < 4$ in Frame 18, namely

$$\frac{z}{(z+2)(z+4)} = \frac{2}{z+4} - \frac{1}{z+2}$$

$$= \cdots + \frac{8}{z^4} - \frac{4}{z^3} + \frac{2}{z^2} - \frac{1}{z} + \frac{1}{2} - \frac{z}{8} + \frac{z^2}{32} - \frac{z^3}{128} + \cdots$$

Here the residue is $a_{-1} = -1$ and so $\oint_c \frac{z\,dz}{(z+2)(z+4)} = 2\pi i(-1) = -2\pi i$

where c lies entirely within the region of convergence.

The Residue theorem extends to the case where the contour contains a finite number of singularities. If $f(z)$ is analytic inside and on the simple closed contour c except at the finite number of points $z_0, z_1, z_2, \ldots$, each with a Laurent series expansion and each with corresponding residues $\overset{(0)}{a}_{-1}$, $\overset{(1)}{a}_{-1}$, $\overset{(2)}{a}_{-1}, \ldots$ then

$$\oint_c f(z)\,dz = 2\pi i\left\{\overset{(0)}{a}_{-1} + \overset{(1)}{a}_{-1} + \overset{(2)}{a}_{-1}\right\} = 2\pi i\{\text{sum of residues inside c}\}$$

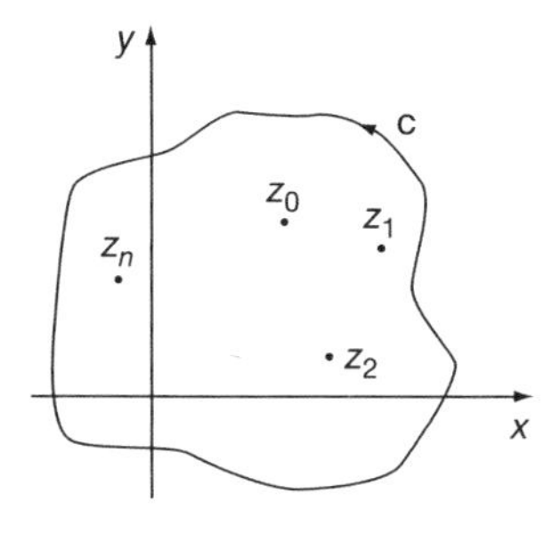

What could be more straightforward?

Next frame

Calculating residues

23

When evaluating these integrals the major part of the exercise is to find the residues, and it would be very tedious if we had to find a Laurent series for each and every singularity. Fortunately there is a simpler method for poles. If $f(z)$ is analytic inside and on the simple closed contour c except at the interior point z_0 at which there is a pole of order n, then

$$a_{-1} = \underset{z \to z_0}{Lim}\left[\frac{1}{(n-1)!}\frac{d^{n-1}}{dz^{n-1}}\left((z-z_0)^n f(z)\right)\right]$$

Example

Find the residues at all the poles of $f(z) = \dfrac{3z}{(z+2)^2(z^2-1)}$.

$f(z)$ has a pole of order 2 (a double pole) at $z = -2$ and two poles of order 1 (simple poles) at $z = \pm 1$.

At $z = -2$ the residue is $a_{-1} = \underset{z \to -2}{Lim}\left[\dfrac{1}{(2-1)!}\dfrac{d^{2-1}}{dz^{2-1}}\left((z+2)^2 f(z)\right)\right]$

$$= \underset{z \to -2}{Lim}\left[\frac{d}{dz}\left(\frac{3z}{z^2-1}\right)\right]$$

$$= \underset{z \to -2}{Lim}\left[\frac{3(z^2-1)-6z^2}{(z^2-1)^2}\right]$$

$$= \frac{3(4-1)-24}{(4-1)^2} = -\frac{5}{3}$$

At $z = 1$ the residue is ....

24

$$\boxed{\frac{1}{6}}$$

Because

At $z = 1$ the residue is $a_{-1} = \underset{z \to 1}{Lim}\left[\dfrac{1}{(1-1)!}\dfrac{d^{1-1}}{dz^{1-1}}\left((z-1)f(z)\right)\right]$

$$= \underset{z \to 1}{Lim}\left[\frac{d^0}{dz^0}\left(\frac{3z}{(z+2)^2(z+1)}\right)\right]$$

The zeroth derivative of an expression is the expression itself

$$= \underset{z \to 1}{Lim}\left[\frac{3z}{(z+2)^2(z+1)}\right]$$

$$= \frac{3}{(3)^2(2)} = \frac{1}{6}$$

At $z = -1$ the residue is ....

25

$$\boxed{\frac{3}{2}}$$

Because

$$\text{At } z = -1 \text{ the residue is } a_{-1} = \underset{z\to -1}{Lim}\left[\frac{1}{(1-1)!}\frac{d^{1-1}}{dz^{1-1}}((z+1)f(z))\right]$$
$$= \underset{z\to -1}{Lim}\left[\frac{d^0}{dz^0}\left(\frac{3z}{(z+2)^2(z-1)}\right)\right]$$
$$= \underset{z\to -1}{Lim}\left[\frac{3z}{(z+2)^2(z-1)}\right]$$
$$= \frac{-3}{(1)^2(-2)}$$
$$= \frac{3}{2}$$

Move to the next frame

Integrals of real functions

26

The Residue theorem can be very usefully employed to evaluate integrals of real functions that cannot be evaluated using the real calculus. Even when an integral is susceptible to evaluation by the real calculus, the use of the residue calculus can often save a great amount of effort. We shall look at three types of real integral and in each case we shall proceed by example.

Integrals of the form $\int_0^{2\pi} F(\cos\theta, \sin\theta)\,d\theta$

Example

Evaluate $\displaystyle\int_0^{2\pi} \frac{1}{4\cos\theta - 5}\,d\theta$.

To evaluate this integral we make use of the exponential representation of a complex number of unit length, namely $z = e^{i\theta}$,and the exponential form of the trigonometric functions

$$\cos\theta = \frac{e^{i\theta} + e^{-i\theta}}{2} = \frac{z + z^{-1}}{2} \text{ and } \sin\theta = \frac{e^{i\theta} - e^{-i\theta}}{2i} = \frac{z - z^{-1}}{2i},$$

and finally $dz = ie^{i\theta}\,d\theta = iz\,d\theta$ so that $d\theta = dz/iz$

▶

Using these relations we can transform the real integral from 0 to 2π into a contour integral in the complex plane where the contour c is the *unit circle centered on the origin*. That is

$$\int_0^{2\pi} \frac{1}{4\cos\theta - 5}\, d\theta = \oint_c \frac{1}{4\dfrac{z+z^{-1}}{2} - 5} \times \frac{dz}{iz}$$

$$= -i\oint_c \frac{1}{2z^2 - 5z + 2}\, dz$$

$$= -i\oint_c \frac{1}{(2z-1)(z-2)}\, dz$$

The complex integrand has two simple poles, one at $z = \dfrac{1}{2}$ which is inside the contour c and another at $z = 2$ which is outside the contour c. Using the Residue theorem

$$-i\oint_c \frac{1}{(2z-1)(z-2)}\, dz = -i \times 2\pi i \times \{\text{residue at } z = 1/2\}$$

The residue at $z = 1/2$ is

$$\lim_{z\to 1/2} \left\{(z - 1/2)\frac{1}{(2z-1)(z-2)}\right\} = \lim_{z\to 1/2} \left\{\frac{1}{2(z-2)}\right\}$$

$$= -\frac{1}{3}$$

so that

$$\int_0^{2\pi} \frac{1}{4\cos\theta - 5}\, d\theta = -i\oint_c \frac{1}{(2z-1)(z-2)}\, dz$$

$$= -i \times 2\pi i \times \{\text{residue at } z = 1/2\}$$

$$= -2\pi/3$$

Now you try one

$$\int_0^{2\pi} \frac{d\theta}{2 + \cos\theta} = \ldots\ldots\ldots\ldots$$

27

$$\boxed{\frac{2\pi}{\sqrt{3}}}$$

Because

$$\int_0^{2\pi}\frac{d\theta}{2+\cos\theta}=\oint_c\frac{dz/iz}{2+\dfrac{z+z^{-1}}{2}}$$ where c is the unit circle centered on the origin.

$$=-i\oint_c\frac{2\,dz}{z^2+4z+1}$$

$$=-i\oint_c\frac{2\,dz}{(z+2-\sqrt{3})(z+2+\sqrt{3})}$$

The integrand has two simple poles, one at $z=-2+\sqrt{3}$ which is inside c and another at $z=-2-\sqrt{3}$ which is outside c. Therefore

$$-i\oint_c\frac{2\,dz}{(z+2-\sqrt{3})(z+2+\sqrt{3})}=-i\times 2\pi i\times\left\{\text{residue at } z=-2+\sqrt{3}\right\}$$

The residue is

$$\lim_{z\to-2+\sqrt{3}}\left\{(z+2-\sqrt{3})\frac{2}{(z+2-\sqrt{3})(z+2+\sqrt{3})}\right\}$$

$$=\lim_{z\to-2+\sqrt{3}}\left\{\frac{2}{(z+2+\sqrt{3})}\right\}=\frac{1}{\sqrt{3}}$$ and so

$$\int_0^{2\pi}\frac{d\theta}{2+\cos\theta}=-i\oint_c\frac{2\,dz}{(z+2-\sqrt{3})(z+2+\sqrt{3})}=-i\times 2\pi i\times\frac{1}{\sqrt{3}}$$

$$=2\pi\times\frac{1}{\sqrt{3}}=\frac{2\pi}{\sqrt{3}}$$

Integrals of the form $\int_{-\infty}^{\infty}F(x)\,dx$

28

Example

Evaluate $\int_{-\infty}^{\infty}\frac{1}{1+x^4}\,dx$.

To evaluate this integral we must consider the integral $\oint_c\frac{1}{1+z^4}\,dz$ where c is the contour shown in the figure, so that

$$\oint_c\frac{1}{1+z^4}\,dz=\int_s\frac{dz}{1+z^4}+\int_{-R}^{R}\frac{dx}{1+x^4}=2\pi i\{\text{sum of residues inside c}\}$$

Notice that along the real axis between $-R$ and R, $z=x$. Provided $R>1$ we can evaluate this integral using the Residue theorem. That is

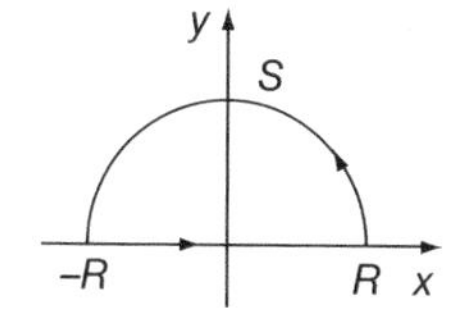

$$\oint_c\frac{1}{1+z^4}\,dz=2\pi i\times\{\text{sum of residues inside c}\}$$

29

$$\boxed{\frac{\pi}{\sqrt{2}}}$$

Because

The integrand $\frac{1}{1+z^4}$ possesses four simple poles at $z = e^{\pi i/4}$, $e^{3\pi i/4}$, $e^{5\pi i/4}$, $e^{7\pi i/4}$ of which only the first two are inside c.

The residue at $z = e^{\pi i/4}$ is $\underset{z \to e^{\pi i/4}}{Lim} \left\{ \left(z - e^{\pi i/4} \right) \times \frac{1}{1+z^4} \right\}$

$$= \underset{z \to e^{\pi i/4}}{Lim} \left\{ \frac{1}{4z^3} \right\} \text{ by L'Hôpital's rule}$$

$$= \frac{e^{-3\pi i/4}}{4}$$

The residue at $z = e^{3\pi i/4}$ is $\underset{z \to e^{3\pi i/4}}{Lim} \left\{ \left(z - e^{3\pi i/4} \right) \times \frac{1}{1+z^4} \right\}$

$$= \underset{z \to e^{3\pi i/4}}{Lim} \left\{ \frac{1}{4z^3} \right\} \text{ by L'Hôpital's rule}$$

$$= \frac{e^{-9\pi i/4}}{4} = \frac{e^{-\pi i/4}}{4}$$

Therefore

$$\oint_c \frac{1}{1+z^4} \, dz = 2\pi i \times \left\{ \frac{1}{4} \left(e^{-3\pi i/4} + e^{-\pi i/4} \right) \right\}$$

Now $e^{-3\pi i/4} = \cos\frac{3\pi}{4} - i \sin\frac{3\pi}{4} = -\frac{1}{\sqrt{2}} - \frac{i}{\sqrt{2}}$ and

$$e^{-\pi i/4} = \cos\frac{\pi}{4} - i \sin\frac{\pi}{4} = \frac{1}{\sqrt{2}} - \frac{i}{\sqrt{2}} \text{ and so}$$

$$\oint_c \frac{1}{1+z^4} \, dz = 2\pi i \times \left\{ \frac{1}{4} \left(\frac{-2i}{\sqrt{2}} \right) \right\} = \frac{\pi}{\sqrt{2}}$$

We now look at the components of this integral in the next frame

30

We now recognize that

$$\oint_c \frac{1}{1+z^4} \, dz = \int_{-R}^{R} \frac{1}{1+x^4} \, dx + \int_S \frac{1}{1+z^4} \, dz$$

because $z = x$ along the real line.

Now we let R increase indefinitely and take limits, so that

$$\underset{R \to \infty}{Lim} \oint_c \frac{1}{1+z^4} \, dz = \int_{-\infty}^{\infty} \frac{1}{1+x^4} \, dx + \underset{R \to \infty}{Lim} \int_S \frac{1}{1+z^4} \, dz = \frac{\pi}{\sqrt{2}}$$

because the value of the contour integral is independent of the value of R. We shall now proceed to show that $\underset{R \to \infty}{Lim} \int_S \frac{1}{1+z^4} \, dz = 0$.

▶

Writing $z = Re^{i\theta}$ so that, on S, $dz = Re^{i\theta}\,d\theta$, the limit of the integral becomes

$$\underset{R\to\infty}{Lim}\int_S \frac{Re^{i\theta}}{1+R^4e^{i4\theta}}\,d\theta = 0$$

Notice that the requirement that ensures that the integral along the semicircle vanishes in the limit is equivalent to the requirement that the degree of the denominator be at least two degrees higher than the numerator.

Now you try one.

$$\int_{-\infty}^{\infty}\frac{x^2\,dx}{(x^2+1)^2} = \ldots\ldots\ldots\ldots$$

31

$$\boxed{\frac{\pi}{2}}$$

Because

Consider the integral $\oint_c \frac{z^2\,dz}{(z^2+1)^2}$ where the contour c is the same semicircular contour as in the previous example. Here the integrand has two double poles at $z = i$ and $z = -i$ but only the pole at $z = i$ is inside the contour. The residue at $z = i$ is

$$\underset{z\to i}{Lim}\left\{\frac{d}{dz}(z-i)^2\frac{z^2}{(z-i)^2(z+i)^2}\right\} = \underset{z\to i}{Lim}\left\{\frac{2z(z+i)^2 - z^2 2(z+i)}{(z+i)^4}\right\}$$
$$= -\frac{i}{4}$$

Therefore

$$\oint_c \frac{z^2\,dz}{(z^2+1)^2} = 2\pi i\left(-\frac{i}{4}\right) = \frac{\pi}{2}$$

Taking limits

$$\underset{R\to\infty}{Lim}\oint_c \frac{z^2\,dz}{(z^2+1)^2} = \int_{-\infty}^{\infty}\frac{x^2\,dx}{(x^2+1)^2} + \underset{R\to\infty}{Lim}\int_S \frac{z^2\,dz}{(z^2+1)^2} = \frac{\pi}{2}$$

Where, in the second integral on the right-hand side, the degree of the denominator is two higher than the degree of the numerator, and so

$$\underset{R\to\infty}{Lim}\int_S \frac{z^2\,dz}{(z^2+1)^2} = 0, \text{ therefore } \int_{-\infty}^{\infty}\frac{x^2\,dx}{(x^2+1)^2} = \frac{\pi}{2}$$

32 Integrals of the form $\int_{-\infty}^{\infty} F(x)\begin{Bmatrix}\sin x\\ \cos x\end{Bmatrix} dx$

These integrals are often referred to as Fourier integrals because of their appearances within Fourier analysis.

Example

Evaluate $\int_{-\infty}^{\infty} \frac{\cos kx}{a^2+x^2}\,dx$ where $a > 0$ and $k > 0$.

To evaluate this integral we consider the contour integral $\oint_c \frac{e^{ikz}}{a^2+z^2}\,dz$ where c is the semicircular contour of the previous problems and whose integrand possesses two simple poles at $z = ai$ and $z = -ia$ of which only the first is inside the contour. Consequently

$$\oint_c \frac{e^{ikz}}{a^2+z^2}\,dz = 2\pi i\{\text{residue at } z = ia\} = \ldots\ldots\ldots\ldots$$

33

$$\boxed{\frac{\pi e^{-ka}}{a}}$$

Because

The residue at $z = ai$ is

$$\underset{z\to ai}{Lim}\left\{(z-ia)\frac{e^{ikz}}{a^2+z^2}\right\} = \underset{z\to ia}{Lim}\left\{\frac{e^{ikz}}{z+ia}\right\} = \frac{e^{ik(ia)}}{2ia} = -\frac{ie^{-ka}}{2a} \text{ and so}$$

$$\oint_c \frac{e^{ikz}}{a^2+z^2}\,dz = 2\pi i\left\{-\frac{ie^{-ka}}{2a}\right\} = \frac{\pi e^{-ka}}{a}$$

Taking limits as $R \to \infty$

$$\underset{R\to\infty}{Lim}\oint_c \frac{e^{ikz}}{a^2+z^2}\,dz = \int_{-\infty}^{\infty} \frac{e^{ikz}}{a^2+z^2}\,dz + \underset{R\to\infty}{Lim}\int_S \frac{e^{ikz}}{a^2+z^2}\,dz = \frac{\pi e^{-ka}}{a}$$

In the second integral on the right-hand side, the degree of the denominator is two higher than the degree of the numerator, and so

$$\underset{R\to\infty}{Lim}\int_S \frac{e^{ikz}}{a^2+z^2}\,dz = 0, \text{ therefore } \int_{-\infty}^{\infty} \frac{e^{ikx}}{a^2+x^2}\,dx = \frac{\pi e^{-ka}}{a}. \text{ That is}$$

$$\int_{-\infty}^{\infty} \frac{\cos kx + i\sin kx}{a^2+x^2}\,dx = \frac{\pi e^{-ka}}{a} = 2\pi i\{\text{residue at } z = ia\}.$$

Consequently

$$\int_{-\infty}^{\infty} \frac{\cos kx}{a^2+x^2}\,dx = \frac{\pi e^{-ka}}{a} = -2\pi\,\mathrm{Im}\{\text{residue at } z = ia\} \text{ and}$$

$$\int_{-\infty}^{\infty} \frac{\sin kx}{a^2+x^2}\,dx = 0 = 2\pi\,\mathrm{Re}\{\text{residue at } z = ia\}$$

▶

Notice that e^{ikz} is easier to use than $\cos kx = (e^{ikx} + e^{-ikx})/2$, and it also gives the solution to the related integral with $\cos kx$ replaced with $\sin kx$.

Finally, to finish off the Program, here is one for you to try.

$$\int_{-\infty}^{\infty} \frac{\cos \pi x}{x^2 + x + 1}\, dx = \ldots\ldots\ldots\ldots$$

34

$$\boxed{0}$$

Because

Consider $\oint_c \frac{e^{i\pi z}}{z^2 + z + 1}\, dz$ where c is the semicircular contour of the previous problem. The integrand is singular at the simple poles $z = (-1 \pm \sqrt{3}i)/2$ where only $z = (-1 + \sqrt{3}i)/2$ is inside the contour. The residue at $z = (-1 + \sqrt{3}i)/2$ is then

$$\begin{aligned}
&\lim_{z \to (-1+\sqrt{3}i)/2} \left\{ \left(z - \left[-1 + \sqrt{3}i\right]/2\right) \frac{e^{i\pi z}}{z^2 + z + 1} \right\} \\
&= \lim_{z \to (-1+\sqrt{3}i)/2} \left\{ \frac{e^{i\pi z}}{z - \left[-1 - \sqrt{3}i\right]/2} \right\} \\
&= \frac{e^{i\pi(-1+\sqrt{3}i)/2}}{\sqrt{3}i} \\
&= \frac{e^{-i\pi/2} e^{-\sqrt{3}\pi/2}}{\sqrt{3}i} \\
&= -\frac{e^{-\sqrt{3}\pi/2}}{\sqrt{3}} \qquad \text{since } e^{-i\pi/2} = -i
\end{aligned}$$

Therefore

$$\oint_c \frac{e^{i\pi z}}{z^2 + z + 1}\, dz = 2\pi i \left\{ \frac{e^{-\sqrt{3}\pi/2}}{\sqrt{3}} \right\} = -i \frac{2\pi e^{-\sqrt{3}\pi/2}}{\sqrt{3}}$$

that is

$$\oint_c \frac{e^{i\pi z}}{z^2 + z + 1}\, dz = \oint_c \frac{\cos \pi z + i \sin \pi z}{z^2 + z + 1}\, dz = -i \frac{2\pi e^{-\sqrt{3}\pi/2}}{\sqrt{3}}$$

and so

$$\oint_c \frac{\cos \pi z}{z^2 + z + 1}\, dz = 0 \text{ and } \oint_c \frac{\sin \pi z}{z^2 + z + 1}\, dz = -\frac{2\pi e^{-\sqrt{3}\pi/2}}{\sqrt{3}}$$

Note that, again, the contribution from the contour integral along the semicircle is zero.

The **Summary** now follows. Check it through in conjunction with the **Can You?** checklist before goint on to the **Test exercise**. The **Further problems** provide additional practice.

Summary

35

1 *Maclaurin series*
The Maclaurin series expansion of a function of a complex variable z is

$$f(z) = f(0) + zf'(0) + z^2\frac{f''(0)}{2!} + z^3\frac{f'''(0)}{3!} + \cdots$$

2 *Ratio test for convergence*
The ratio test for convergence of a series of terms of a complex variable

$$f(z) = a_0(z) + a_1(z) + a_2(z) + a_3(z) + \cdots + a_n(z) + \cdots$$

is that given

$$\underset{n\to\infty}{Lim}\left|\frac{a_{n+1}(z)}{a_n(z)}\right| = L$$

then if $L < 1$ the series converges and so the expansion is valid
$L > 1$ the series diverges and so the expansion is invalid
$L = 1$ the ratio test fails to give a conclusion.

3 *Radius and circle of convergence*
The radius of the circle within which a series expansion is valid is called the *radius of convergence* of the series and the circle is called the *circle of convergence*. The radius of convergence can be found using the ratio test for convergence.

4 *Singular points*
Any point at which $f(z)$ fails to be analytic, that is where the derivative does not exist, is called a *singular point.*

Poles
If $f(z)$ has a singular point at z_0 and for some natural number n

$$\underset{z\to z_0}{Lim}\{(z - z_0)^n f(z)\} = L \neq 0$$

then the singular point (also called a singularity) is called *a pole of order n*.

Removable singularity
If $f(z)$ has a singular point at z_0 but $\underset{z\to z_0}{Lim}\{f(z)\}$ exists then the singular point is called a *removable singularity*.

5 *Circle of convergence*
When an expression is expanded in a Maclaurin series, the *circle of convergence* is always centered on the origin and the *radius of convergence* is determined by the location of the first singular point met as z moves out from the origin.

6 *Taylor's series*
Provided $f(z)$ is analytic inside and on a simple closed curve c, the Taylor series expansion of $f(z)$ about a point z_0 which is interior to c is given as

$$f(z) = f(z_0) + (z - z_0)f'(z_0) + \frac{(z - z_0)^2 f''(z_0)}{2!} + \cdots + \frac{(z - z_0)^n f^{(n)}(z_0)}{n!} + \cdots$$

where, here, the expansion is about the point z_0 which is the center of the circle of convergence. The circle of convergence is given as $|z - z_0| = R$ where R is the radius of convergence. Maclaurin's series is a special case of Taylor's series where $z_0 = 0$.

7 *Laurent's series*
The *Laurent series expansion* provides a series expansion valid within an annular region *centered on the singular point.*

Let $f(z)$ be singular at $z = z_0$ and let c_1 and c_2 be two concentric circles centered on z_0. Then if $f(z)$ is analytic in the annular region between c_1 and c_2 and c is any concentric circle lying within the annular region between c_1 and c_2 we can expand $f(z)$ as a Laurent series in the form

$$f(z) = \cdots + \frac{a_{-2}}{(z - z_0)^2} + \frac{a_{-1}}{(z - z_0)} + a_0 + a_1(z - z_0) + a_2(z - z_0)^2 + \cdots$$

$$= \sum_{n \to -\infty}^{\infty} a_n (z - z_0)^n \text{ where } a_n = \frac{1}{2\pi i}\oint_c \frac{f(z)}{(z - z_0)^{n+1}}\,dz$$

8 *Residues*

In the Laurent series

$$f(z) = \cdots + \frac{a_{-2}}{(z - z_0)^2} + \frac{a_{-1}}{(z - z_0)} + a_0 + a_1(z - z_0) + a_2(z - z_0)^2 + \cdots$$

the coefficient a_{-1} is referred to as the *residue* of $f(z)$.

Residue theorem
Provided $f(z)$ is analytic at all points inside and on the simple closed contour c, apart from the single isolated singularity at z_0 which is interior to c, then

$$\oint_c f(z)\,dz = 2\pi i a_{-1}$$

9 The Residue theorem extends to the case where the contour contains a finite number of singularities. If $f(z)$ is analytic inside and on the simple closed contour c except at the finite number of points $z_0, z_1, z_2, \cdots$ each with a Laurent series expansion and each with corresponding residues $\overset{(0)}{a}_{-1}, \overset{(1)}{a}_{-1}, \overset{(2)}{a}_{-1}, \cdots$ then

$$\oint_c f(z)\,dz = 2\pi i\left\{\overset{(0)}{a}_{-1} + \overset{(1)}{a}_{-1} + \overset{(2)}{a}_{-1} + \cdots\right\}$$

▶

10 *Calculating residues*

$$a_{-1} = \lim_{z \to z_0} \left[\frac{1}{(n-1)!} \frac{d^{n-1}}{dz^{n-1}} \left((z - z_0)^n f(z) \right) \right]$$

11 *Real integrals*

The Residue theorem can be very usefully employed to evaluate integrals of real functions.

Integrals of the form $\int_0^{2\pi} F(\cos\theta, \sin\theta)\, d\theta$

Use $z = e^{i\theta}$ and the exponential form of the trigonometric functions $\cos\theta = \dfrac{e^{i\theta} + e^{-i\theta}}{2} = \dfrac{z + z^{-1}}{2}$, $\sin\theta = \dfrac{e^{i\theta} - e^{-i\theta}}{2i} = \dfrac{z - z^{-1}}{2i}$ and $dz = ie^{i\theta}d\theta = iz\, d\theta$ so that $d\theta = dz/zi$. Convert the integral into a contour integral around the unit circle centered on the origin and use the Residue theorem.

Integrals of the form $\int_{-\infty}^{\infty} F(x)\, dx$ and $\int_{-\infty}^{\infty} F(x) \left\{ \begin{matrix} \sin x \\ \cos x \end{matrix} \right. dx$

Consider integrals of the form $\oint_c F(z)\, dz$ and $\oint_c F(z) e^{iz}\, dz$ respectively, where the contour c is a semicircle with the diameter lying along the real axis. The principle is that the integral can be evaluated by the Residue theorem and then the contour can be expanded to cover the required extent of the real axis, the integration along the semicircle giving a zero contribution.

☑ Can You?

36 Checklist 9

Check this list before and after you try the end of Program test

On a scale of 1 to 5 how confident are you that you can: **Frames**

- Expand a function of a complex variable about the origin in a Maclaurin series? 1 to 7

  Yes ☐ ☐ ☐ ☐ ☐ *No*

- Determine the circle and radius of convergence of a Maclaurin series expansion? 8 to 10

  Yes ☐ ☐ ☐ ☐ ☐ *No*

- Recognize singular points in the form of poles of order *n*, removable and essential singularities? 11

  Yes ☐ ☐ ☐ ☐ ☐ *No*

- Expand a function of a complex variable about a point in the complex plane in a Taylor series, transforming the coordinates with a shift of origin?
  Yes ☐ ☐ ☐ ☐ ☐ *No*

12 to 14

- Expand a function of a complex variable about a singular point in a Laurent series?
  Yes ☐ ☐ ☐ ☐ ☐ *No*

15

- Recognize the principal and analytic parts of the Laurent series and link the form of the principal part to the type of singularity?
  Yes ☐ ☐ ☐ ☐ ☐ *No*

16 to 20

- Recognize the residue of a Laurent series and state the Residue theorem?
  Yes ☐ ☐ ☐ ☐ ☐ *No*

21 and 22

- Calculate the residues at the poles of an expression without resort to deriving the Laurent series?
  Yes ☐ ☐ ☐ ☐ ☐ *No*

23 to 25

- Evaluate certain types of real integrals using the Residue theorem?
  Yes ☐ ☐ ☐ ☐ ☐ *No*

26 to 34

Test exercise 9

37

1 Expand each of the following in a Maclaurin series and determine the radius and the circle of convergence in each case.

(a) $f(z) = e^z$

(b) $f(z) = \ln(1 + 4z)$.

2 Determine the location and nature of the singular points in each of the following.

(a) $f(z) = \dfrac{3z}{(z+1)^5}$

(b) $f(z) = z^{10}e^{1/z}$

(c) $f(z) = z\sin(1/z)$

(d) $f(z) = \dfrac{1 - \cos z}{z^2}$

3 Expand $f(z) = \sin z$ in a Taylor series about the point $z = \pi/4$ and determine the radius of convergence.

▶

4 Expand each of the following in a Laurent series. In (a) and (c) determine the nature of the singularity from the principal part of the series.

(a) $f(z) = (5 - z)\cos\dfrac{1}{z+3}$ about the point $z = -3$

(b) $f(z) = \dfrac{2z}{(z+1)(z+3)}$ valid for $1 < |z| < 3$

(c) $f(z) = \dfrac{1}{z^3(z-2)^2}$ about the point $z = 2$.

5 Calculate the residues at each of the singularities of

$$f(z) = \frac{3z-1}{z^2(z+1)^2(z-1)}.$$

6 Evaluate each of the following integrals.

(a) $\displaystyle\int_0^{2\pi} \frac{d\theta}{5\cos\theta - 13}$

(b) $\displaystyle\int_{-\infty}^{\infty} \frac{dx}{x^2 + x + 1}$

(c) $\displaystyle\int_{-\infty}^{\infty} \frac{\cos 3x}{x^4 + 2x^2 + 1}\, dx$

Further problems 9

38

1 For each of the following find the Maclaurin series expansion and determine the radius of convergence.

(a) $\sinh z$

(b) $\tan z$

(c) $\ln\left(\dfrac{1+z}{1-z}\right)$

(d) a^z, where $a > 0$

(e) $\dfrac{15z^2}{(5-3z)^3}$.

2 By using the appropriate Maclaurin series expansions, show that

(a) $(\cos z)' = -\sin z$

(b) $\cos z = \dfrac{e^z + e^{-z}}{2}$

(c) $(e^z)' = e^z$.

3 Given the series expansion for $(1+z)^{-1}$

(a) show by integration that this is compatible with the series expansion for $\ln(1+z)$

(b) by differentiation find $\sum_{n=1}^{\infty}(-1)^n n z^n$ and $\sum_{n=1}^{\infty}(-1)^n n^2 z^n$.

4 Use the ratio test to test each of the following for convergence.

(a) $\sum_{n=0}^{\infty} \frac{(2n)!}{(n!)^2} z^n$

(b) $\sum_{n=0}^{\infty} \frac{z^n}{1-3n}$

(c) $\sum_{n=0}^{\infty} \frac{n^2 z^n}{1-3n}$

(d) $\sum_{n=0}^{\infty} \frac{(\cos n\pi) z^n}{2n-1}$

(e) $\sum_{n=1}^{\infty} \frac{(-1)^n z^n}{(n+1)!}$

5 Find the Taylor series about the point indicated of each of the following.

(a) e^z about the point $z = 2$

(b) $\cos z$ about the point $z = \pi/6$

(c) $(z-3)\sin(z+3)$ about the point $z = 3$

(d) $(2z-5)^{-1}$ about the point $z = 1/3$

(e) $(2z-5)^{-1}$ about the point $z = 3$.

6 Find the series expansion of $z \ln z$ valid for $|z-1| < 1$.

7 Find the circle of convergence of each of the following when expanded in a Taylor series about the point indicated.

(a) $e^{-z}\cos(z-2)$ about the point $z = 1$

(b) $\frac{z^3}{(z^2+6)}$ about the point $z = 0$

(c) $\frac{z-2}{(z-6)(z-4)}$ about the point $z = 5$

(d) $\frac{z^2}{(e^z+1)}$ about the point $z = 0$.

8 Locate and classify all of the singularities of each of the following.

(a) $\frac{(z-1)^3}{z^2(z^2-1)^2}$

(b) $z^{-2}e^{-1/z}$.

9 Find the Laurent series about the point indicated of each of the following.

(a) $\frac{1}{z}\sin\left(\frac{1}{z}\right)$ about the point $z = 0$

(b) $\frac{1}{2z-3}$ about the point $z = 3/2$

(c) $\frac{z}{(z-2)(z-3)}$ about the point $z = 3$.

10 Find the Laurent series of $\frac{z-1}{(z+2)(z+5)}$ that is valid for

(a) $2 < |z| < 5$

(b) $|z| > 5$

(c) $|z| < 2$.

11 Evaluate each of the following integrals.

(a) $\int_0^{2\pi} \frac{d\theta}{2+\sin\theta}$

(b) $\int_0^{2\pi} \frac{d\theta}{\alpha+\beta\sin\theta}$ for $\alpha > |\beta|$

(c) $\int_0^{2\pi} \frac{d\theta}{1+\alpha^2-2\alpha\cos\theta}$ where $0 < \alpha < 1$

(d) $\int_0^{2\pi} \frac{\sin^2\theta d\theta}{5-4\cos\theta}$

(e) $\int_0^{2\pi} \frac{d\theta}{5-3\cos\theta}$

(f) $\int_{-\infty}^{\infty} \frac{dx}{x^2+6x+13}$

(g) $\int_{-\infty}^{\infty} \frac{x^2 dx}{x^4+6x^2+13}$

(h) $\int_{-\infty}^{\infty} \frac{x^2 dx}{(x^2+4)^2}$

(i) $\int_{-\infty}^{\infty} \frac{x^2+x+1}{x^4+x^2+1} dx$

(j) $\int_{-\infty}^{\infty} \frac{dx}{x^6+1}$

(k) $\int_{-\infty}^{\infty} \frac{x^2 \sin\pi x\, dx}{x^4+6x^2+13}$

(l) $\int_{-\infty}^{\infty} \frac{\sin\pi x}{x^4+x^2+1} dx$.

Answers

Test exercise 1 (page 48)

1 (a) does not, (b) and (c) do

2 (a) $h(x) = \dfrac{2}{x-2} - 3x + 3$; domain $2 < x < 3$, range $-4 < x < \infty$

(b) $k(x) = -\dfrac{3}{5(x-2)(x-1)}$; domain $2 < x < 3$, range $-\infty < x < -0.3$

4 (a) $f(x) = 5(\sqrt{x}+3)^4$ (b) $f(x) = 25\sqrt{x}$ (c) $f(x) = ((x+3)^4+3)^4$

5 $f(x) = a[c(a[b(x)])]$ where $a(x) = x - 4$, $b(x) = 5x$ and $c(x) = x^3$

$f^{-1}(x) = b^{-1}[a^{-1}(c^{-1}[a^{-1}(x)])]$: $a^{-1}(x) = x + 4$, $b^{-1}(x) = x/5$, $c^{-1}(x) = x^{\frac{1}{3}}$

$f^{-1}(x) = \dfrac{(x+4)^{\frac{1}{3}} + 4}{5}$

6 (a) 0.6428 (b) 3.5495 (c) 0 **7** (a) period $2\pi/7$, amplitude 4, phase 0
(b) period π, amplitude 2, phase 0 relative to $\cos 2\theta$
(c) period $2\pi/3$, amplitude ∞, phase 4/3 **8** (a) 3 (b) 4.5 (c) 0 **9** (a) $\pm n\pi$
(b) $\pm 35.26° \pm n \times 360°$, $\pm 30° \pm n \times 360°$ (c) $1.381 \pm 2n\pi$ radians **10** (a) 0
(b) 1/3 (c) 3.9949 (d) 625 (e) 4.2618 (f) 4.6052 **11** (a) 2
(b) 0.6931, 1.0986 (c) 6 (d) $\pm 32/81$ (e) 1/9 **12** $f_e(x) = \dfrac{a^x + a^{-x}}{2}$ and

$f_o(x) = \dfrac{a^x - a^{-x}}{2}$ **13** (a) 0 (b) 6 (c) $-5/28$ (d) $-1/9$ (e) $\pi/4$

Further problems 1 (page 49)

1 $x = \sqrt{3}$ **2** (a) domain $-1 < x < 1$, range $-\infty < f(x) < \infty$

3 the two straight lines $y = x/3$ and $y = -x/3$

6 yes; consider $a = \sqrt{2}$ and $x = 2$ as an example **7** (a) $f^{-1}(x) = \left(\dfrac{x}{6} + 2\right)^{\frac{1}{3}}$

(b) $f^{-1}(x) = \left(x^{\frac{1}{3}} + 2\right)^{\frac{1}{3}}$ (c) $f^{-1}(x) = \left(\dfrac{(x+2)^{\frac{1}{3}}}{6} + 2\right)^{\frac{1}{3}}$

10 (b) $f'(x) = \begin{cases} -1 & \text{if } x < 0 \\ 1 & \text{if } x > 0 \end{cases}$ (c) $f'(0)$ does not exist

13 (a) -5.702 (b) 0.8594 (c) -2.792 **14** (a) 0 (b) -4 (c) 1/2

Test exercise 2 (page 78)

1 (a) $-i$ (b) i (c) 1 (d) -1 **2** (a) $29 - 2i$ (b) $-2i$ (c) $111 + 56i$ (d) $1 + 2i$

3 $x = 10.5$, $y = 4.3$ **4** (a) $5.831\underline{|59°3'}$ (b) $6.708\underline{|153°26'}$ (c) $6.403\underline{|231°24'}$

5 (a) $-3.5355(1+i)$ (b) $3.464 - 2i$ **6** (a) $10e^{0.650i}$
(b) $10e^{-0.650i}$, $2.303 + 0.650i$, $2.303 - 0.650i$ **7** ie

Further problems 2 (page 78)

1 (a) $115 + 133i$ (b) $2.52 + 0.64i$ (c) $\cos 2x + i \sin 2x$ **2** $(22 - 75i)/41$

3 $0.35 + 0.17i$ **4** 0.7, 0.9 **5** $-24.4 + 22.8i$ **6** $1.2 + 1.6i$ **8** $x = 18$, $y = 1$

9 $a = 2$, $b = -20$ **10** $x = \pm 2$, $y = \pm 3/2$ **12** $a = 1.5$, $b = -2.5$ **13** $\sqrt{2}e^{2.3562i}$

14 2.6 **16** $R=(R_2C_3-R_1C_4)/C_4$, $L=R_2R_4C_3$ **18** $E=(1811+1124i)/34$
20 $2+3i$, $-2+3i$

Test exercise 3 (page 105)

1 $5.831\angle 210°58'$ **2** (a) $-1.827+0.813i$ (b) $3.993-3.009i$
3 (a) $36\angle 197°$ (b) $4\angle 53°$ **4** $8\angle 75°$ **5** $2\angle 88°$, $2\angle 208°$, $2\angle 328°$;
principal root $=2\angle 328°$

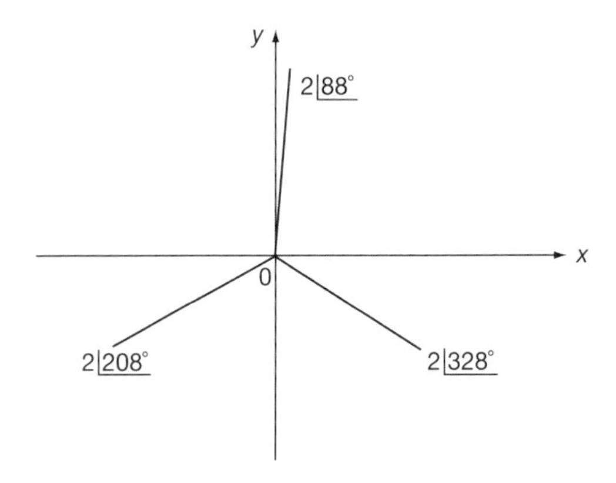

6 $\sin 4\theta = 4\sin\theta\cos\theta - 8\sin^3\theta\cos\theta$ **7** $\cos^4\theta = \frac{1}{8}[\cos 4\theta + 4\cos 2\theta + 3]$

8 (a) $x^2+y^2-8x+7=0$ (b) $y=\dfrac{x+2}{\sqrt{3}}$

Further problems 3 (page 105)

1 $x=0.27$, $y=0.53$ **2** $-3+\sqrt{3}i$; $-2\sqrt{3}i$
3 $3.606\angle 56°19'$, $2.236\angle 296°34'$; $24.2-71.6i$; $75.6e^{-1.244i}$
4 $1.336(\angle 27°, \angle 99°, \angle 171°, \angle 243°, \angle 315°)$; $1.336(e^{0.4712i}, e^{1.7279i}, e^{2.9845i}, e^{-2.0420i}, e^{-0.7854i})$
5 $2.173+0.899i$, $2.351e^{0.392i}$ **6** $\sqrt{2}(1+i)$, $\sqrt{2}(-1+i)$, $\sqrt{2}(-1-i)$, $\sqrt{2}(1-i)$
7 $1\angle 36°$, $1\angle 108°$, $1\angle 180°$, $1\angle 252°$, $1\angle 324°$; $e^{0.6283i}$ **8** $x=-4$ and $x=2\pm 3.464i$
9 $1\angle 102°18'$, $1\angle 222°18'$, $1\angle 342°18'$; $0.953-0.304i$
11 $1.401(\angle 58°22', \angle 130°22', \angle 202°22', \angle 274°22', \angle 346°22')$
principal root $=1.36-0.33i=1.401e^{-0.2379i}$
12 $-0.36+0.55i$, $-1.64-2.55i$ **13** $-ie$, i.e. $-2.718i$ to 3 dp
14 $\sin 7\theta = 7s-56s^3+112s^5-64s^7$ ($s\equiv\sin\theta$)
15 $\frac{1}{32}[10-15\cos 2x+6\cos 4x-\cos 6x]$
16 $x^2+y^2+\frac{20}{3}x+4=0$; center $\left(-\frac{10}{3},0\right)$, radius $8/3$
17 $x^2+y^2-(1+\sqrt{3})x-(1+\sqrt{3})y+\sqrt{3}=0$, center $\left(\frac{1+\sqrt{3}}{2},\frac{1+\sqrt{3}}{2}\right)$, radius $\sqrt{2}$
18 $x^2+y^2=16$ **19** (a) $2x^2+2y^2-x-1=0$ (b) $x^2+y^2+2x+2y=0$
20 (a) $x^2+y^2-4x=0$ (b) $x^2+y^2+x-2=0$ **22** (a) $y=3$ (b) $x^2+y^2=4k^2$

Test exercise 4 (page 131)

1

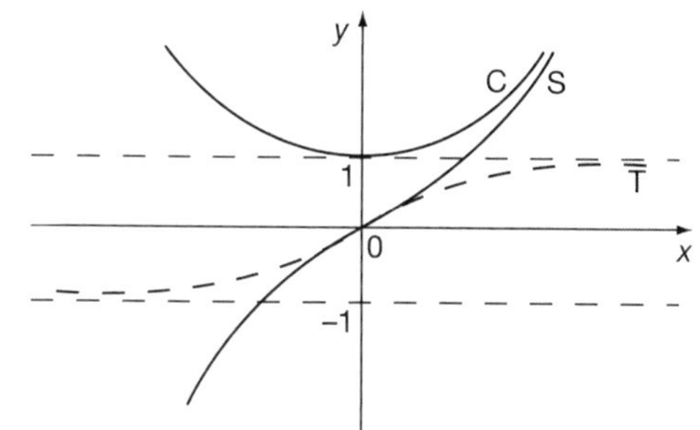

2 67.25 **3** 19.40 **4** (a) 1.2125 (b) ± 0.6931 **5** $x = 0.3466$
6 (a) $y = 224$ (b) $x = \pm 48.12$ **7** $-\coth A$ **8** $\sin x \cosh y - i \cos x \sinh y$

Further problems 4 (page 131)

2 $x = 0, x = 0.549$ **5** (a) 0.9731 (b) ± 1.317 **7** (a) $0.9895 + 0.2498i$
(b) $0.4573 + 0.5406i$ **10** $x = 0, x = \frac{1}{2}\ln 2$ **12** $x = 0.3677$ or -1.0986
14 $1.528 + 0.427i$ **17** 1.007

Test exercise 5 (page 170)

1 (a) $w = 6 - 2i$ (b) $w = 3 - 2i$ (c) $w = 3i$ (d) $w = 2$
2 Magnification = 2.236; rotation = 63° 26′; translation = 1 unit to right, 3 units downwards

3

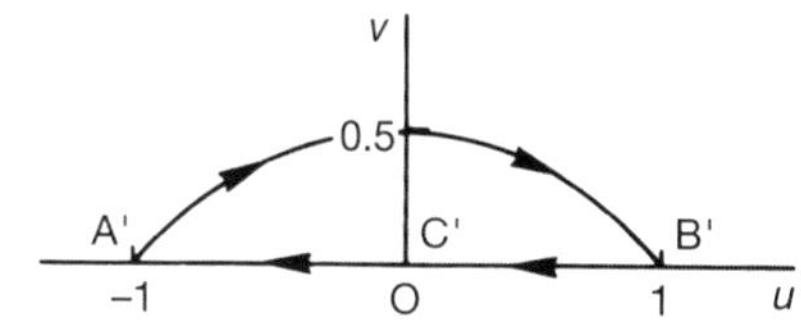

$$v = \frac{1}{2}(1 - u^2)$$

4

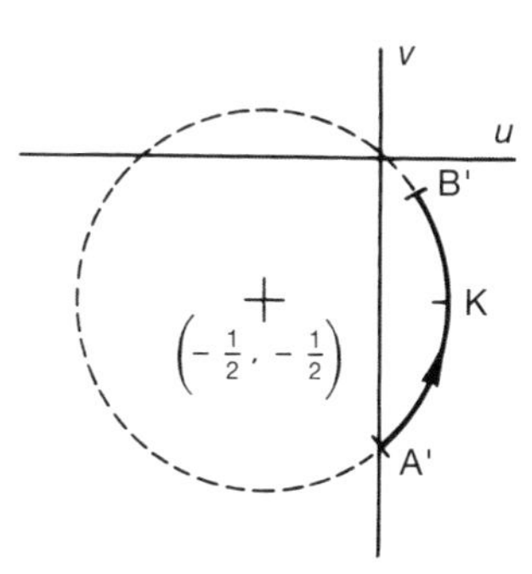

Minor arc of circle, center $\left(-\frac{1}{2}, -\frac{1}{2}\right)$, radius $\frac{1}{\sqrt{2}} = 0.7071$, between A′ $(-i)$ and B $(0.12 - 0.16i)$

5 (a) center $\left(u = 0, v = \frac{2}{3}\right)$ (b) radius $\frac{1}{3}$
6 center $\left(u = \frac{2}{3}, v = 0\right)$; radius $\frac{2}{3}$

Further problems 5 (page 171)

1 Triangle A′B′C′ with A′ $(-1+2i)$, B′ $(5+2i)$, C′ $(2+5i)$

2 (a) A′ $(-8+9i)$; B′ $(23+14i)$
(b) Magnification $= \sqrt{29} = 5.385$; rotation $= 68°12'$; translation = nil

3 Straight line joining A′ $(5-7i)$ to B′ $(-3-i)$; magnification $= 3.162$; rotation $= 161°34'$ counter-clockwise; translation = 2 to right, 4 upwards

4

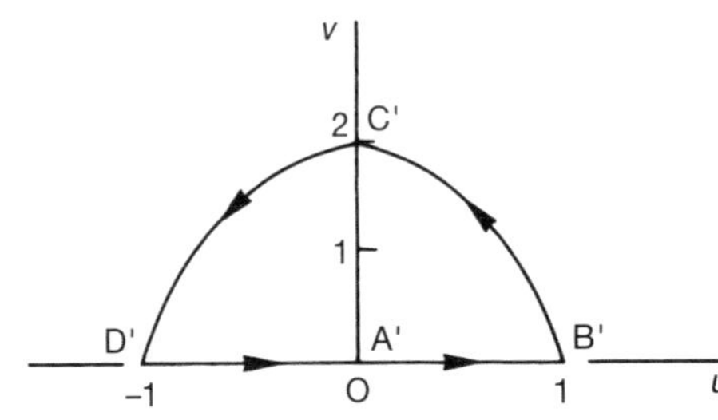

A′ B′: $v = 0$

B′C′: $u = 1 - \dfrac{v^2}{4}$

C′D′: $u = \dfrac{v^2}{4} - 1$

D′A′: $v = 0$

5

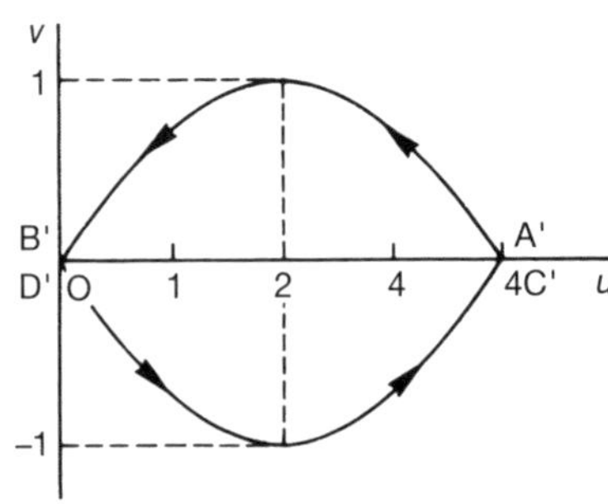

A′B′ and C′D′: $v = \dfrac{1}{4}(4u - u^2)$

B′C′ and D′A′: $v = \dfrac{1}{4}(u^2 - 4u)$

6 A′ $(1-2i)$; B′ $(-23+10i)$; C′ $(1-8i)$ A′B′: $u = 2 - \dfrac{v^2}{4}$;

B′C′: $v = \dfrac{(u-1)^2}{32} - 8$; C′A′: $u = 1$ **7** circle, center $\left(\dfrac{1}{2} - \dfrac{2}{3}i\right)$, radius $\dfrac{7}{6}$

8 (a) circle, center $\left(\dfrac{1}{3} - 0i\right)$, radius $\dfrac{2}{3}$ (b) region outside the circle in (a)

9 circle, center $\left(\dfrac{3}{2} + 0i\right)$, radius 1; clockwise development

10 circle, $u^2 + v^2 - \dfrac{22u}{5} + \dfrac{8}{5} = 0$, center $\left(\dfrac{11}{5} + 0i\right)$, radius $\dfrac{9}{5}$

11 circle, $u^2 + v^2 - \dfrac{u}{2} = 0$, center $\left(\dfrac{1}{4} + 0i\right)$, radius $\dfrac{1}{4}$; region inside this circle

12 circle, center $\left(-\dfrac{7}{3} + 0i\right)$, radius $\dfrac{5}{3}$

13 (a) circle, center $\left(\frac{3}{5}, 0\right)$, radius $\frac{2}{5}$, developed clockwise

(b) region outside the circle in (a)

14 $v = -\frac{u}{3}$

Test exercise 6 (page 194)

1 (a) $\frac{\partial z}{\partial x} = 12x^2 - 5y^2$ $\frac{\partial z}{\partial y} = -10xy + 9y^2$ $\frac{\partial^2 z}{\partial x^2} = 24x$ $\frac{\partial^2 z}{\partial y^2} = -10x + 18y$

$\frac{\partial^2 z}{\partial y \cdot \partial x} = -10y$ $\frac{\partial^2 z}{\partial x \cdot \partial y} = -10y$ (b) $\frac{\partial z}{\partial x} = -2\sin(2x + 3y)$ $\frac{\partial z}{\partial y} = -3\sin(2x + 3y)$

$\frac{\partial^2 z}{\partial x^2} = -4\cos(2x + 3y)$ $\frac{\partial^2 z}{\partial y^2} = -9\cos(2x + 3y)$ $\frac{\partial^2 z}{\partial y \cdot \partial x} = -6\cos(2x + 3y)$

$\frac{\partial^2 z}{\partial x \cdot \partial y} = -6\cos(2x + 3y)$ (c) $\frac{\partial z}{\partial x} = 2xe^{x^2 - y^2}$ $\frac{\partial z}{\partial y} = -2ye^{x^2 - y^2}$

$\frac{\partial^2 z}{\partial x^2} = 2e^{x^2 - y^2}(2x^2 + 1)$ $\frac{\partial^2 z}{\partial y^2} = 2e^{x^2 - y^2}(2y^2 - 1)$ $\frac{\partial^2 z}{\partial y \cdot \partial x} = -4xye^{x^2 - y^2}$

$\frac{\partial^2 z}{\partial x \cdot \partial y} = -4xye^{x^2 - y^2}$ (d) $\frac{\partial z}{\partial x} = 2x^2\cos(2x + 3y) + 2x\sin(2x + 3y)$

$\frac{\partial^2 z}{\partial x^2} = (2 - 4x^2)\sin(2x + 3y) + 8x\cos(2x + 3y)$

$\frac{\partial^2 z}{\partial y \cdot \partial x} = -6x^2\sin(2x + 3y) + 6x\cos(2x + 3y)$ $\frac{\partial z}{\partial y} = 3x^2\cos(2x + 3y)$

$\frac{\partial^2 z}{\partial y^2} = -9x^2\sin(2x + 3y)$ $\frac{\partial^2 z}{\partial x \cdot \partial y} = -6x^2\sin(2x + 3y) + 6x\cos(2x + 3y)$

2 (a) $2V$ **3** P decreases 375 W **4** $\pm 2.5\%$

Further problems 6 (page 195)

10 $\pm 0.67E \times 10^{-5}$ approx. **12** $\pm(x + y + z)\%$ **13** y decreases by 19% approx.
14 $\pm 4.25\%$ **16** 19% **18** $\delta y = y\{\delta x \cdot p\cot(px + a) - \delta t \cdot q\tan(qt + b)\}$

Test exercise 7 (page 245)

1 (a) $dz = 4x^3\cos 3y\,dx - 3x^4\sin 3y\,dy$ (b) $dz = 2e^{2y}\{2\cos 4x\,dx + \sin 4x\,dy\}$

(c) $dz = xw^2\{2yw\,dx + xw\,dy + 3xy\,dw\}$ **2** (a) $z = x^3y^4 + 4x^2 - 5y^3$

(b) $z = x^2\cos 4y + 2\cos 3x + 4y^2$ (c) not exact differential

3 9 square units **4** (a) 278.6 (b) $\pi/2$ (c) 22.5 (d) 48 (e) -21

(f) -54π **5** Area $= \frac{5}{12}$ square units **6** (a) 2 (b) 0

Further problems 7 (page 246)

1 14 **2** 1.6 **3** $\frac{\pi}{36}\{9 - 4\sqrt{3}\}$ **4** $\frac{1}{2}\{\pi^4 + 4\}$ **5** $\frac{9\pi}{256}$ **6** $\frac{1}{2} \cdot \ln 2$

7 $2 - \pi/2$ **8** $\frac{1}{8}$ **9** 14 **10** (a) 39.24 (b) 0 **11** $\frac{2}{3}$

Test exercise 8 (page 293)

1 (a) regular at all points (b) $z = -5$ (c) regular at all points

(d) $z = -1$ and $z = 4$ (e) $z = 0$, where $z = x + iy$

2 (a) $v(x, y) = \cosh x\cos y + C$ (b) $v(x, y) = 6(y^2 - x^2) - 4x + C$ **4** $4\pi i$

5 (a) $z = 0$ (b) $z = \pm 1$ (c) no critical point (d) $z = \pm\sqrt{2}$ (e) $z = 0$
(f) no critical point **6** $w = \cosh\dfrac{\pi z}{4}$; D′: $w = 1$

Further problems 8 (page 294)

3 circle, center (5, −2), radius $\sqrt{2}$ **4** circle, center $\left(-\dfrac{1}{3}, 0\right)$, radius $\dfrac{2}{3}$, counter-clockwise **5** (a) $v(x, y) = 2y(x-1) + C$
(b) $v(x, y) = 3x^2y - y^3 - 2xy + y + C$ (c) $v(x, y) = x^2 - 2x - y^2 + C$
(d) $v(x, y) = e^{x^2-y^2}\sin 2xy + C$ **6** (a) $10\pi i$ (b) $6\pi i$ **7** (a) 0 (b) $4\pi i$
(c) $10\pi i$ **9** $2\pi i$ **10** $10\pi i$ **11** (a) (1) $z = 0$ (2) $z = \pm 1$
(b) ellipse, center (0, 0), semi major axis $\frac{5}{2}$, semi minor axis $\frac{3}{2}$
12 (a) $u^2 + v^2 = 1$ (b) $u^2 + (v-1)^2 = 2$; $\theta = 45°$. **13** Unit circle becomes the real axis on the w-plane. Region within the circle maps onto the upper half plane **14** $w = \sin\dfrac{z\pi}{2a}$

Test exercise 9 (page 323)

1 (a) $f(z) = 1 + z + \dfrac{z^2}{2!} + \dfrac{z^3}{3!} + \ldots + \dfrac{z^n}{n!} + \ldots$ valid for $|z| < \infty$
(b) $f(z) = 4z - \dfrac{(4z)^2}{2} + \dfrac{(4z)^3}{3} - \ldots + \dfrac{(-1)^{n+1}(4z)^n}{n} + \ldots$ valid for $|z| < 1/4$ **2** (a) pole of order 5 at $z = -1$ (b) essential singularity at $z = 0$ (c) essential singularity at $z = 0$ (d) removable singularity at $z = 0$

3 $f(z) = \dfrac{1}{\sqrt{2}}\left\{1 + (z - \pi/4) - \dfrac{(z-\pi/4)^2}{2!} - \dfrac{(z-\pi/4)^3}{3!} + \dfrac{(z-\pi/4)^4}{4!} + \dfrac{(z-\pi/4)^5}{5!} - \ldots\right\}$; valid for $|z| < \infty$

4 (a) $f(z) = -(z+3) + 8 + \dfrac{1}{2(z+3)} - \dfrac{4}{(z+3)^2} - \dfrac{1}{24(z+3)^3} + \dfrac{1}{3(z+3)^4} + \ldots$; essential singularity
(b) $f(z) = \dfrac{3}{z+3} - \dfrac{1}{z+1} = \ldots - \dfrac{1}{z^3} + \dfrac{1}{z^2} - \dfrac{1}{z} + 1 - \dfrac{z}{3} + \dfrac{z^2}{9} - \dfrac{z^3}{27} + \ldots$
(c) $f(z) = \dfrac{1}{8(z-2)^2} - \dfrac{3}{16(z-2)} + \dfrac{3}{16} - \dfrac{5(z-2)}{32} + \dfrac{15(z-2)^2}{64} + \ldots$; pole of order 2 **5** double pole at $z = 0$; residue −4, double pole at $z = -1$, residue 7/2, single pole at $z = 1$, residue 1/2 **6** (a) $-\pi/6$ (b) $2\pi/\sqrt{3}$
(c) $2\pi e^{-3}$

Further problems 9 (page 324)

1 (a) $z + \dfrac{z^3}{3!} + \dfrac{z^5}{5!} + \ldots + \dfrac{z^{2n+1}}{(2n+1)!} + \ldots$, $|z| < \infty$

(b) $z + \dfrac{z^3}{3} + \dfrac{2z^5}{15} + \dfrac{17z^7}{315} + \ldots$, $|z| < \pi/2$

(c) $2\left\{z + \dfrac{z^3}{3} + \dfrac{z^5}{5} + \ldots + \dfrac{z^{2n+1}}{2n+1} + \ldots\right\}$, $|z| < 1$

(d) $1 + z\ln a + \dfrac{z^2(\ln a)^2}{2!} + \dfrac{z^3(\ln a)^3}{3!} + \ldots + \dfrac{z^n(\ln a)^n}{n!} + \ldots,\ |z| < \infty$

(e) $\dfrac{3z^2}{25} + \dfrac{27z^3}{125} + \dfrac{162z^4}{625} + \dfrac{810z^5}{3125} + \ldots,\ |z| < 5/3;$

$-\dfrac{5}{9z} - \dfrac{25}{9z^2} - \dfrac{250}{27z^3} - \dfrac{6250}{243z^4} - \ldots,\ |z| > 5/3$ **3** (b) $-\dfrac{z}{(z+1)^2},\ \dfrac{z(z-1)}{(z+1)^3}$

4 (a) convergent for $|z| < \infty$ (b) convergent for $|z| < 1$
(c) convergent for $|z| < 1$ (d) convergent for $|z| < 1$ (e) convergent for $|z| < \infty$

5 (a) $e^2\left\{1 + (z-2) + \dfrac{(z-2)^2}{2!} + \dfrac{(z-2)^3}{3!} + \ldots + \dfrac{(z-2)^n}{n!} + \ldots\right\}$

(b) $\dfrac{\sqrt{3}}{2} - \dfrac{(z-\pi/6)}{2} - \dfrac{\sqrt{3}(z-\pi/6)^2}{2\times 2!} + \dfrac{(z-\pi/6)^3}{2\times 3!} + \dfrac{\sqrt{3}(z-\pi/6)^4}{2\times 4!} + \ldots$

(c) $(z-3)\sin 6 + (z-3)^2\cos 6 - \dfrac{(z-3)^3\sin 6}{2!} - \dfrac{(z-3)^4\cos 6}{3!}$

$+\dfrac{(z-3)^5\sin 6}{4!} + \ldots$ (d) $-\left\{\dfrac{3}{13} + 2\left(\dfrac{3}{13}\right)^2(z-1/3)\right.$

$\left. + 4\left(\dfrac{3}{13}\right)^3(z-1/3)^2 + \ldots + 2^n\left(\dfrac{3}{13}\right)^{n+1}(z-1/3)^n + \ldots\right\}$

(e) $1 - 2(z-3) + 4(z-3)^2 + \ldots + (-2)^n(z-3)^n + \ldots$

6 $(z-1) + \dfrac{(z-1)^2}{1\times 2} - \dfrac{(z-1)^3}{2\times 3} + \dfrac{(z-1)^4}{3\times 4} - \dfrac{(z-1)^5}{4\times 5} + \ldots$ **7** (a) $z = \infty$
(b) $|z| = \sqrt{6}$ (c) $|z-5| = 1$ (d) $z = \infty$ **8** (a) poles of order 2 at $z = 0$ and $z = -1$, removable singularity at $z = \pm 1$ (b) essential singularity at $z = 0$

9 (a) $\dfrac{1}{z^2} - \dfrac{1}{z^4 3!} + \dfrac{1}{z^6 5!} - \dfrac{1}{z^8 7!} + \ldots,\ |z| > 0$

(b) $\dfrac{1}{2}\left(z - \dfrac{3}{2}\right)^{-1},\ |2z-3| > 0$

(c) $\dfrac{3}{z-3} - 2\{1 - (z-3) + (z-3)^2 - (z-3)^3 + \ldots\},\ 0 < |z-3| < 1$

10 (a) $\ldots + \dfrac{8}{z^4} - \dfrac{4}{z^3} + \dfrac{2}{z^2} - \dfrac{1}{z} + \dfrac{2}{5} - \dfrac{2z}{25} + \dfrac{2z^2}{125} - \dfrac{2z^3}{625} + \ldots$

(b) $\dfrac{1}{z} - \dfrac{8}{z^2} + \dfrac{46}{z^3} - \dfrac{242}{z^4} + \ldots$ (c) $-\dfrac{1}{10} + \dfrac{17z}{100} - \dfrac{109z^2}{1000} + \dfrac{593z^3}{10000} - \ldots$

11 (a) $2\pi/\sqrt{3}$ (b) $\dfrac{2\pi}{\sqrt{\alpha^2 - \beta^2}}$ (c) $\dfrac{2\pi}{|\alpha^2 - 1|}$ (d) $\pi/4$ (e) $\pi/2$ (f) $\pi/2$

(g) $\pi\sqrt{\sqrt{13}/8 - 3/8}$ (h) $\pi/4$ (i) $2\pi/\sqrt{3}$ (i) $2\pi/3$ (k) 0 (l) 0

Index